MRE

Materials Research and Engineering
Edited by B. Ilschner and N. J. Grant

C. R. Boër · N. Rebelo
H. Rydstad · G. Schröder

Process Modelling of Metal Forming and Thermomechanical Treatment

With 195 Figures

Springer-Verlag
Berlin Heidelberg NewYork Tokyo

CLAUDIO R. BÖER, Ph. D.

HANS A.B. RYDSTAD, Tekn. Dr.

GÜNTHER SCHRÖDER, Dr.-Ing.

BBC Brown, Bovery & Company, Limited
Research Center
CH-5401 Baden, Switzerland

NUNO M.R.S. REBELO
MARC Analysis Research Corp.
Palo Alto, Calif., U.S.A.

ISBN 3-540-16401-4 Springer-Verlag Berlin Heidelberg New York Tokyo
ISBN 0-387-16401-4 Springer-Verlag New York Heidelberg Berlin Tokyo

Library of Congress Cataloging-in-Publication Data
Process modelling for metal forming and thermomechanical treatment.
(Materials research and engineering)
Includes bibliographies.
1. Metal-work--Mathematical models. 2. Metal-work--Simulation methods.
I. Boër, C. R. (Claudio R.) II. Series.
TS213.P68 1986   671.3   86-3771
ISBN 0-387-16401-4 (U.S.)

Printing: Color-Druck, G. Baucke, Berlin
Binding: Helm, Berlin
2161/3020-543210

Editor's Preface

It is the objective of the series "Materials Research and Engineering" to publish information on technical facts and processes together with specific scientific models and theories.

Fundamental considerations assist in the recognition of the origin of properties and the roots of processes. By providing a higher level of understanding, such considerations form the basis for further improving the quality of both traditional and future engineering materials, as well as the efficiency of industrial operations.

In a more general sense, theory helps to integrate facts into a framework which ties relations between physical equilibria and mechanisms on the one hand, product development and economical competition on the other. Aspects of environmental compatibility, conservation of resources and of socio-cultural interaction form the final horizon - a subject treated in the first volume of this series, "Materials in World Perspective".

The four authors of the present book endeavor to present a comprehensive picture of process modelling in the important field of metal forming and thermomechanical treatment. The reader will be introduced to the rapidly-growing new field of application of computer-aided numerical methods to the quantitative simulation of complex technical processes. Extensive use is made of the state of scientific knowledge related to materials behavior under mechanical stress and thermal treatment.

This book not only demonstrates a high degree of originality, but also presents an unusually large amount of original research results which were obtained by the authors and their co-workers at the Brown Boveri Corporate Research Center at Baden, Switzerland, in the last 5 years. It may be expected, therefore, that this book, along with the others in the "MRE" series, will attract attention and render valuable service to the community of those who, in research and development, strive

for a better performance based on new methodic approaches and
on profound understanding of clearly defined facts.

                                              B. Ilschner

Lausanne, November 1985

Foreword

In metal forming processes, the controlled change of the work-piece geometry in the deformation zone is influenced by various parameters e.g. plastic properties of workpiece materials, tribological conditions in the contact zone between workpiece and tool, tool geometry, tool material, process temperature and velocity. The optimum layout of a metal forming process therefore requires a deep and broad knowledge of all of the above-mentioned parameters. In particular, it requires a good understanding of the interaction between these parameters, including the influence of the machine tool involved with regard to process-time relations.

As soon as the first steps were taken in the development of metal forming technology from shop floor experience level to increasingly scientifically determined high technology level, the modelling of metal forming processes was born. Process modelling as a design procedure, as presented in this book, is based, of course, on theoretical and experimental process analysis. The process simulation, which is still closely connected with process analysis, follows as a logical step. In addition to mathematical modelling, which is performed by a large variety of methods of the theory of plasticity from elementary analysis by "slab method" to large plastic strain finite element analysis, physical modelling is being used to a large extent for various processes, taking advantage of model materials under application of the similarity theory. Special modelling methods and techniques were developed, particularly after the broad introduction of powerful and fast computers. Today computer-aided process modelling plays an increasingly important role for metal forming process modelling.

This book aims to be a presentation on the state of the art in process modelling in metal forming, based on intense research work in the Research Center of a leading industrial company. In its first part, the book provides a condense introduction into the existing methods of mathematical modelling and physical modelling and in its second part, presents process adapted modelling of forging, rolling, drawing and several thermomechanical treatment methods. It demonstrates very clearly, that process modelling following systematic process analysis may be considered as the key to improved product quality and safer production by metal forming in future.

I wish this book every success and I would like to take this opportunity to congratulate the authors on the excellent presentation of the contents of this comprehensive and, at the same time, systematic and clear discussion of their subject.

Kurt Lange

Stuttgart, November 1985

Contents

1 Preface

The expression "process model" as used here refers to a mathematical model which has been developed to a level at which it can quantiatively describe the essential characteristics of a process and which, when implemented as a computer program, permits the stepwise simulation of the process. The practical aim of a computer process model is always the generation of quantitative statements concerning the process in question, whereby these should be faster, cheaper or more extensive than can be obtained by laboratory experiments or tests during actual fabrication.

Clearly the need for the development of process models is particularly great in situations where it is technically difficult to measure the process parameters or materials properties which are altered during the process. The same applies where the costs of operational tests are high. As a result, it appears that, in addition to casting and welding, various forming processes such as rolling, extrusion and forging and heat-treatment procedures such as annealing and quenching are important areas of application for process models.

A new dimension is provided to the interest in process models by the increased use of computers in design and development (CAE: Computer-Aided Engineering; CAD: Computer-Aided Design). Process models should be looked upon as the technology modules within a CAE/CAD system which permit an engineering treatment of the process-related problems.

Only within the last 50 years have scientific methods been extensively applied to manufacturing technologies. In the last 25 years, it has gradually become clear that the further development of the technologies must be based upon an engineering approach which can rely on computer process models and modern measuring techniques. In the last 10 years, powerful FEM process models have been developed. These permit more extensive statements to be made concerning properties of the workpieces and process limits than could be obtained previously by testing.

The present book draws upon work on process models for forming procedures and for special heat-treatment methods which was carried out between 1980 and 1985 at the Brown, Boveri Research Center, Dättwil, Switzerland. The central aim was the application of forming procedures, in particular isothermal forging, to the manufacture of highly-developed turbomachinery components, such as radial impellers and turbine blades. Since this was also in conjunction with the application of new materials, such as oxide-dispersion strengthened superalloys, the aims could only be achieved by interdisciplinary co-operation between materials scientists and manufacturing engineers. The decision to include specialists for process modelling in this team has been shown to have been extremely helpful.

The present book is based on the following concept: the main part comprises process models developed in the course of our work and applications from industrial manufacturing practice and recent research projects. Introductory chapters provide the necessary information in terms of basic theoretical principles of plasticity theory, materials laws, solution procedures and the determination of characteristic physical properties. In fact, the book presents the limited activities of a team as an illustration of the present state of work with process models. It cannot be considered a textbook in the classical sense of the word and even less a monograph which attempts to present the entire state of knowledge in this rapidly-moving field.

The book is intended for engineers in the manufacturing industry who are either active in the field of technology development or who have the task of deciding on the medium-term introduction of CAD/CAE projects. Students and scientists in the fields of manufacturing technology, materials science and information science will be provided with a modern introduction to an advanced interdisciplinary area.

The book is based on concepts arising from research work performed at the Corporate Research Center, Dättwil, of Brown, Boveri and Co., Ltd., Switzerland. We would like to thank Professor R. Schnörr of the Board of Directors and Professor A.P. Speiser and Dr. R.W. Meier, Directors of the Research Center, for their generous support of our work in the field and for their en-

couragement during the writing of this book. In the course of
the work on process models, the authors drew on the knowledge
and experience of teachers, colleagues and friends in the scientific community.

During the development of the process models and the solution
of problems, we were provided with valuable support in the form
of criticism, suggestions and discussions with various people
within the company. In particular, we would like to thank:
Dr. J. Albrecht who contributed Chapter 7.5, Dr. B. Eliasson,
Dr. G.H. Gessinger, Dr. P. Gudmundson, P.-O. Larsson, R. Lüthi,
W. Kuhn, Dr. H. Riegger who assisted with the preparation of
Chapter 5, Dr. W. Schneider, Dr. R. Singer, R. Tièche and Dr. C.
Wüthrich.

We should also like to thank Dr. J.C. Nagtegaal from MARC Analysis Research Corporation, U.S.A., for assistance in reviewing
the Finite Element elastic-plastic formulation, the Department
of Mechanical Engineering, University of Porto, for providing one
of the authors with facilities for the writing of the manuscript
and W.D. Webster Jr., GMI Engineering & Management, USA, who
contributed Chapter 6.3.

Various members of the BBC Corporate Research Center contributed
to the production of the book. In particular, we would like to
mention Mrs. B. Nowatzek, who typed the various versions of the
manuscripts with great professional knowledge and accuracy and
considerable patience and forbearance. Also Mrs. W. Scarlin,
who undertook the task of unifying the styles of the contributions from the various authors into a consistent form.

The drawings were produced with care and style by Mrs. M. Zamfirescu and Mr. S. Guglielmino. Photographs were made by Mr. E.
Schönfeld and Mr. R. Baumann, to whom we convey our thanks.

Last, but not least, we would like to express our particular
thanks to our families and especially Victoria J. Santos for
their help and understanding.

2 Mathematical Modelling

(other symbols are defined in the text)

c_p	specific heat
E	Young's modulus
f_i	surface traction
F_x, F_y, F_z	forces
h	heat transfer coefficient or height of element
H	instantaneous height
J_1, J_2, J_3	deviatoric stress invariants
k	shear flow stress
k_{ijkl}	elasticity constants
L	length of bar element
m	friction factor $0 \leq m \leq 1$
n_j	normal unit vector
N_1, N_2	shape function or interpolations function
p	hydrostatic pressure or applied pressure
p_u, p_l	applied pressure in upper and lower dies
P	total load

q_i	body loads or heat flux
t	time
T_i	surface loads or temperatures
$u_{i,j}$	displacement
$du_{i,j}$	displacement increments
v_o	axial velocity of die
Δv	velocity discontinuity
v_{t_1}, v_{t_2}	tangential velocities
α	thermal expansion coefficient (or angle)
$\bar{\varepsilon}$	effective strain
$\dot{\bar{\varepsilon}}$	effective strain rate
$\varepsilon_x, \varepsilon_y, \varepsilon_z, \gamma_{xy}, \gamma_{xz}, \gamma_{yz}$	generalized state of strain
ε_{ij}	infinitesimal deformation
$d\varepsilon_{ij}^*$	admissable strain increment tensor
μ	friction factor $\quad 0 \leq \mu \leq 0.577$
ρ	density
$\bar{\sigma}$	equivalent or effective stress
$\sigma_x, \sigma_y, \sigma_z, \tau_{xy}, \tau_{xz}, \tau_{yz}$	generalized state of stress
$\sigma_z, \sigma_r, \sigma_t$	axisymmetric state of stress $\quad \} \quad$ σ_z axial stress, σ_r radial stress, σ_t hoop stress

σ_f flow stress in a uniaxial test

σ_m average stress

$\sigma_i = \sigma_1, \sigma_2, \sigma_3$ principal stresses ($\sigma_1 > \sigma_2 > \sigma_3$)

$\sigma_i' = \sigma_1', \sigma_2', \sigma_3'$ deviatoric stresses
(or s_1, s_2, s_3)

σ_{ij}^* admissable stress tensor

σ_Y yield stress

τ friction shear stress or
tangential shear stress

τ_{max} maximum shear stress

τ_Y shear yield stress

2.1 Introduction

At the very heart of metal forming analysis is the theory of plasticity. Initially proposed last century by Tresca and Mohr, it reached a stabilized form in the twenties as a result of work done by von Mises and Hencky. In the early sixties it was completed for infinitesimal analysis [2.1]. A brief description of this theory follows in section 2.2. The theory for large deformations is still being developed today.

The _infinitesimal theory of plasticity_, simplified by discarding the elastic part of the deformation (also known as rigid-plastic flow theory), has produced several approximate methods of analysis which have proved very useful in metal forming. We will present the basic assumptions that lead to the slab method, the upper-bound method and finite element method.

The availability of computers with relatively inexpensive, large facilities for data processing fostered the development of the Finite Element method. Originally used in structural analysis, it rapidly expanded into other fields, and has been applied to metal forming analysis since the early seventies. This is the area in which most of the recent work in this field has been done. Two schools have developed in the area of Finite Element analysis. The first one is based on the rigid-plastic approach (the flow theory), for which the infinitesimal theory of plasticity has been sufficient. It leads to relatively simple formulations which have allowed the difficult problems specific to metal forming applications to be attacked. The second one is based on the more complete elasto-plastic approach which almost always requires a _large deformation theory of plasticity_. The formulations are more complicated and have followed, if not actually led to, developments in the theory itself. Today it can simulate forming processes as intricate as the flow approach.

2.2 Infinitesimal Theory of Plasticity

For a complete, consistent, and elegant presentation of the infinitesimal theory of plasticity, the reader is referred to reference [2.1]. This paper shows that all the theory can be derived from three basic assumptions:

1) At any time, t, the infinitesimal deformation $\varepsilon_{ij}(\underset{\sim}{x}, t)$ can
   be decomposed as a sum of an elastic part and of a plastic
   part:

$$\varepsilon_{ij}(\underset{\sim}{x}, t) = \varepsilon_{ij}{}^{e} + \varepsilon_{ij}{}^{p}$$

2) A yield function (surface) in stress space exists.

3) Material is stable, i.e., any increase in deformation will
   not produce a decrease in the flow stress.

In the following pages, the same theory will be described, but in
a less formal manner and with special emphasis placed upon the
physical considerations behind it.

To begin with, there are several physically-observed phenomena
that are usually ignored in the theory, such as:

a. Influence of strain rate on the elastic limit

b. Creep

c. Bauschinger effect

d. Hysteresis loops

e. Anisotropy. This is usually introduced only for sheet metal
   forming analysis, see Hill [2.2].

f. Size effects

g. In all rigid-plastic analysis, the elastic part of the deformation.

2.2.1 Yield Criteria

We are searching for those stress states which will make the material yield. If we consider that these stress states define a
yield surface in stress space, it means that any stress state inside the surface corresponds to an elastic state, and any stress
state on the surface corresponds to a plastic state.

It is known that in metals, plastic deformation is produced by dislocations, which implies that materials are incompressible during plastic deformation, and any hydrostatic state of stress does not produce yielding. Therefore, the yield surface must be cylindrical, with an axis defined by $\sigma_x = \sigma_y = \sigma_z$.

It is also concluded that only shear stresses produce yielding. Hence the most simple yielding criterion that can be devised states that the material will yield when the maximum shear stress reaches a certain value:

$$\tau_{max} = \frac{\sigma_1 - \sigma_3}{2} = \pm k$$

or $\quad \sigma_1 - \sigma_3 = 2k$

σ_1 = maximum principal stress

σ_3 = minimum principal stress
k = shear flow stress

This is the Tresca yield criterion, proposed in 1864. The value of the constant in a tensile or compression test can be determined as follows:

$$\sigma_2 = \sigma_3 = 0$$

therefore:

$$\sigma_1 = \sigma_f = 2k \tag{2.1}$$

σ_f = flow stress in a uniaxial test

In a torsion test:

$$\sigma_1 = -\sigma_3$$
$$\sigma_2 = 0$$

therefore:

$$\sigma_1 - \sigma_3 = \sigma_f = 2\tau_{max} = 2k = \bar{\sigma}$$

If all stress components in the yield criterion are to be included, the following should be taken into consideration: flow stress is specific to the material only. This means it should not depend on the chosen axis, and is thus a function of the stress tensor invariants only.

As hydrostatic stress does not cause yielding, the deviatoric stresses may be considered as:

$$\sigma_i' = \sigma_i - \sigma_m \qquad \text{with} \qquad \sigma_m = \frac{1}{3}(\sigma_x + \sigma_y + \sigma_z) = -p$$

p = hydrostatic pressure

and the invariants:

$$J_1 = \sigma_x' + \sigma_y' + \sigma_z'$$

$$J_2 = -(\sigma_x'\sigma_y' + \sigma_y'\sigma_z' + \sigma_z'\sigma_x') + \tau_{xy}^2 + \tau_{yz}^2 + \tau_{zx}^2$$

$$J_3 = \det \begin{vmatrix} \sigma_x' & \tau_{xy} & \tau_{xz} \\ \tau_{xy} & \sigma_y' & \tau_{yz} \\ \tau_{xz} & \tau_{yz} & \sigma_z' \end{vmatrix}$$

and the flow stress as a function of its invariants J_1, J_2, J_3, where $J_1 = 0$. The yield surface should then be defined as:

$$f(J_2, J_3) = \text{constant}$$

The most simple conceivable form is:

$$J_2 = \text{constant}$$

which is the von Mises yield criterion (flow rule) proposed in 1913. This can be written in the general form:

$$\sqrt{\frac{1}{2}\left[(\sigma_x - \sigma_y)^2 + (\sigma_y - \sigma_z)^2 + (\sigma_z - \sigma_x)^2\right] + 6\,(\tau_{xy}^{\,2} + \tau_{yz}^{\,2} + \tau_{zx}^{\,2})} = \bar{\sigma} = \text{constant} \qquad (2.2)$$

The left-hand side of equation (2.2) is commonly represented by $\bar{\sigma}$ and known as effective stress or equivalent stress. For the case of principal axes x, y, z (τ_{xy}, τ_{yz} τ_{zx} = 0):

$$\sqrt{\frac{1}{2}\left[(\sigma_1 - \sigma_2)^2 + (\sigma_2 - \sigma_3)^2 + (\sigma_3 - \sigma_1)^2\right]} = \bar{\sigma}$$

Again in a tensile test with principal axes σ_1, σ_2, σ_3:

$$\sigma_2 = \sigma_3 = 0$$

$$\sigma_1 = \sigma_f \qquad \text{for the beginning of flow.}$$

According to the von Mises yield criterion, plastic flow starts if $\bar{\sigma}$ in the expression (2.2) reaches the value of flow stress σ_f.

$\bar{\sigma}$ is the combination of the stress components that indicates how close the stress state is to the yield surface, or that which is equivalent to the uniaxial test.

In a torsion test with principal axes σ_1, σ_2, σ_3:

$$\sigma_1 = -\sigma_2 \;; \qquad \sigma_3 = 0$$

$$\sqrt{3} \cdot \sigma_1 = \sqrt{3}\,\tau_{max} = \sigma_f \qquad (2.3)$$

What the expressions (2.1) and (2.4) mean, is that if we determine the yield surface, with, for example, a tensile test, then the two criteria will not predict yielding at the same point in torsion, or vice-versa. The difference between the two criteria is at most about 15 %, which is for most cases within the scatterband of experimental flow stress determination.

Assuming that the material is stable, then the yield surface must be convex. Otherwise it would be possible to have changes in plastic state (plastic deformation) with a decrease in the equivalent stress.

If the yield surfaces in a principal stress space, seen from the $\sigma_1 = \sigma_2 = \sigma_3$ axis, are represented, the following diagram (Figure 2.1) is obtained:

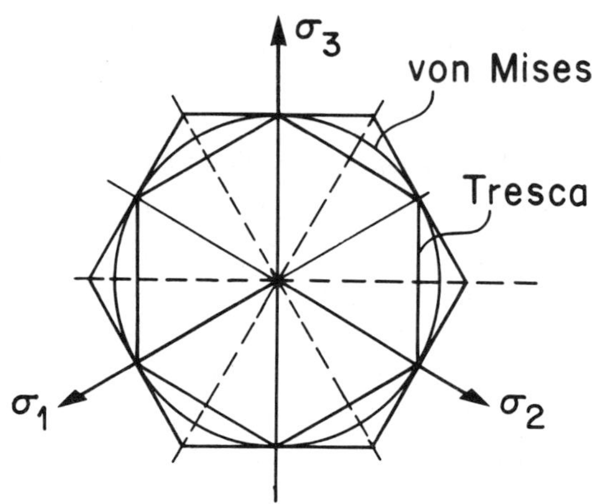

Figure 2.1: Yield surfaces in principal stress space.

The Tresca yield surface is the inner hexagon, and the von Mises yield surface is the circle.

Isotropy and absence of the Bauschinger effect impose symmetry with respect to the six lines represented. Convexity requires that all admissible yield surfaces lie between the Tresca hexagon and the outer hexagon. The von Mises yield criterion follows experiments fairly well, and there is no point in further complicating the theory.

2.2.2 Work Hardening

Consider that after yielding, there was a small amount of plastic deformation $d\underset{\sim}{\varepsilon}^p$. Upon unloading and loading again, the yield surface may have been altered (hardening), therefore:

$$\sigma_Y = g\,(d\underset{\sim}{\varepsilon}^p).$$

As the plastic deformation occurs at constant volume, g would be expected to be a function of the shear components of deformation.

It has been shown experimentally that various things can happen to the yield surface upon plastic deformation. Material parameters play a role, the path of deformation is very important, and anisotropy usually develops in the material after it has been plastically deformed. There are several theories in the literature as to how to take the various phenomena into account. The general ones are usually hard to handle in practical applications, and the great majority of the analysis is made with so-called isotropic hardening. This means that the effect of plastic deformation on the yield surface is its isotropic expansion.

As the effective stress becomes the scalar "measure" of the stress tensor, a similar effective strain increment can be defined as an isotropic combination of shear-like terms in the strain increments:

$$d\bar{\varepsilon} = \frac{\sqrt{2}}{3}\left[(d\varepsilon_x^p - d\varepsilon_y^p)^2 + (d\varepsilon_y^p - d\varepsilon_z^p)^2 + (d\varepsilon_z^p - d\varepsilon_x^p)^2 + \right.$$

$$\left. + \frac{3}{2}(d\gamma_{xy}^{p\,2} + d\gamma_{yz}^{p\,2} + d\gamma_{zx}^{p\,2})\right]^{1/2} \tag{2.4}$$

The factor $\frac{\sqrt{2}}{3}$ was conveniently chosen so that $d\bar{\varepsilon} = d\varepsilon$ in a uniaxial test. Also, the incremental amount of plastic work done per unit volume becomes $dW^p = \bar{\sigma}d\bar{\varepsilon}$.

For isotropic hardening:

$$\sigma_Y = g\,(d\bar{\varepsilon}) \qquad \text{is given.}$$

As the plastic deformation proceeds, $\bar{\varepsilon}$ cannot be calculated based on the components of $\underset{\sim}{\varepsilon}^p$. In fact, any original shape can always be recovered after some plastic deformation by imposing another appropriate plastic deformation. Therefore, the components of $\underset{\sim}{\varepsilon}^p$

have no meaning other than an overall change in geometry. The
real amount of plastic deformation is measured by integrating $d\bar{\varepsilon}$
which is always positive.

$$\bar{\varepsilon} \;=\; \int d\bar{\varepsilon}$$

After a finite amount of deformation:

$$\sigma_f \;=\; g\,(\bar{\varepsilon}).$$

The yield surface is defined then by:

$$\bar{\sigma} \;=\; g\,(\bar{\varepsilon})$$

with its expansion following the uniaxial stress-strain curve as
shown in Figure 2.2

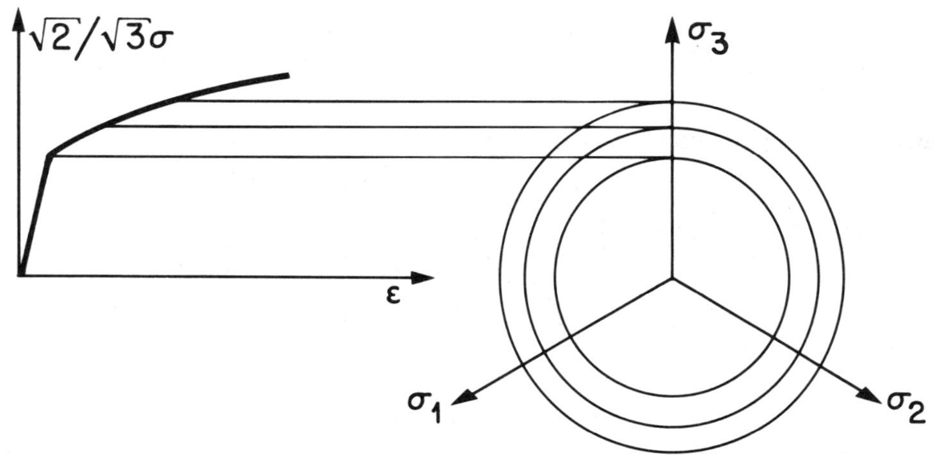

Figure 2.2: Expansion of yield surface with work hardening.

2.2.3 Plastic Deformation

The permanent character of plastic deformation, and the fact
that, unlike in elasticity, a certain configuration can be obtained via an infinite variation of the amounts of plastic deformation, lead us to conclude that a state of stress provides

information only for the increment of strain which it produces. In order to obtain the components of such strain increments, the two following physically acceptable assumptions should be taken as a starting point:

a) The principle stress directions coincide with the principal strain increment directions;

b) The principal shear components of strain increment are proportional to the principal shear stress components.

It can be concluded that:

$$\frac{d\varepsilon_x^p}{\sigma'_x} = \frac{d\varepsilon_y^p}{\sigma'_y} = \frac{d\varepsilon_z^p}{\sigma'_z} = \frac{d\gamma_{xy}^p/2}{\tau_{xy}} = \frac{d\gamma_{xz}^p/2}{\tau_{xz}} = \frac{d\gamma_{yz}^p/2}{\tau_{yz}} = d\lambda$$

By inserting these six equations in the definition of $d\bar{\varepsilon}$ (2.4), the following equation is obtained:

$$d\bar{\varepsilon} = \frac{2}{3} d\lambda \, \bar{\sigma}$$

and the 6 equations become:

$$d\varepsilon_x^p = \frac{d\bar{\varepsilon}}{\bar{\sigma}} \left[\sigma_x - \frac{1}{2} (\sigma_y + \sigma_z) \right] \tag{2.5}$$

$$d\varepsilon_y^p = \frac{d\bar{\varepsilon}}{\bar{\sigma}} \left[\sigma_y - \frac{1}{2} (\sigma_x + \sigma_z) \right] \tag{2.6}$$

$$d\varepsilon_z^p = \frac{d\bar{\varepsilon}}{\bar{\sigma}} \left[\sigma_z - \frac{1}{2} (\sigma_x + \sigma_y) \right] \tag{2.7}$$

$$d\gamma_{xy}^p = \frac{d\bar{\varepsilon}}{\bar{\sigma}} 3\tau_{xy} \tag{2.8}$$

$$d\gamma_{yz}^p = \frac{d\bar{\varepsilon}}{\bar{\sigma}} 3\tau_{yz} \tag{2.9}$$

$$dy^p_{xz} = \frac{d\bar{\varepsilon}}{\bar{\sigma}} \, 3\tau_{xz} \tag{2.10}$$

With the aid of equation (2.3) the yield surface equation can be written in the following fashion:

$$h\,(\underset{\sim}{\sigma}) = \bar{\sigma} - \sigma_Y = 0$$

or

$$\frac{1}{\sqrt{2}} \, [(\sigma_x - \sigma_y)^2 + (\sigma_y - \sigma_z)^2 + (\sigma_z - \sigma_x)^2 +$$

$$+ 6 \, (\tau_{xy}^2 + \tau_{yz}^2 + \tau_{xz}^2)]^{1/2} - \sigma_Y = 0$$

It is now possible to calculate:

$$\frac{\partial h}{\partial \sigma_x} = \frac{1}{\sqrt{2}} \, \frac{1}{2} \, [(\sigma_x - \sigma_y)^2 + (\sigma_y - \sigma_z)^2 + (\sigma_z - \sigma_x)^2 +$$

$$+ 6 \, (\tau_{xy}^2 + \tau_{yz}^2 + \tau_{xz}^2)]^{-1/2} \cdot [2\,(\sigma_x - \sigma_y) -$$

$$- 2\,(\sigma_z - \sigma_x)] = \frac{1}{\bar{\sigma}} \, [\sigma_x - \frac{1}{2}\,(\sigma_y + \sigma_z)] \tag{2.11}$$

A comparison between (2.11) and (2.5) gives:

$$d\varepsilon^p_x = d\bar{\varepsilon} \, \frac{\partial h}{\partial \sigma_x} \tag{2.12}$$

and similarly:

$$d\varepsilon^p_y = d\bar{\varepsilon} \, \frac{\partial h}{\partial \sigma_y} \tag{2.13}$$

$$d\varepsilon^p_z = d\bar{\varepsilon} \, \frac{\partial h}{\partial \sigma_z} \tag{2.14}$$

$$dy^p_{xy} = d\bar{\varepsilon} \, \frac{\partial h}{\partial \tau_{xy}} \tag{2.15}$$

$$d\gamma^p_{yz} = d\bar{\varepsilon} \, \frac{\partial h}{\partial \tau_{yz}} \tag{2.16}$$

$$d\gamma^p_{xz} = d\bar{\varepsilon} \, \frac{\partial h}{\partial \tau_{xz}} \tag{2.17}$$

Equations (2.13 - 2.17) show that the plastic strain increments are normal to the yield surface. In fact, this result can be mathematically derived from the stability assumptions, from where assumptions a) and b) in this section become natural consequences.

When the theory is to be extended to viscoplasticity, or other types of material behavior considered, equations such as (2.12), which explain the existence of a "plastic potential", become a common point of departure.

2.3 Problem Solution

Any metal forming problem is a boundary value problem, in which
we have a volume V of material, bound by a surface S, as in Figure 2.3. Compatibility equations and equilibrium equations apply
within the volume. In order to simplify the equations, a tensor
notation will be used. The first equations mean that for any load
producing an infinitesimal strain increment field $d\varepsilon_{ij}$, it must
be possible to derive such a field from a displacement increment
field du_i through:

$$d\varepsilon_{ij} = \frac{1}{2}(du_{i,j} + du_{j,i}) \qquad (2.18)$$

where a comma denotes a derivative with respect to the axis following it.

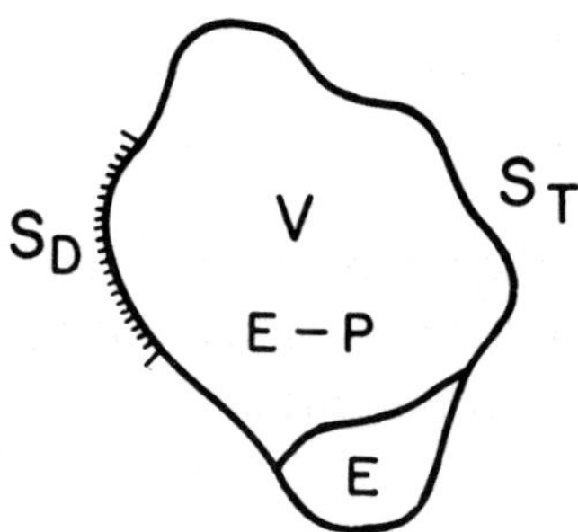

<u>Figure 2.3</u>: Solid mechanics boundary value problem.

The plastic part of strain increments satisfy incompressibility:

$$d\varepsilon_{ii}^{p} = 0 \qquad (2.19)$$

The second equations mean that for any load producing a stress
field σ_{ij} in which body loads, q_i, are present, the following
applies:

$$\sigma_{ij,j} + q_i = 0 \qquad (2.20)$$

where a summation convention is implied. Furthermore, the surface can be subdivided into two parts. One is S_T on which forces/unit area - T_i - are applied and on which force boundary conditions have to be satisfied:

$$\sigma_{ij} n_j = T_i \qquad (2.21)$$

where n_j are the components of a unit vector, normal to the surface.

On the other part, S_D, displacement increments dU_i are applied, and the following conditions have to be satisfied:

$$du_i = dU_i \qquad (2.22)$$

The stresses and strain increments produced by a certain load are related by constitutive equations. The body may be divided in two parts; one, E, in which yielding does not take place, and where elasticity equations apply:

$$d\varepsilon_{ij} - \alpha \Delta T \delta_{ij} = K_{ijkl}\, d\sigma_{ij} \qquad (2.23)$$

with K_{ijkl} being elasticity constants, and α the thermal expansion coefficient. In the other part, E-P, where yielding has occurred, the yield criterion applies:

$$\bar{\sigma} = \sigma_f \qquad (2.24)$$

as well as elasticity and plasticity stress-strain relations:

$$d\varepsilon_{ij} - d\varepsilon_{ij}^p - \alpha \Delta T \delta_{ij} = K_{ijkl}\, d\sigma_{ij} \qquad (2.25)$$

$$d\varepsilon_{ij}^p = d\bar{\varepsilon}\, \frac{\partial h}{\partial \sigma_{ij}} \qquad (2.26)$$

It becomes obvious that it is very difficult to obtain closed form solutions that satisfy all these requirements. In fact, there are only a few complete solutions in the literature, all for very simple problems.

Even if rigid-plastic materials are considered, equations 2.18, 2.19, 2.20, 2.21, 2.22, 2.24 and 2.26 have to be satisfied, with the risk of having part of volume E becoming rigid, with an undetermined stress field. The following table gives all the requirements for a rigid-plastic analysis.

Given the difficulties in satisfying all the equations, several approximate numerical solutions have been developed through the years, even before the theory was completely defined.

TABLE 2.1

	MECHANICS OF DEFORMATION	
STRESS REQUIREMENTS		**DEFORMATION REQUIREMENTS**
1. EQUILIBRIUM EQUATIONS 2. YIELD CRITERION 3. MATERIAL FLOW PROPERTIES 4. STRESS BOUNDARY CONDITIONS	FLOW RULE (STRESS-STRAIN RELATIONS)	1. COMPATIBILITY EQUATIONS 2. INCOMPRESSIBILITY 3. DISPLACEMENT BOUNDARY CONDITIONS

A metal forming system can be modelled at different levels of detail. It is clear that the more complete the model, the larger will also be the amount of input data required, the algorithms necessary to describe the process will be more complex, and the amount of output data will be considerable and therefore more difficult to interpret.

At the present state of the art in computer-aided modelling of metal forming systems, we can say that there exist, basically, two main groups of modelling approach:

- system approach
- detailed mechanics approach

The properties and integrity of the final formed product is a function of all the components of the metal forming system. Therefore, if a large number of these variables is taken into consideration, the simulation will be better able to predict the true system behavior. Furthermore, in the system approach simulation, there is generally a need to collect a large amount of information about the behavior of the system in a relatively short time, to pin down the range of values of the parameters that optimize the process. In this way the number of iterations with a more detailed, but also more costly, simulation will be considerably reduced.

2.3.1 System Approach

The system simulation procedure is illustrated in Figure 2.4 following an application of forging of complex parts made of a nickel-base alloy. It should be noticed that the computer simulation is always followed by an experimental simulation to verify both the results of the computer simulation and to determine the parameters that cannot be predicted by the theoretical analysis (such as microstructure).

This double-track simulation is given, for example, in reference [2.3] and reference [2.4]. The amount of experimentation is according to the confidence level reached with the analytical model: the confidence will increase with the amount of verification done. The experimental techniques can also be computer-aided

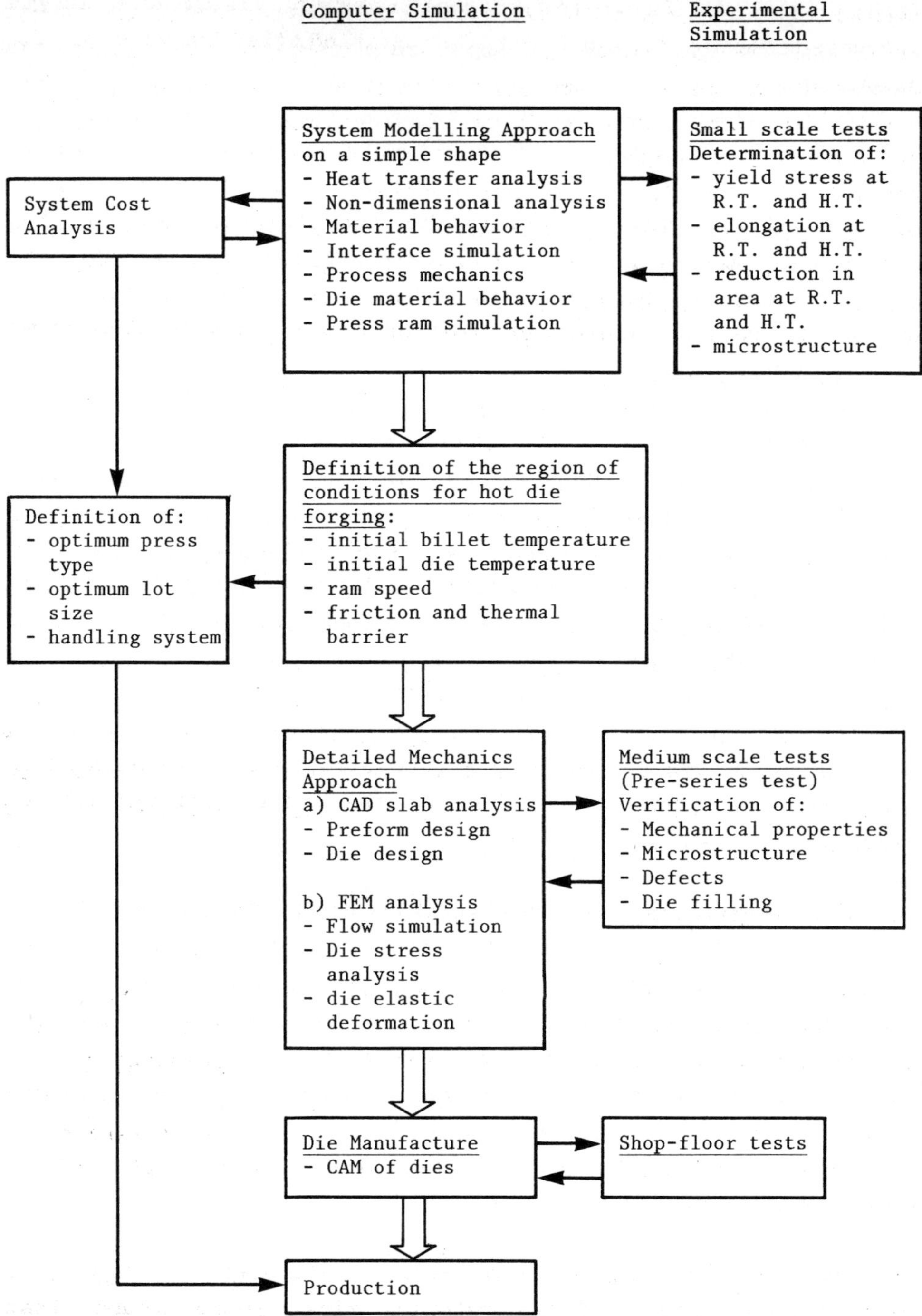

Figure 2.4: Simulation Procedure of Forging of Parts.

[2.5], [2.6] in order to increase accuracy, allowing a larger collected amount of data and facilitating the analysis of the output data.

Parallel to the computer and experimental simulation of the physical processing system, it is necessary to determine the economical viability. This has to be done parallel to design of the system, at various stages, to ensure that the various modifications imposed on the system in order to meet the necessary requirements do not exceed the expected costs. This is also shown in Figure 2.4.

It is possible to develop various methods to allow such cost analysis and it is very useful if it is possible to start right from the beginning of the system design process as, for example, from the product design [2.7]. In fact, it is often the case that a product will be designed and forged without enough consideration having been given to the effect of various features on overall forging costs.

The methods that are mainly used for process modelling of metal forming at a system level are the slab method or elementary analysis, the upper-bound method and the Finite-Difference technique.

The slab method considers the stresses on a plane perpendicular to the metal flow direction. A slab, straight or curved, of infinitesimal thickness is selected in this plane at any arbitrary point in the deformed metal. The forces on the slab are balanced and this will result in a differential equation of static equilibrium. By analytical or numerical integration of the differential equation and with the introduction of the boundary conditions, it is possible to determine the forming forces and to obtain other information.

The starting point of the slab method (elementary theory) is a major simplification of the material flow. It is assumed that the velocity along the slab cutting surfaces is constant, which means that the slab cutting surfaces will be planes during forming.

The method lends itself to an easy discretization, dividing the deformation zone into planes and directions of metal flow as shown in Figure 2.5 [2.8] and therefore computer-aided implementation. The solution can thus be rapidly calculated [2.9].

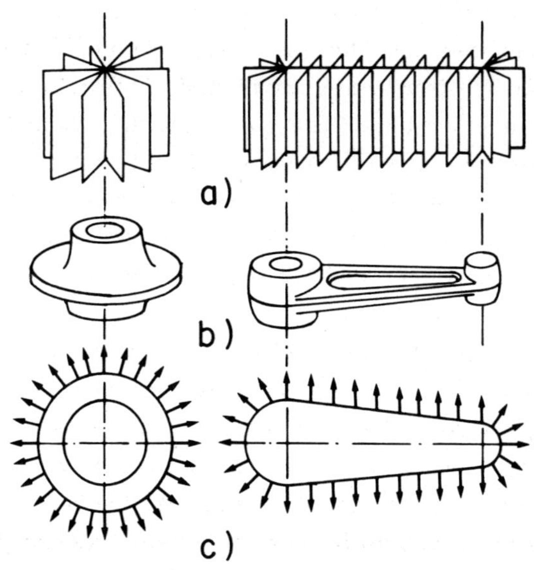

<u>Figure 2.5</u>: Planes and directions of metal flow during forging of two simple shapes: (a) planes of flow, (b) finished forged shapes, (c) directions of flow [2.5].

The slab method is particularly suitable for interactive computer simulation allowing the designer to examine several forging systems in a relatively short time. A typical example is the forging of turbine blades. In view of the large number of parameters given to the designer, it is of importance to have a simple model that gives quick interactive simulations [2.10]. This allows a solution that can then be checked by a fully flow computation to be obtained.

In Chapter 4.1.2 an example is given of interactive simulation of forging of an airfoil section.

The upper-bound method is used for determining the velocity, strain rate, and strain distributions in a deforming metal. A velocity field which satisfies the incompressibility of metals and the velocity boundary conditions is considered. The total forming energy and also the forming load are obtained from the computed deformation, shear and friction energies based on the above velocity field. The value of the forming load, derived from such a velocity field, is necessarily higher than the actual load required by the process, because the real velocity field needs the minimum of energy and, therefore, represents an upper-bound to the actual forming load. By including one or more variable parameters in the considered velocity field, it is possible to determine an optimized upper-bound velocity field. The values of the parameters are determined by minimizing the total forming energy with respect to these parameters [2.11].

The Finite-Difference technique is mainly used in metal forming analysis for the calculation of temperature distribution. The heat transfer problem is non steady-state in a moving incompressible medium with heat sources. An approximate numerical method of solution based on Finite-Difference techniques can be used [2.12]. This method approximates the heat generation and the simultaneous heat transfer as taking place in two consecutive steps. The heat generation and heat transportation take place at the beginning of a time interval Δt, associated with a small step of deformation. Then the heat transfer takes place as for a stationary medium during Δt. The repetition of these two steps simulates numerically the deformation process and gives the temperature distribution as a function of time. The results for strip rolling [2.8] give an example of a steady-state deformation process. The heat is transported into the deformation zone, heat is generated by deformation in the material and by friction at the roll-workpiece interface, heat is conducted to the work rolls, and heat is carried away with the deformed material. The entire strip and the rolls within the arc of contact were divided into a grid system.

The same approach can be used to simulate the temperature distribution during non steady-state deforming processes [2.5].

2.3.2 Detailed Mechanics Approach

When the region where the optimum conditions for a metal forming operation have been determined with a system modelling approach, it is possible to simulate in more detail the deformation process. In Figure 2.4 it is shown that there are basically two steps at this level of modelling: first the preform and dies are designed with a "slab type" of analysis that allows a quick and interactive determination of several solutions; secondly the metal flow is accurately simulated with a FEM (Finite Element Method) approach to verify the die filling for complex geometrical parts as well as strain and strain rate distribution for control of the microstructure. Furthermore, the FEM can be used to analyze the die stresses and elastic deformations.

At this stage, sub-scale or medium scale tests can also be done to verify the mechanical properties and microstructure of the final product. It is also possible to check for defects and die filling.

In Figure 2.6 a block diagram is shown for process design and control taken from Ref. [2.13]. The deformation mechanics has a central role in the design and control of metal forming processes and a more detailed simulation is therefore justified. One of the most powerful theoretical tools in simulating metal flow during deformation processes is the <u>Finite Element Method</u>. In recent years, the use of this method of analysis has been a major topic in metal forming research [2.13].

Techniques other than the Finite Element method have been developed in recent years to investigate in detail the plastic deformation zone. Some of them, having potential if further developed, are here briefly summarized.

<u>Combined Numerical and Algebraic Numerical Computer Processing</u> [2.14] is an approach based on symbolic computer processing of formulae expressions. The method permits the treatment of complex analytical solutions to forming problems. As an example, a solution was obtained for flat-bar rolling with spread. The approach uses the computer language LISP, which is widely adopted in the

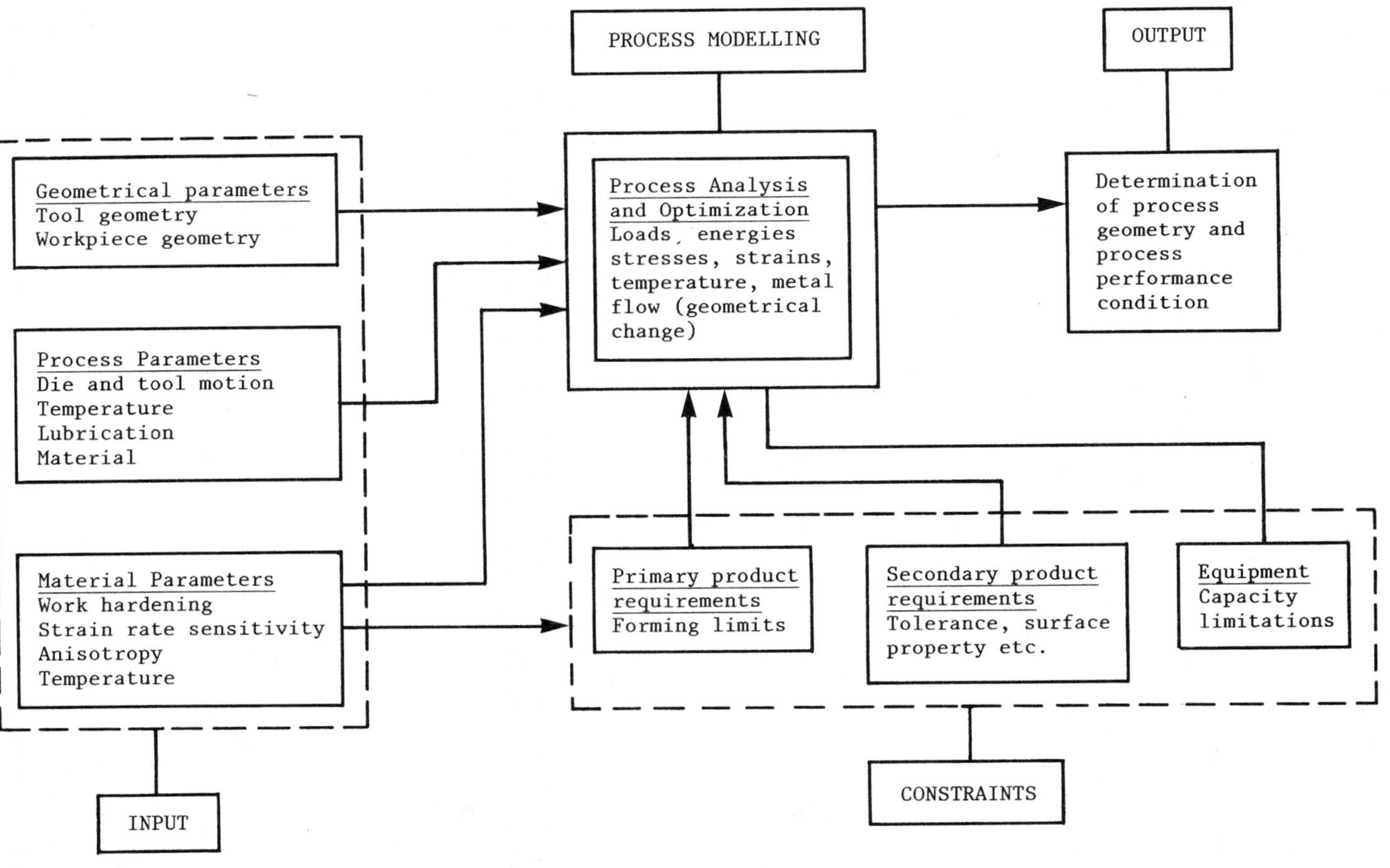

<u>Figure 2.6:</u> Block diagrams for process design and control [2.13].

field of Artificial Intelligence.

<u>Upper-Bound Elemental Technique</u> (UBET) [2.15], [2.16] is based on the upper-bound technique widely used to estimate loads and stresses in forming operations. The deforming body is divided into unit regions and the total velocity field over the workpiece is defined, satisfying the boundary conditions and the rate of energy dissipation. The advantage of the method is mainly the possibility of using a small computer and easy data input.

<u>Matrix Operator Technique for Slip Line Field Construction</u> [2.17] The technique permits a reduction of some of the drawbacks of the slip line approach, such as a trial-and-error procedure when the boundary conditions are such that no initial slip line is evident. The method can have potential if it is further developed to bring it into use in industrial problems.

2.4 Elementary Analysis or "Slab Method"

The slab method assumes that a square grid placed in the deformation zone would be distorted uniformly into rectangular elements as shown from a) to c) in Figure 2.7. A practical deformation pattern is like a) to b) (Figure 2.7) when friction at the interface is present.

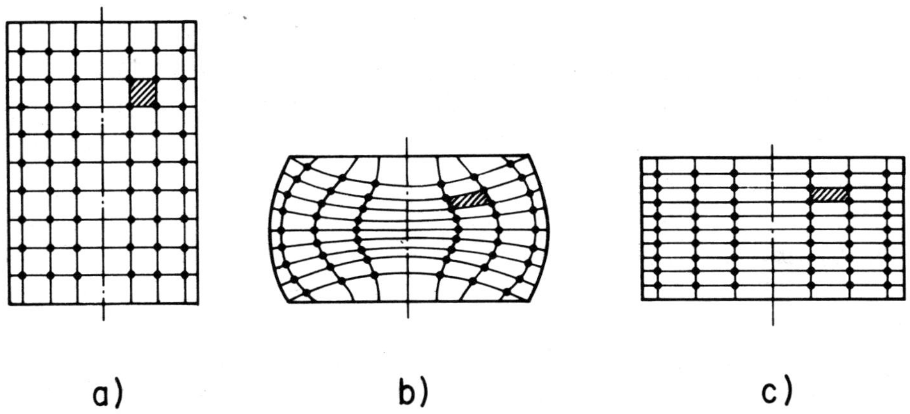

Figure 2.7: Distortion of a square grid (a) in a real deformation pattern when friction is present (b) and as it is assumed in the slab method (c) when there is no friction.

For the calculation of the deformation stresses and force, the deformation zone is divided into <u>elements</u> such as slabs, slices, tubes (see Figure 2.8) with infinitesimal dimensions. Integrating

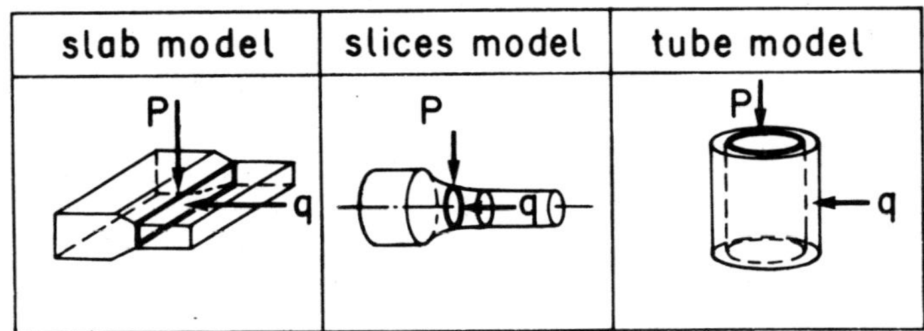

Figure 2.8: Different elements used in the slab method for the calculation of stresses and forces.

over the deformation zone and taking into account the boundary
conditions, it is possible to approximately calculate the stress
distribution and the forces. Figure 2.9 shows a typical example.

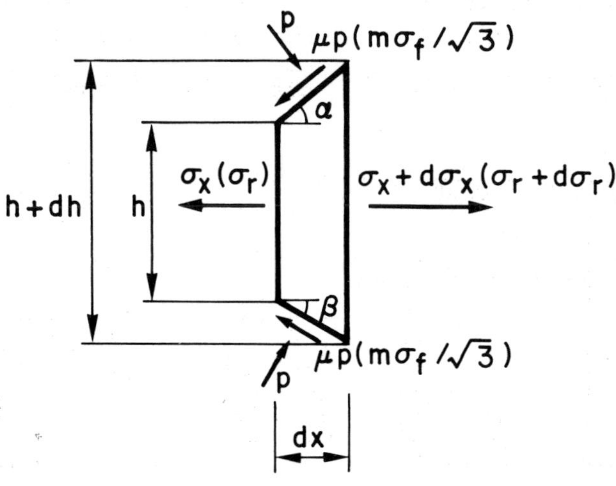

<u>Figure 2.9</u>: Slab element for plane strain or axisymmetric con-
dition.

Friction is usually described by the coefficient μ - tangential
stress = μ times normal stress p, or by a friction factor m -
tangential stress = m times the shear yield stress of adjacent
material. Note that the sign of the friction stresses depends on
the direction in which the material is flowing.

The plane surface assumption implies that the directions x, y in
plane strain or r, z for an axisymmetrical problem are principal
stress and strain directions. The von Mises yield criterion is
written thus:

$$\sigma_y - \sigma_x = -\frac{2}{\sqrt{3}}\bar{\sigma} \qquad \text{plane strain} \qquad (2.27)$$

$$\sigma_z - \sigma_r = -\bar{\sigma} \qquad \text{axisymmetric} \qquad (2.28)$$

It is also assumed that the flow stress $\bar{\sigma}$ is constant in each
slab, although it may change from slab to slab.

2.4.1 The General Case

With the previous assumptions, the equilibrium equations for each
slab (Figure 2.9) along direction x (consider plane strain for
example) can be written:

$$(\sigma_x + d\sigma_x)(h + dh) - \sigma_x h + \frac{pdx}{\cos\alpha} \sin\alpha + p \frac{dx}{\cos\beta} \sin\beta -$$

$$- \tau \frac{dx}{\cos\alpha} \cos\alpha - \frac{\tau dx}{\cos\beta} \cos\beta = 0 \qquad (2.29)$$

neglecting second order terms:

$$hd\sigma_x + \sigma_x dh + pdx\,(tg\alpha + tg\beta) - 2\tau dx = 0$$

$$dx = \frac{dh}{tg\alpha + tg\beta}$$

$$hd\sigma_x + \sigma_x dh + pdh - 2\frac{\tau \quad dh}{(tg\alpha + tg\beta)} = 0 \qquad (2.30)$$

assuming that:

$$\sigma_y \approx -p \qquad (2.31)$$

the yield criterion becomes:

$$\sigma_x - \sigma_y = \frac{2}{\sqrt{3}} \bar{\sigma} \approx$$

$$\sigma_x + p = \frac{2}{\sqrt{3}} \bar{\sigma}$$

$$d\sigma_x = d\sigma_y$$

therefore:

$$d\sigma_y + \left(\frac{2}{\sqrt{3}} \bar{\sigma} - \frac{2 \tau}{(tg\alpha + tg\beta)} \right) \frac{dh}{h} = 0 \qquad (2.32)$$

In general, τ is used according to the type of friction considered. There may be slight variations in the geometrical simplifications according to the external definition of the problem.

The procedure for solving a problem is always to divide the deforming area into parts so that an equation such as that given above can be determined. The equations of each part can then be integrated, starting with a part that has a known boundary condition, such as free surface normal to the x direction:

$$\sigma_x = 0 \qquad \text{for} \quad x = b$$

$$\sigma_y = -\frac{2}{\sqrt{3}}\,\bar{\sigma}$$

If a rigid zone forms, it is possible to separate it from deforming zones by shear lines, whose position can be optimized during the solution. The load is obtained by integrating σ_y along the surface. Incremental deformations for a given configuration are obtained by using the flow rule and incompressibility. The geometry can then be updated for a new configuration, and the calculation restarts. In this way, for instance, it is possible to simulate a forging process.

Table 2.1 shows that the slab method uses only the stress requirements and approximations are made at the equilibrium equations level in order to obtain a solution. The deformation part is derived from that solution, and there are no guarantees of its being correct.

The whole solution procedure can be automated in computer programs, which, due to the simplicity of the calculations, are most suitable for coupling with interactive graphical displays and serve as a computer aid to designers or operators of metalworking processes.

2.4.2   The Plane Strain Situation

2.4.2.1   Frictionless Strip Drawing

The following example of drawing a strip without friction in plane strain (see Figure 2.10) was taken. The height of strip h is much smaller than the width w. Taking a slice or a slab of length dx and determining the <u>equilibrium of forces</u>:

a) $\Sigma F_x = 0$ $\quad\quad (\sigma_x + d\sigma_x)(h + dh)\, w - \sigma_x\, h\, w\; +$

$$+\ 2p\ (w\ \frac{dx}{\cos\ \alpha})\ \sin\ \alpha\ =\ 0 \quad\quad\quad (2.33)$$

b) $\Sigma F_y = 0$ $\quad\quad \sigma_y\ dx\ w + p\ \cos\ \alpha\ (\frac{dx}{\cos\ \alpha})\ w\ =\ 0 \quad (2.34)$

simplifying:

$$\sigma_x\ dh + h d\sigma_x + 2p\ \tan\ \alpha\ dx\ =\ 0$$

$$\sigma_y + p\ =\ 0$$

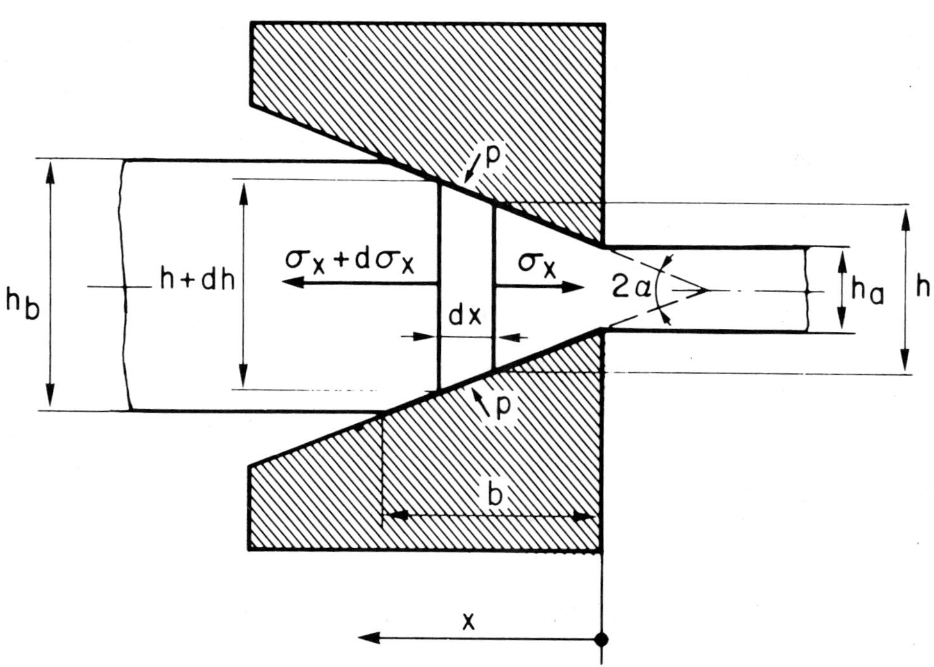

Figure 2.10: Frictionless strip drawing (plane strain).

The <u>yield criterion</u> for plane strain:

$$\sigma_x + p = \frac{2\bar{\sigma}}{\sqrt{3}} \qquad (2.35)$$

is substituted in the force equilibrium equation (2.33):

$$\sigma_x \, dh + h\,d\sigma_x + \frac{4\bar{\sigma}}{\sqrt{3}} \tan \alpha \, dx - 2\,\sigma_x \tan \alpha \, dx = 0 \qquad (2.36)$$

It is:

$$dh = 2 \tan \alpha \cdot dx$$

$$\frac{dh}{2} = \tan \alpha \, dx$$

Therefore:

$$h\,d\sigma_x + \frac{2\bar{\sigma}}{\sqrt{3}} \, dh = 0 \qquad (2.37)$$

The <u>boundary conditions</u> are:

$$\text{at} \quad x = b \qquad h = h_b \qquad \sigma_x = 0$$

$$x = 0 \qquad h = h_a \qquad \sigma_x = \sigma_{xa}$$

therefore:

$$\int_0^{\sigma_{xa}} d\sigma_x = \frac{2\bar{\sigma}}{\sqrt{3}} \int_{h_b}^{h_a} \frac{dh}{h}$$

$$\sigma_{xa} = \frac{2\bar{\sigma}}{\sqrt{3}} \ln \frac{h_a}{h_b} \qquad (2.38)$$

2.4.2.2 Lateral Flow between Parallel Dies

When upsetting in plane strain between parallel dies, the stress distribution σ_y obtained due to the lateral flow is given by (following Figure 2.11):

$$\frac{d\sigma_y}{dx} + \frac{2\tau}{h} = 0 \tag{2.39}$$

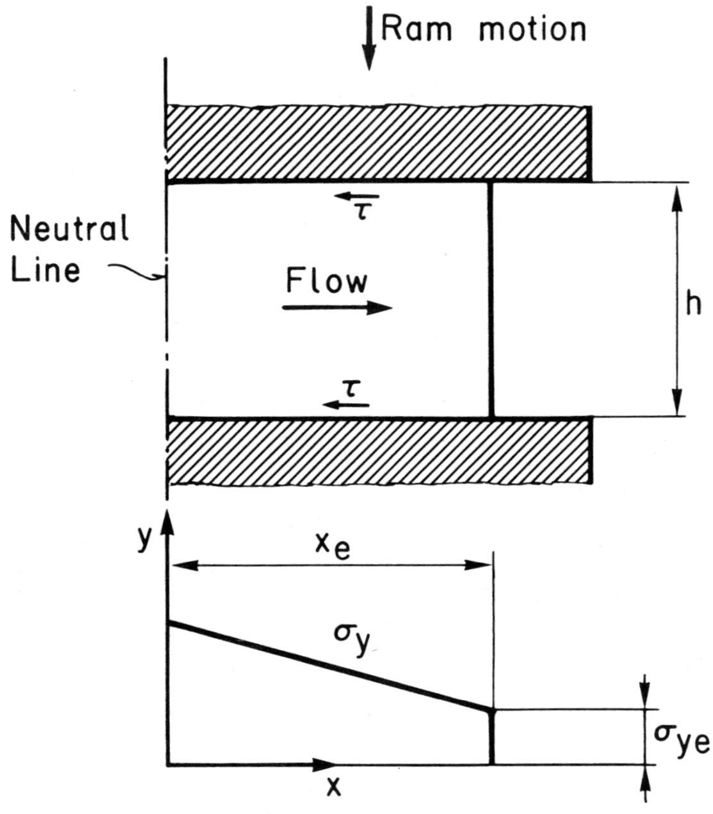

<u>Figure 2.11</u>: Plane strain lateral flow and stress σ_y distribution between flat parallel dies.

With the boundary conditions $\sigma_y = \sigma_{ye}$ at $x = x_e$ and integrating:

$$\sigma_y = \frac{2\tau}{h} (x_e - x) + \sigma_{ye} \qquad \text{is obtained.} \tag{2.40}$$

Integrating over the deformation zone, the load per unit length:

$$P = \int_0^{x_e} \sigma_y \, dx = x_e \left(\sigma_{ye} + \frac{\tau \cdot x_e}{h} \right) \quad \text{is obtained.} \quad (2.41)$$

2.4.2.3 Plane Strain Lateral Flow between Inclined Dies

A general deformation zone ABCD as shown in Figure 2.12 with all stresses acting upon it is considered. If the static equilibrium of forces in the x direction is expressed:

$$\sigma_x \cdot h - (\sigma_x + d\sigma_x)(h + dh) - \tau \frac{dx}{\cos \alpha} \cos \alpha -$$

$$- \tau \frac{dx}{\cos \beta} \cos \beta + p_u \frac{dx}{\cos \alpha} \cdot \sin \alpha +$$

$$+ p_l \frac{dx}{\cos \beta} \sin \beta = 0 \qquad (2.4.2)$$

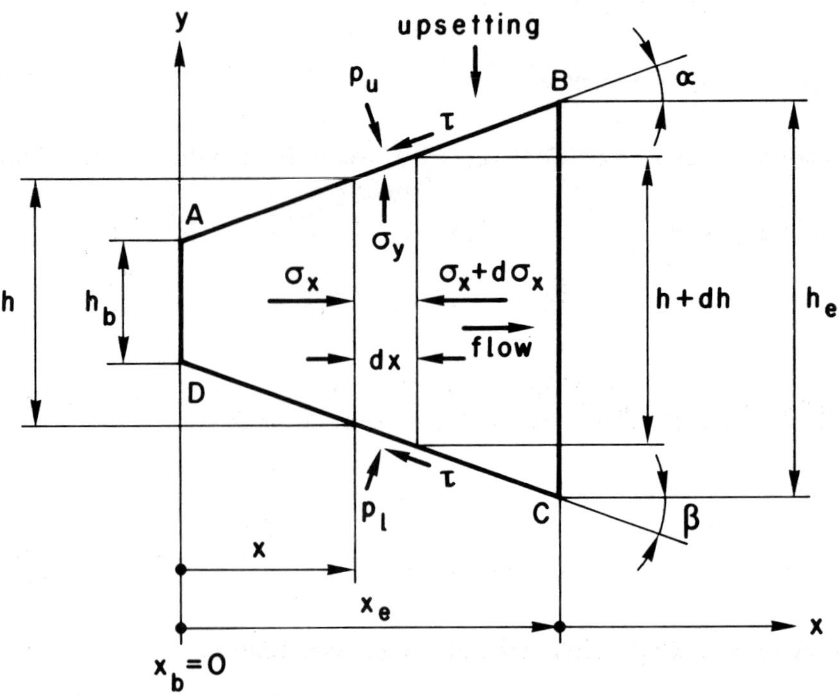

Figure 2.12: Slab method principle: general plain strain case.

Ignoring higher order differentials:

$$- \sigma_x dh - h d\sigma_x - \tau dx - \tau dx + dx \, (p_u \tan \alpha +$$

$$+ p_1 \tan \beta) \; = \; 0 \qquad (2.43)$$

where the subscripts "u" and "l" refer to upper and lower plates, respectively. The subscripts "b" and "e" refer to the beginning and end of the deformation zone considered, respectively.

The static equilibrium in the y direction gives, at the upper interface:

$$\sigma_y \, dx - p_u \frac{dx}{\cos \alpha} \cos \alpha - \tau \frac{dx}{\cos \alpha} \sin \alpha \; = \; 0 \qquad (2.44)$$

or:

$$p_u \; = \; \sigma_y - \tau \tan \alpha \qquad (2.45)$$

and at the lower interface:

$$p_1 \; = \; \sigma_y - \tau \tan \beta \qquad (2.46)$$

Introducing 2.46 and 2.45 in 2.43, the following is obtained:

$$- h d\sigma_x - \sigma_x dh - 2\tau dx + dx \, [\sigma_y \tan \alpha - \tau \tan^2 \alpha +$$

$$+ \sigma_y \tan \beta - \tau \tan^2 \beta] \; = \; 0 \qquad (2.47)$$

From the geometry of the deformation element it is:

$$dh \; = \; dx \, (\tan \alpha + \tan \beta) \qquad (2.48)$$

$$h \; = \; h_b + x \, (\tan \alpha + \tan \beta) \qquad (2.49)$$

and from material flow according to von Mises:

$$\sigma_y - \sigma_x \; = \; \frac{2}{\sqrt{3}} \, \bar{\sigma} \qquad (2.50)$$

and:

$$d\sigma_y = d\sigma_x \tag{2.51}$$

Therefore equation 2.47 becomes, after simplification:

$$[- h_b - x (\tan \alpha + \tan \beta)] \, d\sigma_y + dx (\tan \alpha + \tan \beta) (\sigma_y - \sigma_x) - [2\tau + \tau (\tan^2 \alpha + \tan^2 \beta)] \, dx = 0 \tag{2.52}$$

If:

$$K_1 = \tan \alpha + \tan \beta \tag{2.53}$$

$$K_2 = - \frac{2}{\sqrt{3}} \bar{\sigma} K_1 + \tau (2 + \tan^2 \alpha + \tan^2 \beta) \tag{2.54}$$

is introduced, this gives:

$$\frac{d\sigma_y}{dx} = - \frac{K_2}{h_b + K_1 x} \tag{2.55}$$

The boundary conditions are:

$$\sigma_y = \sigma_{ye} \qquad \text{at} \qquad x = x_e$$

and integrating equation 2.55:

$$\sigma_y = \int_o^{x_e} \left(\frac{- K_2}{h_b + K_1 x} \right) dx$$

resulting in:

$$\sigma_y = \left[- K_2 \left(\frac{1}{K_1} \right) \ln (h_b + K_1 x) + C \right]_o^{x_e} \tag{2.56}$$

Applying the boundary conditions and knowing that:

$$h_e = h_b + K_1 x_e \tag{2.57}$$

the constant of integration:

$$C = \sigma_{ye} + \frac{K_2}{K_1} \ln h_e \qquad \text{is arrived at.} \tag{2.58}$$

Therefore the final solution:

$$\sigma_y = \frac{K_2}{K_1} \ln \left(\frac{h_e}{h_b + K_1 x} \right) + \sigma_{ye} \qquad \text{is obtained.} \tag{2.59}$$

Equation (2.59) gives the σ_y stress distribution over the deformation zone in Figure 2.12. This distribution may increase or decrease from the end to the beginning of the deformation zone in accordance with the value of the angles α and β (see Figures 2.13, 2.14 and equations 2.53, 2.54 and 2.55) and with the magnitude of the friction shear stress τ.

Integrating the axial stress σ_y over the deformation zone gives the total compression load per unit length:

$$P = \int_{x_b=0}^{x_e} \sigma_y \, dx \tag{2.60}$$

Solving (2.60):

$$P = -\frac{K_2}{K_1} \int_{x_b=0}^{x_e} \ln (h_b + K_1 x) \, dx +$$

$$+ \int_{x_b=0}^{x_e} \left(\sigma_{ye} + \frac{K_2}{K_1} \ln h_e \right) dx$$

and finally:

$$P = -\frac{K_2}{K_1} \cdot \frac{1}{K_1} \left[h_e (\ln h_e - 1) - h_b (\ln h_b - 1) \right] +$$

$$+ \left(\sigma_{ye} + \frac{K_2}{K_1} \ln h_e \right) x_e \qquad\qquad (2.61)$$

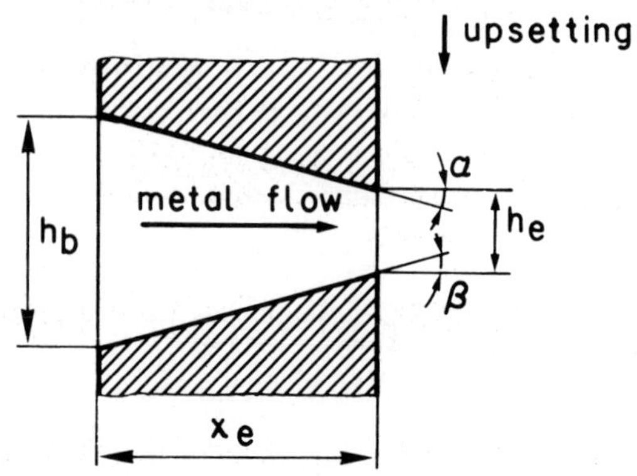

a) Converging Flow $(\alpha < 0,\ \beta < 0)$

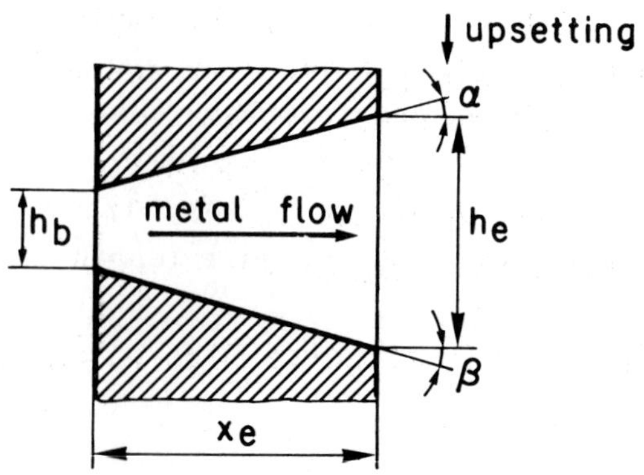

b) Diverging Flow $(\alpha > 0,\ \beta > 0)$

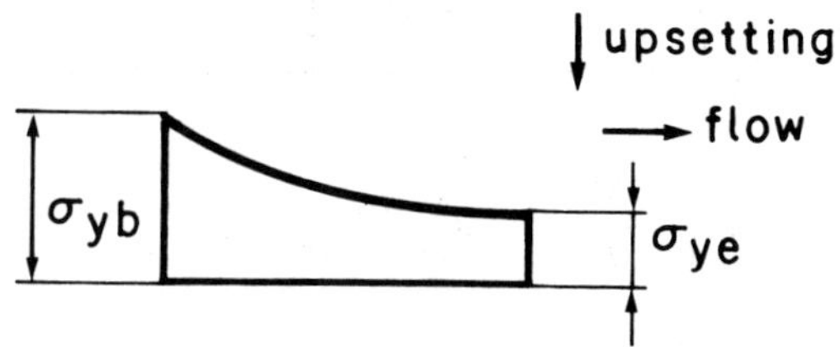

c) Stress Distribution

Figure 2.13: Deformation units for upset forging between inclined plates for predominantly horizontal (converging or diverging) metal flow.

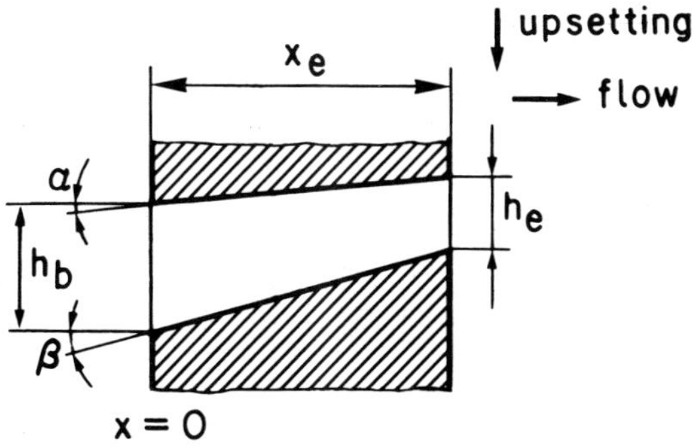

a) Ascending Flow $(\alpha > 0, \beta < 0)$

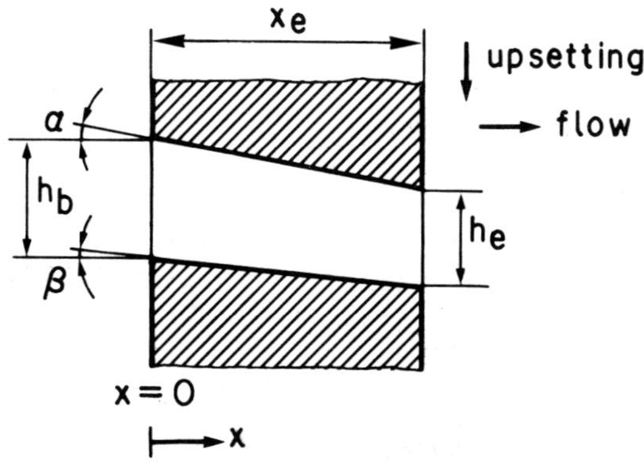

b) Descending Flow $(\alpha < 0, \beta > 0)$

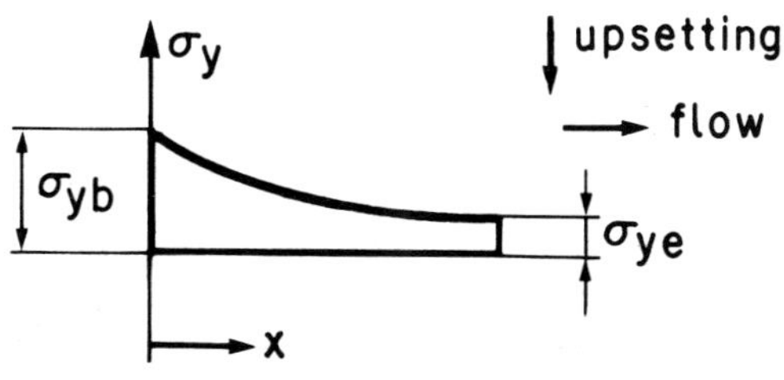

c) Stress Distribution

Figure 2.14: Deformation units for upset forging between inclined plates for ascending and descending metal flow.

2.4.2.4 <u>Flow into a Rib</u>

The geometric variables and the stress distribution are given in
Figure 2.15.

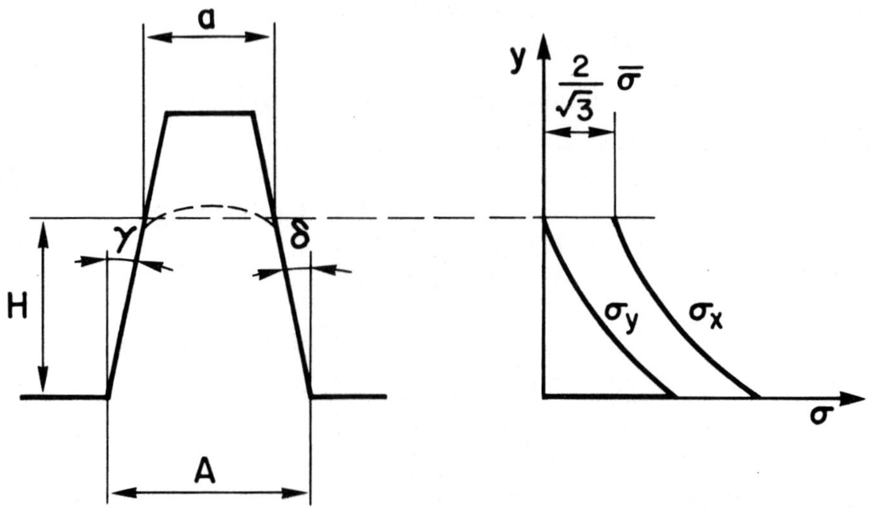

<u>Figure 2.15</u>: Flow into a rib: geometric variable and stress
distribution.

The stresses σ_x and σ_y in the rib can be expressed as:

$$\sigma_x = \frac{K_2}{K_1} \ln \left(\frac{a}{A + K_1 y} \right) + \sigma_{xe} \tag{2.62}$$

$$\sigma_y = \sigma_x - \frac{2}{\sqrt{3}} \bar{\sigma} \tag{2.63}$$

where:

$$K_1 = -(\tan \gamma + \tan \delta) \tag{2.64}$$

$$K_2 = -\frac{2}{\sqrt{3}} \bar{\sigma} K_1 + \mu\bar{\sigma} (2 + \tan^2 \gamma + \tan^2 \delta) \tag{2.65}$$

o   For $y = H$ it is $\sigma_{ye} = 0$

therefore: $\sigma_{xe} = \dfrac{2}{\sqrt{3}} \bar{\sigma}$

o   For $y = 0$ it is:

$$\sigma_x = \frac{K_2}{K_1} \ln \left(\frac{a}{A}\right) + \sigma_{xe}$$

but $a = A + K_1 H$ therefore:

$$\sigma_x = \frac{K_2}{K_1} \ln \left(\frac{A + K_1 H}{A}\right) + \frac{2}{\sqrt{3}} \bar{\sigma} \tag{2.66}$$

Equation 2.66 can also be used to determine the pressure distribution normal to the rib and therefore could be used to calculate the stress distribution in the die.

Substituting (2.66) in (2.63) gives:

$$\sigma_y = \frac{K_2}{K_1} \ln \left(\frac{A + K_1 H}{A}\right) \tag{2.67}$$

2.4.3   The Axisymmetric Situation

2.4.3.1   Stress Distribution for the Flat Axisymmetric Case

Following Figure 2.16, the equations that establish the forces equilibrium for an angle $d\gamma$ are derived. The equilibrium in the r direction can be written:

$$\sigma_r \, r \, h \, d\gamma - (\sigma_r + d\sigma_r)(r + dr) \, d\gamma \cdot h +$$

$$+ \, 2\sigma_t \sin \frac{d\gamma}{2} \, h \, dr - 2 \, \tau \, d\gamma \, r \, dr \; = \; 0 \tag{2.68}$$

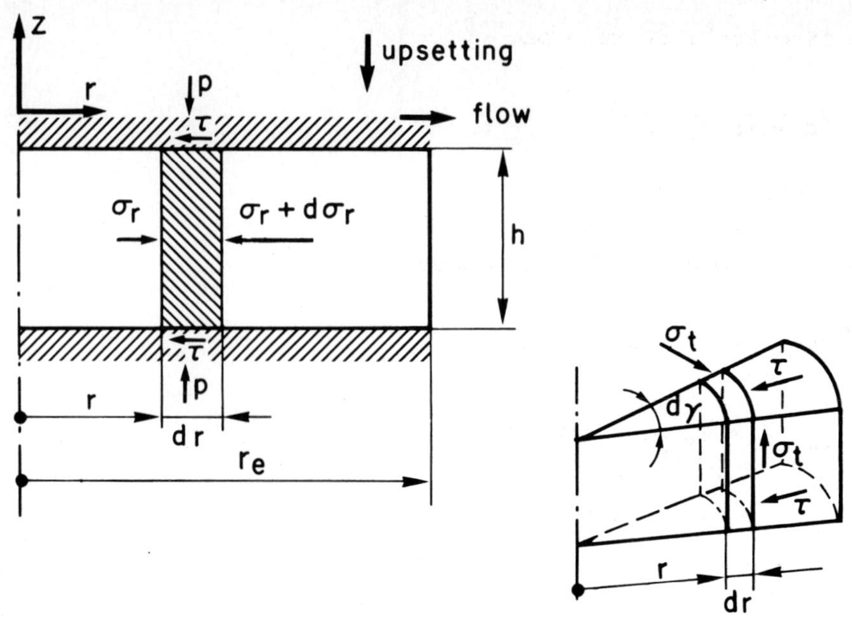

<u>Figure 2.16</u>: Axisymmetric element with flat parallel dies.

With $\sin \frac{d\gamma}{2} \simeq \frac{d\gamma}{2}$ and for uniform deformation:

$$\varepsilon_z \text{ (axial strain)} = -2\varepsilon_r \text{ (radial strain)} =$$

$$= -2\varepsilon_t \text{ (hoop strain)}$$

and therefore:

$$d\varepsilon_r = d\varepsilon_t \qquad \text{or} \qquad \sigma_r = \sigma_t.$$

Therefore equation (2.68) becomes:

$$\sigma_r \, r \, h \, d\gamma - \sigma_r \, r \, h \, d\gamma - \sigma_r \, dr \, d\gamma \, h - d\sigma_r \, r \, d\gamma \, h -$$

$$- d\sigma_r \, dr \, d\gamma \, h + 2 \, \sigma_t \frac{d\gamma}{2} \, h \, dr - 2 \, \tau \, d\gamma \, r \, dr = 0$$

Simplifying:

$$- h \, d\sigma_r - 2\tau \, dr = 0 \qquad\qquad (2.69)$$

44

but from the flow rule is:

$$\sigma_z - \sigma_r = -\bar{\sigma}$$

or: $\quad d\sigma_z = d\sigma_r$

Therefore equation (2.69) becomes:

$$h d\sigma_z = -2\tau dr$$

$$\frac{d\sigma_z}{dr} = -\frac{2\tau}{h}$$

Integrating:

$$\int_{\sigma_{ze}}^{\sigma_z} d\sigma_z = -\int_{r_e}^{r} \frac{2\tau}{h} dr$$

$$\sigma_z - \sigma_{ze} = -\frac{2\tau}{h} (r - r_e)$$

$$\sigma_z = \frac{2\tau}{h} (r_e - r) + \sigma_{ze} \tag{2.70}$$

For: $\quad r = 0 \quad$ is $\quad \sigma_z = \sigma_{max}$

$$\sigma_{max} = \sigma_{ze} + \frac{2\tau}{h} r_e$$

but: $\quad \tau = \mu \bar{\sigma} \quad$ and $\quad \sigma_{ze} = \bar{\sigma} \quad$ with $\quad 0 \leq \mu \leq 0.577$

$$\sigma_{max} = \bar{\sigma} + \frac{2 \mu \bar{\sigma}}{h} r_e$$

$$\sigma_{max} = \bar{\sigma} \left[1 + \mu \frac{D}{h} \right] \tag{2.71}$$

2.4.3.2 <u>Stress Distribution for Converging Outward Flow</u>

Following Figure 2.17, the static <u>equilibrium in the radial direction</u> gives:

$$(\sigma_r + d\sigma_r)\ r\ d\gamma\ (h + dh) - \sigma_r\ (r + dr)\ d\gamma\ h +$$

$$+ 2\sigma_t\ \sin\frac{d\gamma}{2}\ (h + \frac{dh}{2})\ dr - \tau\ \frac{d\gamma\ r\ dr}{\cos\alpha}\ \cos\alpha -$$

$$- \tau\ \frac{d\gamma\ r\ dr}{\cos\beta}\ \cos\beta - p_u\ \frac{r\ d\gamma\ dr}{\cos\alpha}\ \sin\alpha -$$

$$- p_l\ \frac{r\ d\gamma\ dr}{\cos\beta}\ \sin\beta\ =\ 0 \tag{2.72}$$

with: $\qquad \sin\dfrac{d\gamma}{2} \approx \dfrac{d\gamma}{2}\qquad$ and with:

(axial strain) $\varepsilon_z = - 2\varepsilon_r$ (radial strain) $= - 2\varepsilon_t$ (hoop strain)

or: $\quad d\varepsilon_r\ =\ d\varepsilon_t \qquad$ or $\qquad \sigma_r\ =\ \sigma_t$

and by ignoring the higher order differentials, equation (2.72) becomes:

$$\sigma_r\ dh + h\ d\sigma_r - 2\tau\ dr - dr\ (p_u\ \tan\alpha +$$

$$+ p_l\ \tan\beta)\ =\ 0 \tag{2.73}$$

Note that:

$$dh\ =\ dr\ (\tan\alpha + \tan\beta) \tag{2.74}$$

$$h\ =\ h_b - (r - r_b)\ (\tan\alpha + \tan\beta) - dh \tag{2.75}$$

$$1 + \tan^2\alpha\ =\ 1/\cos^2\alpha \tag{2.76}$$

$$1 + \tan^2\beta\ =\ 1/\cos^2\beta \tag{2.77}$$

and from the flow rule:

$$\sigma_z - \sigma_r = -\bar{\sigma} \qquad \text{and} \qquad d\sigma_z = d\sigma_r. \tag{2.78}$$

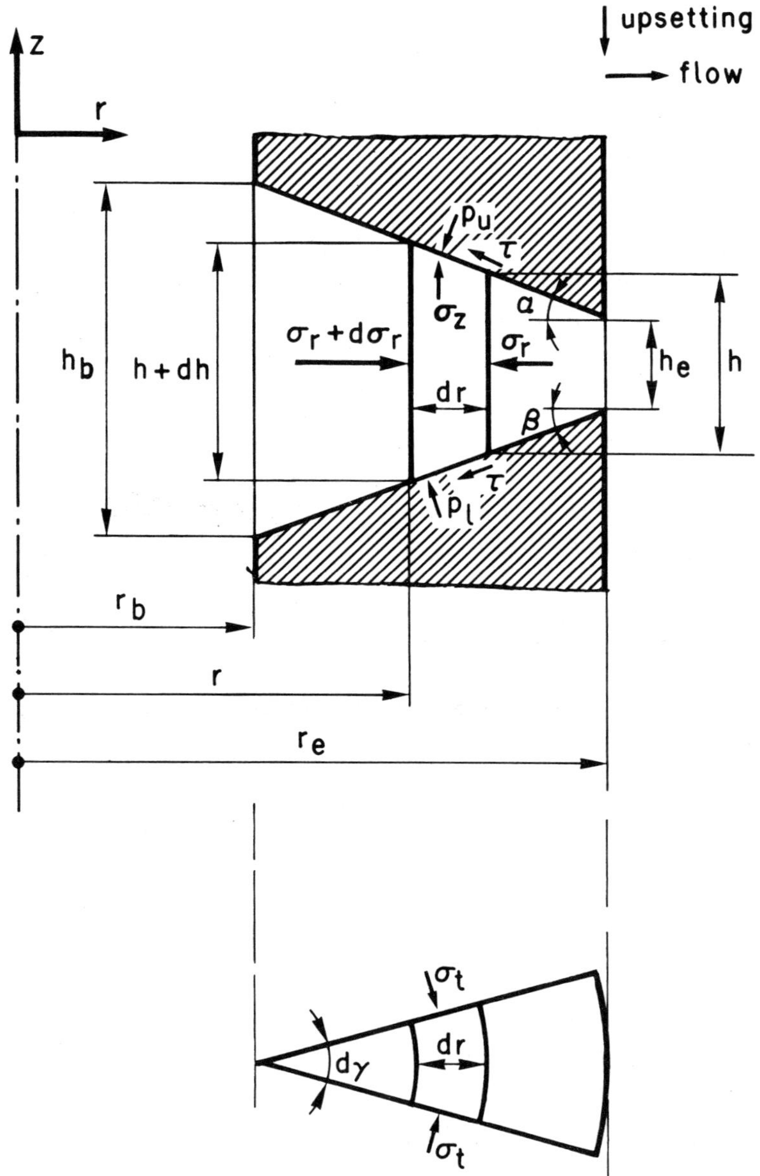

Figure 2.17: Axisymmetric element with inclined dies.

The <u>equilibrium in the z-direction</u> gives:

$$\sigma_z \, dr \, d\gamma \, r - p_u \, \frac{r \, d\gamma \, dr}{\cos \alpha} \, \cos \alpha \, +$$

$$+ \tau \, \frac{r \, d\gamma \, dr}{\cos \alpha} \, \sin \alpha \; = \; 0 \qquad\qquad (2.79)$$

Therefore at the upper interface, equation (2.79) gives:

$$p_u \; = \; \sigma_z + \tau \, \tan \alpha \qquad\qquad (2.80)$$

and at the lower interface:

$$p_1 \; = \; \sigma_z + \tau \, \tan \beta \qquad\qquad (2.81)$$

Introducing equations (2.74) to (2.77) and equations (2.80) and (2.81) in (2.73) produces:

$$\bar{\sigma} \, dr \, (\tan \alpha + \tan \beta) + \sigma_z \, dr \, (\tan \alpha + \tan \beta) + [h_b +$$

$$+ \, r_b \, (\tan \alpha + \tan \beta)] \, d\sigma_z - r \, (\tan \alpha + \tan \beta) \, d\sigma_z -$$

$$- \, dr \, (\tan \alpha + \tan \beta) \, d\sigma_z - 2\tau \, dr - dr \, \sigma_z \, (\tan \alpha +$$

$$+ \, \tan \beta) - dr \, \tau \, (\tan^2 \alpha + \tan^2 \beta) \; = \; 0 \qquad\qquad (2.82)$$

If it is:

$$K_1 \; = \; - \, (\tan \alpha + \tan \beta) \qquad\qquad (2.83)$$

equation (2.82) becomes:

$$- \, \bar{\sigma} \, K_1 \, dr + (h_b - r_b \, K_1) \, d\sigma_z + r \, K_1 \, d\sigma_z -$$

$$- \, dr \, \tau \, (1 + \tan^2 \alpha + 1 + \tan^2 \beta) \; = \; 0$$

If it is:

$$K_2 = -\bar{\sigma} K_1 - \tau \left(\frac{1}{\cos^2 \alpha} + \frac{1}{\cos^2 \beta} \right) \qquad (2.84)$$

$$K_3 = h_b - r_b K_1 \qquad (2.85)$$

then:

$$\frac{d\sigma_z}{dr} = -\frac{K_2}{K_3 + rK_1} \qquad \text{is obtained.} \qquad (2.86)$$

With $\sigma_z = \sigma_{ze}$ at $r = r_e$, the integration gives:

$$\sigma_z = \frac{K_2}{K_1} \ln \left(\frac{h_e}{K_3 + K_1 r} \right) + \sigma_{ze} \qquad (2.87)$$

The load can be obtained as:

$$\frac{P}{2\pi} = \int_{r_b}^{r_e} \sigma_z \, r \, dr \qquad (2.88)$$

2.5 Upper-Bound Method

A kinematically admissible displacement increment field is one
that satisfies the deformation requirements of Table 2.1. A
statically admissible stress field is one that satisfies the
stress requirements of Table 2.1.

Consider now for a material point the stress tensor σ_{ij} and the
strain increment tensor $d\varepsilon_{ij}$ solution of a problem, and the
stress tensor σ_{ij}^* obtained through equations (2.26) from a kinematically admissible strain increment tensor $d\varepsilon_{ij}^*$, Figure 2.18.

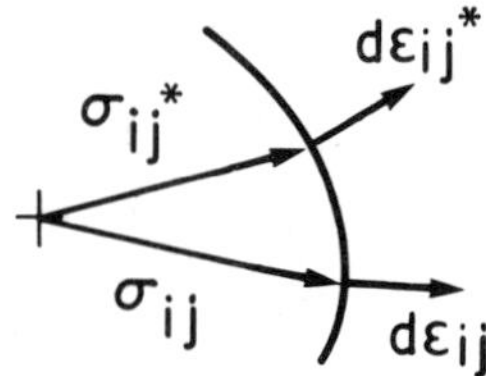

<u>Figure 2.18</u>:   Correct and kinematically admissible strain incre-
ments.

Convexity of yield surface gives $(\sigma_{ij}^* - \sigma_{ij})\, d\varepsilon_{ij}^* \geq 0$ and integrating over the volume:

$$\int_V (\sigma_{ij}^* - \sigma_{ij})\, d\varepsilon_{ij}^* \geq 0 \qquad\qquad (2.89)$$

Consider next, equilibrium equations without body forces (normal
situation in metal forming):

$$\sigma_{ij,j} = 0 \qquad\qquad (2.90)$$

and a kinematically admissible displacement increment field
du_i^*. By multiplying this displacement vector at any point by
equations (2.90) the scalar:

$$\sigma_{ij,j}\, du_i^* = 0 \qquad\qquad \text{is obtained,}$$

which integrated over the volume yields:

$$\int_V \sigma_{ij,j} \, du_i^* \, dV = 0$$

$$\int_V [(\sigma_{ij} du_i^*)_{,j} - (\sigma_{ij} \, du_{i,j}^*)] \, dV = 0$$

Applying the divergence theorem and symmetry of the stress tensor:

$$\int_S \sigma_{ij} \, du_i^* \, n_j \, dS - \int_V \sigma_{ij} \, du_{i,j}^* \, dV = 0$$

$$\int_S T_i \, du_i^* \, ds - \int_V \sigma_{ij} \, d\varepsilon_{ij}^* \, dV = 0$$

The meaning of this expression is that the work done by the stress field on a kinematically admissible displacement increment field equals the work done by the external loads on the same displacement field.

A feature of kinematically admissible displacement increment fields is that they allow for tangential discontinuities along surfaces. In such cases, the volume integral is interpreted as the summation of the volumes limited by external surfaces and discontinuity surfaces, and to the surface integral are added such discontinuity surfaces as:

$$\int_S T_i \, du_i^* \, dS - \int_{S_S} \tau \, |du^*| \, dS_S - \int_V \sigma_{ij} \, d\varepsilon_{ij}^* \, dV = 0 \qquad (2.91)$$

where τ is the tangential shear stress along the discontinuity, and $|du^*|$ the relative displacement increment.

When inequality (2.89) is used, (2.91) becomes:

$$\int_S T_i \, du_i^* \leq \int_V \sigma_{ij}^* \, d\varepsilon_{ij}^* \, dV + \int_{S_S} k \, |du^*| \, dS_S$$

where k is the shear yield stress of the material.

S is now separated into S_T and S_D, where $du_i = du_i^*$:

$$\int_{S_D} T_i \, du_i \, dS \leq \int_V \sigma^*_{ij} \, d\varepsilon^*_{ij} \, dV + \int_{S_S} \tau_Y \, |du^*| \, dS_S -$$

$$- \int_{S_T} T_i \, du^*_i \, dS_T \tag{2.92}$$

The physical meaning of expression (2.92) is as follows: for all possible kinematically admissible displacement fields, the work done by the associated stress field in the volume and surface discontinuities, minus the work done by the known forces along the stress boundary, equate to an amount of work that is greater than, or equal to, the work done by the unknown forces along the displacement boundary.

The derivations were made here in a manner consistent with the material presented before. Both sides of equation (2.92) can be divided by dt and the displacement increment field becomes a velocity field. The strain increments become strain rates. That is the form under which problems are typically solved:

$$\int_{S_D} T_i \, v_i \, dS \leq \int_V \sigma^*_{ij} \, \dot{\varepsilon}^*_{ij} \, dv + \int_{S_S} k \, |\Delta v^*| \, dS_S -$$

$$- \int_{S_T} v^*_i \, dS_T \tag{2.93}$$

It should also be noted that

$$\sigma^*_{ij} \, \dot{\varepsilon}^*_{ij} = \bar{\sigma}^* \, \dot{\bar{\varepsilon}}^* \tag{2.94}$$

This shows that for an <u>ideally plastic</u> material it is sufficient to know the velocity field (strain rate field) to compute the power in the deformation zone.

In terms of energy rates (2.93) becomes:

$$\dot{U}_{\text{unknown external}} \leq \dot{U}_{\text{internal}} + \dot{U}_{\text{redundant (shear)}} + \tag{2.95}$$

$$+ \, \dot{U}_{\text{known external}}$$

where the last term is typically friction.

To understand the redundant work better, a boundary of velocity
discontinuity as shown in Figure 2.19 must be introduced.

The normal components of velocities have to be constant (conservation of volume).

The velocity discontinuity is:

$$\Delta v \; = \; v_{t2} - v_{t1} \qquad\qquad (2.96)$$

which gives rise to shear within the formed metal, that is, to
shear stress:

$$k \; = \; \frac{\bar{\sigma}}{\sqrt{3}} \qquad\qquad (2.97)$$

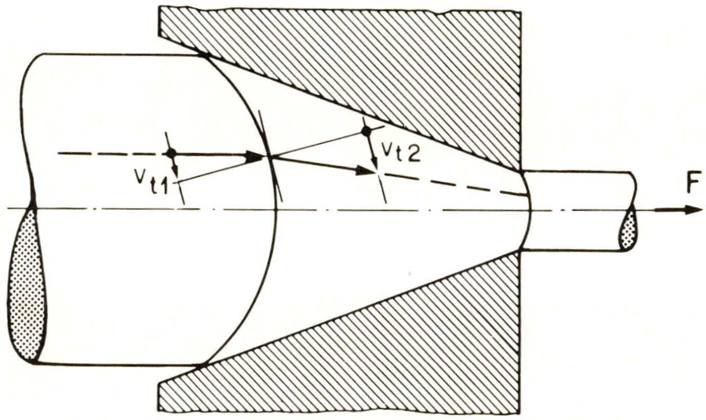

<u>Figure 2.19</u>: Velocity discontinuity at the boundary of metal
deformation zone.

and therefore:

$$\dot{U}_{red} \; = \; \int_{S} k|\Delta v|dS_{S} \qquad\qquad (2.98)$$

This upper-bound theorem is widely applied in metal forming to determine a load estimate to perform a certain operation. This load estimate is always greater than the real one. Families of kinematically admissible velocity fields, defined by a finite number of parameters left as variables are considered. For each, the right-hand side of inequality (2.93) is calculated, and a minimum is found with respect to the above-mentioned variables. This minimum gives the closest load estimate within such families. These estimates (and the particular velocity fields associated with them) are more or less close to reality depending on the "quality" of the fields initially proposed as solutions, and then depend a lot on the engineering "feeling" and experience of the user. Therefore, this method, which is very handy and relatively easy to apply, can only be implemented using separate computer programs for different problems (configurations) where families of fields have already been created by the programmer.

Table 2.1, shows that this method, which is based on the deformation requirements only, satisfies the stress boundary conditions and uses the flow rule to compute stresses. There are no guarantees that these stresses are the correct ones.

In upsetting of cylindrical preforms, the following velocity field, which shows the use of parameters, can be used [2.11]:

$$\text{axial velocity} \qquad v = -2Fz\,(1 - \beta z^2/3) \qquad (2.99)$$

$$\text{radial velocity} \qquad u = F\,(1 - \beta z^2)\,r \qquad (2.100)$$

$$\text{where} \qquad F = \frac{v_o}{2H\,(1 - \beta\,\frac{H^2}{12})} \qquad (2.101)$$

$$v_o = \text{axial velocity of the die}$$

$$H = \text{Instantaneous height of the forging}$$

The parameter β is determined by minimizing the total energy required for incremental forging.

Another example of the use of parameters for optimization of velocity field can be found in the first known solution of a non-axisymmetric velocity field for drawing of round section to square section with a corner radius [2.18].

2.6 Finite Element Analysis

2.6.1 Introduction and Historical Perspective

The Finite Element method, called as such by Clough [2.19], was a part of the overall effort of both engineers and mathematicians to discretize continuum problems. Early applications were in structural analysis, in which field original developments were made. As in many other fields, once engineers began developments, mathematicians moved in to explain and provide a framework of behavior from a theoretical point of view. Shortly thereafter, the mathematical foundations of this numerical method for the solution of field problems led the way to the expansion into other fields of solid mechanics, fluid mechanics, heat transfer, and so on.

The development of the FEM followed. It is a natural consequence of the availability of the digital computer and of reasonably large amounts of memory space. In fact, while hand (or table) calculations require models that provide more or less closed form solutions - even if at the expense of extremely large and complex, but few, mathematical expressions - the computer can handle an extremely large number of simple operations. Therefore, discretization methods which consider a large number of values, calculated from as many relatively simple relations between them, flourished with the new tool available.

Another important point of view is that while closed form solutions are obtained for a specific geometry and usually allow for a limited number of variables, discretization procedures, once established, permit the solution of a much broader class of problems and configuration variations. Of course, not all is on the bright side. Parametric studies are harder to perform. Because all calculations are made with actual figures, one solution always corresponds to a specific case.

The Finite Element method appeared in the metal forming field in the early seventies. Structural analysts started to analyze the plastification stages. Infinitesimal elasto-plastic formulations suitable for small deformations were available. Naturally, the first applications followed this route [2.20, 2.21], but soon it became evident that metal forming required more sophistication. In fact, not only were there very large amounts of deformation, but, in most cases, very large displacements and rotations of the material particles. In such cases, an infinitesimal deformation formulation could not work because the elastic part of the deformation needed to be referred to an original configuration. On the other hand, the constitutive equations produced first order differential equations that posed problems in the time step for their time integration. At this point, two approaches were taken which have since then evolved considerably and independently to achieve the same goals.

The first proceeded with an elasto-plastic analysis in which large deformation formulations were developed. To this day, there are still theoretical developments being made due to the complexity of the variables involved. However, the updated Lagrangean forms using appropriate time integration schemes today produce remarkably good solutions which are no longer overly expensive.

The second, following the tradition of other approximate solutions in metal forming, simply discarded the elastic deformations as being negligible when compared with the plastic ones and considered the material a rigid (thermo-visco) plastic one. This formulation, in exchange for elastic deformations, residual stresses, and a certain indetermination in the definition of rigid zones (and stresses inside them), obtained simple equations in terms of instantaneous velocities and strain rates which posed no problems in updating through time steps and has proved to be very well behaved in extremely harsh deformation conditions. This has liberated its users to address earlier problems specific to the metal forming area, and therefore reach faster solutions, more interesting from the industrial point-of-view.

It should be mentioned that today, from the purely mechanical point-of-view, the solutions obtained by both of the above schools can go equally far, therefore the updated Lagrangean formulation is more complete. It is in the interface with related fields - thermal analysis, die analysis, materials, and CAD/CAM -

that the rigid-plastic approach has a lead because of its earlier achievements.

The authors of this book have been involved with the rigid-plastic formulation over a long period of time. Thus, the applications that are presented and analyzed here have been obtained with programs implementing this approach.

This section (2.6) contains a description of the ingredients used in Finite Element Process Modelling today. First a very succinct background is given on what finite elements are. Then the mathematical formulations used are presented, followed by the Finite Element equations they lead to. Next, a series of details necessary for the success of process modelling applications are emphasized. Finally, some comments are made on where these techniques interface with the design and production process.

2.6.2   Finite Element Background

The Finite Element Method is a series of numerical techniques used in solving boundary value problems, initial value problems and eigenvalue problems. A full book would be necessary to completely describe the theory and the developments that have been made to date, and have made possible the type of analysis described later in this book. Fortunately, books on the subject are abundant, of which [2.22-2.24] are examples. The reader is referred to them to clarify the theoretical background and standard procedures. This work attempts to highlight as many standard features as possible in the process of hand working an example. It is hoped that this will provide enough of a base for the reader as the specific formulations for metal forming analysis are entered into.

The basic process in finite elements is to divide the domain into pieces or elements, approximate the equations governing the problem for each element as a function of selected values, put all elements together, apply boundary conditions to some of those selected values, and solve for such values. Once this has been done, the element equations are referred to in order to obtain the approximate solution variables of the problem.

Assume that with finite elements, an analysis is made of the one-dimensional infinitesimal boundary value problem of stretching a bar made of a linear elastic material by imposing a load L at one end, and fixing the other end, Figure 2.20. For simplicity's sake, the bar is divided into two elements, Figure 2.21. Each element contains two nodes. Figure 2.21 shows the overall configuration and the configuration of the two local elements.

Figure 2.20: Bar stretching.

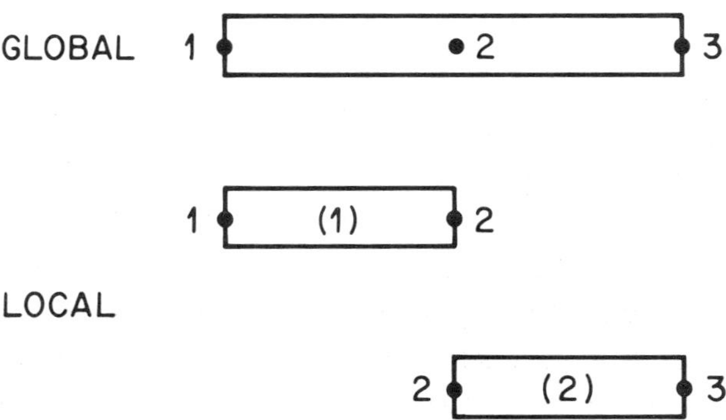

Figure 2.21: Bar division into elements.

The essence of the Finite Element theory is that everything can be described in terms of what happens at the nodes. Firstly a basic variable - in this case, a displacement u - is selected. When its field is known, strains and, through constitutive equations, stresses are then calculated. Surface integration of stresses produces loads. It is, therefore, possible to determine u, and, if this has been correctly done, everything else is derivable. Instead of dealing with the infinite number of u fields possible (kinematically admissible), the infinite number of vectors u_i, where the subscript i denotes node i (thus, $\underset{\sim}{u}$ is a three-dimensional vector), can now be dealt with.

A definition is then made of a field within each element that is uniquely defined by $\underset{\sim}{u}^{(j)}$-vector of nodal variables belonging to element j, in this case, a linear one, such as:

$$u^{(j)} = u_2^{(j)} \frac{x - x_1^{(j)}}{x_2^{(j)} - x_1^{(j)}} + u_1^{(j)} \frac{x_2^{(j)} - x}{x_2^{(j)} - x_1^{(j)}} \qquad (2.103)$$

Equation (2.103) is usually written in the form:

$$u^{(j)} = [N_1 \ N_2]^{(j)} \left\{ \begin{matrix} u_1 \\ u_2 \end{matrix} \right\}^{(j)} \qquad (2.104)$$

where N_1 and N_2 are called shape functions or interpolation functions. These functions are shown at the element level in Figure 2.22; at the global level, in the present case, they are shown in Figure 2.23. The strain is calculated at the element level by:

$$\varepsilon = u_{x,x} = [N_{1,x} \ N_{2,x}]^{(j)} \left\{ \begin{matrix} u_1 \\ u_2 \end{matrix} \right\}^{(j)} \qquad (2.105)$$

$$= \frac{u_2^{(j)} - u_1^{(j)}}{x_2^{(j)} - x_1^{(j)}} \qquad \text{in which } N_{2,x} = -N_{1,x}$$

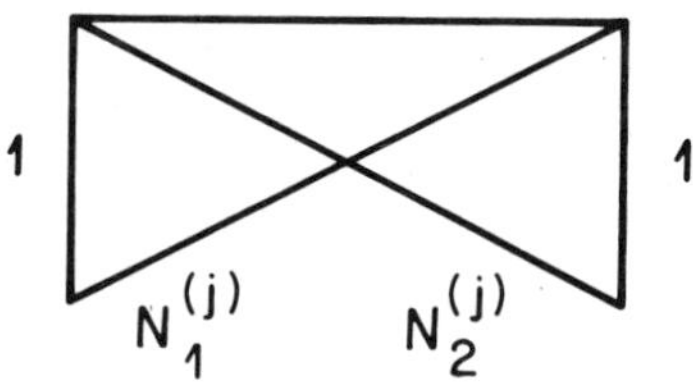

Figure 2.22: Element shape functions.

In a one-dimensional case, Hooke's law produces:

$$\sigma = E \varepsilon \qquad (2.106)$$

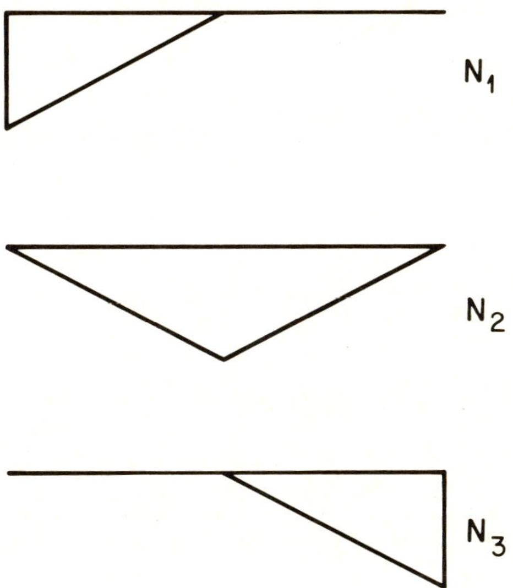

Figure 2.23: Global shape functions.

and the force on the node (2) of element (j) is:

$$F_2^{(j)} = A^{(j)} \sigma \qquad \text{A - cross-sectional area}$$

Equilibrium of the element requires that:

$$F_1^{(j)} + F_2^{(j)} = 0$$

therefore:

$$F_1^{(j)} = - A^{(j)} \sigma$$

or:

$$\left\{ \begin{matrix} F_1 \\ F_2 \end{matrix} \right\}^{(j)} = A^{(j)} E \left[\begin{matrix} -N_{1,x} & -N_{2,x} \\ +N_{1,x} & +N_{2,x} \end{matrix} \right] \left\{ \begin{matrix} u_1 \\ u_2 \end{matrix} \right\}^{(j)}$$

By making $k^{(j)} = A^{(j)} E N_{2,x}$:

$$\left\{ \begin{matrix} F_1 \\ F_2 \end{matrix} \right\}^{(j)} = \left[\begin{matrix} k & -k \\ -k & k \end{matrix} \right]^{(j)} \left\{ \begin{matrix} u_1 \\ u_2 \end{matrix} \right\}^{(j)}$$

or: $\underset{\sim}{F}^{(j)} = \underline{K}^{(j)} \underset{\sim}{u}^{(j)}$

This gives the equations of the two elements:

$$F_1^{(1)} = k^{(1)} u_1^{(1)} - k^{(1)} u_2^{(1)}$$

$$F_2^{(1)} = -k^{(1)} u_1^{(1)} + k^{(1)} u_2^{(1)}$$

$$F_1^{(2)} = k^{(2)} u_1^{(2)} - k^{(2)} u_2^{(2)}$$

$$F_2^{(2)} = -k^{(2)} u_1^{(2)} + k^{(2)} u_2^{(2)}$$

and in order to put everything together, or to assemble it, the following equations are imposed:

$$u_2^{(1)} = u_1^{(2)}$$

$$F_2^{(1)} + F_1^{(2)} - F_2 = 0 \qquad F_1^{(1)} - F_1 = 0 \qquad F_2^{(2)} - F_3 = 0$$

where F_i is an externally applied force at node i. In this case: $F_2 = 0$, $F_3 = L$, and F_1 is unknown, producing:

$$\begin{Bmatrix} F_1 \\ 0 \\ F_3 \end{Bmatrix} = \begin{bmatrix} k^{(1)} & -k^{(1)} & 0 \\ -k^{(1)} & k^{(1)} + k^{(2)} & -k^{(2)} \\ 0 & -k^{(2)} & k^{(2)} \end{bmatrix} \cdot \begin{Bmatrix} u_1 \\ u_2 \\ u_3 \end{Bmatrix}$$

or $\underline{K} \cdot \underset{\sim}{u} = \underset{\sim}{F}$

As u_1 is known and equal to zero its equation can be eliminated as follows:

$$\begin{bmatrix} 1 & 0 & 0 \\ 0 & k^{(1)} + k^{(2)} & -k^{(2)} \\ 0 & -k^{(2)} & k^{(2)} \end{bmatrix} \begin{Bmatrix} u_1 \\ u_2 \\ u_3 \end{Bmatrix} = \begin{Bmatrix} 0 \\ 0 + k^{(1)} \cdot 0 \\ F_3 - 0 \cdot 0 \end{Bmatrix}$$

Note that the unknown force corresponded to a known displacement. The two known forces correspond to unknown displacements. Any degree of freedom will need to have specified either displacement or force. Note also that the system of equations obtained is symmetric. There are a number of numerical procedures to solve such a system of equations based on the Gaussian elimination. The most-used today are the skyline method and the frontal method, both computationally very effective. After $\underset{\sim}{u}$ has been established then:

$$\underline{K} \cdot \underset{\sim}{u} = \underset{\sim}{F} \tag{2.107}$$

gives the unknown forces. Of course, it should also reproduce the imposed forces. Equation (2.105) and (2.106) produce the strains and stresses at the elements.

This haphazard way of proceeding is a natural approach and was, in fact, the process used in the early sixties. In less simple problems, it may not be so straightforward to proceed. To do that, there are other mathematical ways of looking into finite elements.

Consider a functional:

$$J = \int (\tfrac{1}{2} E A \varepsilon^2) \, dx - L \, u_{|x_3} \tag{2.108}$$

which happens to represent the total potential energy of the bar. The first term is the elastic strain energy; the second, the externally applied force potential energy. Considering the invariance of such a functional with respect to the displacement field u, by making its first variation equal to zero and integrating by parts:

$$\delta J = 0$$

$$\delta J = \int EA\varepsilon \, \delta\varepsilon \, dx - L \, \delta u_{|x_3} = 0$$

$$= \int A \, \sigma \, \delta u,_x \, dx - L \, \delta u_{|x_3} = 0$$

$$= A \, \sigma \, \delta u_{|x_3} - A \, \sigma \, \delta u_{|x_1} - L \, \delta u_{|x_3} - \int A \, \sigma,_x \, \delta u \, dx = 0$$

This gives the field u which produces a stationary value of J if, for any arbitrary variation δu, the first variation of J, δJ is zero. From the arbitrariness of δu, the following equations are obtained:

$$A\,\sigma,_x = 0 \qquad\qquad \text{along the bar} \qquad\qquad (2.109)$$

$$A\,\sigma = L \qquad\qquad \text{at } x = x_3 \qquad\qquad (2.110)$$

$$A\,\sigma = 0 \qquad\qquad \text{at } x = x_1 \qquad\qquad (2.111)$$

Equation (2.109) is the equilibrium equation, and equation (2.110) is the boundary condition $F_3 = L$. Therefore, the stationary situation of the potential energy almost solves the problem. The missing point was at the boundary $x = x_1$. However, if all possible displacement fields are restricted to the kinematically admissible ones, then $u = 0$ at $x = x_1$ and $\delta u = 0$ at $x = x_1$. Equation (2.111) disappears then and $A\,\sigma$ will be whatever it will have to be. This is the variational approach to the BVP: if it is possible to construct an appropriate functional of the field variable that satisfies some lower order boundary conditions (also called essential boundary conditions), the value of the variable that makes the functional stationary will satisfy the differential equation over the domain and, automatically, the higher order boundary conditions (also called suppressible boundary conditions). The Finite Element approximation consists of substituting all possible field variables u by a combination of variables defined by the nodal values:

$$u = \underline{N} \cdot \underset{\sim}{u}$$

The functional $J\,(u)$ becomes a function of nodal values $J_1\,(\underset{\sim}{u})$. The stationary situation of J_1 is, hopefully, an approximation to the stationary situation of J. Such is obtained by algebra:

$$dJ_1 = \frac{\partial J_1}{\partial u_i}\,du_i = 0$$

or a set of equations:

$$\frac{\partial J_1}{\partial\,\underset{\sim}{u}} = \underset{\sim}{0}\;. \qquad\qquad (2.112)$$

In our specific case:

$$J_1 = \int \frac{1}{2} E A \, \underset{\sim}{u}^T \, \underline{N}^T,_x \, \underline{N}'_x \, \underset{\sim}{u} dx - L \, \underline{N} \, \underset{\sim}{u} \big|_{x_3} \qquad (2.113)$$

and it is left to the reader to find that $dJ_1 = 0$ yields as before:

$$\underline{K} \, \underset{\sim}{u} = \underset{\sim}{F} \qquad (2.114)$$

To make sure that it is possible to calculate the equations at the element level and then add them, it must be ensured that in J_1 the total integral is the sum of element integrals. It is noted that equilibrium equation corresponds to a second derivative of u, and the problem is said to be of the order of 2p with p = 1. The functional has derivatives of the first order, or p. The shape functions must have continuous derivatives at least to order p within an element and continuous derivatives at least to order p-1 through the boundaries. The first requirement is referred to as completeness; the second, compatibility. In our particular case, compatibility means to make sure the forces applied at each node are well-defined.

Another way of looking at Finite Element approximations is through a weighted residuals method, in particular, a Galerkin one. Assuming again that are considered, displacement fields that satisfy essential boundary conditions, an approximate field then produces an error within the volume:

$$\varepsilon_I = \sigma,_x \qquad (2.115)$$

which must be minimized in some way. With the Galerkin method, a small variation δu produced by a variation of the nodal values is considered:

$$\delta u = N_i \cdot \delta u_i$$

Such a small variation will produce a variation in the error $\delta \varepsilon_I$, and the best approximation when that error becomes zero on the whole volume is considered:

$$\int \varepsilon_I N_i \, \delta u_i \, dV = 0 \qquad (2.116)$$

which produces a set of equations equal to the number of nodal
variables, as each can vary independently.

Substituting (2.115) for (2.116):

$$\int A\, \sigma,_x\, N_i\, \delta u_i\, dx\ =\ 0\ .$$

In order to reduce the order of the derivatives the following can
be integrated by parts:

$$\int A\, \sigma\, N_{i,x}\, \delta u_i\ dx\ -\ A\, \sigma\, N_i\, \delta u_i\Big|_{x_3}\ =\ 0$$

(note that $\delta u_i = 0$ at $x = x_1$)

or using (2.105) and (2.106):

$$\int A\, E\, \delta \underset{\sim}{u}^T\, \underline{N},_x^T\, \underline{N},_x\, \underset{\sim}{u}\ dx\ -\ L\, \delta \underset{\sim}{u}^T\, \underline{N}^T\Big|_{x_3}\ =\ 0\ .$$

Now the arbitrariness in $\delta \underset{\sim}{u}$ gives the set of equations:

$$\int A\, E\ \underline{N},_x^T\ \underline{N},_x \underset{\sim}{u}\, dx\ -\ L\, \underline{N}^T\Big|_{x_3}\ =\ \underset{\sim}{0}$$

which is equal to:

$$\underline{K}\, \underset{\sim}{u}\ =\ \underset{\sim}{F} \qquad\qquad\qquad (2.107)\ \text{rep.}$$

as before. The main advantage of looking at the problem from this
perspective is that there is no need to worry about finding a
functional. In fact, for many problems, it does not exist.

It has already been stated that provided completeness and compatibility are satisfied, work can be done at the element level,
all necessary integrals calculated and then added up or as-
sembled. In the example described, the integrals are very easy
to calculate. However, in most applications, they are very difficult, if not impossible, to obtain in closed form, hence the
common procedure of numerical integration. Simply stated, the
integral by a weighted sum of function values:

$$\int f(x) \ dx \ = \ \sum_{i=1}^{N} W_i \ f(p_i) \tag{2.117}$$

where N is the number of integration points, W_i are the weights, and p_i are the locations of the integration points. Numerical analysis textbooks provide these locations and weights, as well as the accuracy obtainable with the procedure. An added advantage of numeric integration is that if all the equations are expressed in dimensionless local (natural) coordinates, it is very easy to program and automate the integrations.

2.6.3 Stress-Strain Formulations

As far as deformation is concerned in process modelling, a solid mechanics boundary value problem represented in general terms in Figure 2.3 is being dealt with. Here there is a body of volume V in which compatibility and equilibrium equations are satisfied, part of which, E, may be stressed in the elastic domain, and the rest of which, E-P, may be stressed in the plastic domain, having, therefore, different constitutive relations between stress and strain. Such volume is limited by a boundary surface S, in part of which, S_D, a motion, may be imposed, and in the rest of which, S_F, loads may be imposed. Only quasi-static problems will be dealt with. Therefore, no dynamic considerations will be taken into account.

In order to solve this problem, consider a configuration during the loading process and, as in most forming problems, no distributed body loads. Equilibrium equations are, in tensor form, written as:

$$\sigma_{ij,j} \ = \ 0 \tag{2.118}$$

σ_{ij} being the Cauchy stress tensor.

Multiplying by an arbitrary variation of the velocity field, and integrating over the volume, they become equivalent to:

$$\int_V \sigma_{ij,j} \ \delta v_i \ dv \ = \ 0 \tag{2.119}$$

By using the divergence theorem, and imposing $\delta v_i = 0$ on S_D, which means considering velocity fields kinematically admissible, the common statement of the virtual power principle:

$$\int_V \sigma_{ij}\, \delta v_{i,j}\, dv - \int_{S_f} f_i\, \delta v_i\, ds_f = 0 \tag{2.120}$$

is obtained where:

$$f_i = \sigma_{ij}\, n_j \tag{2.121}$$

is the surface traction, with n_j the components of the surface normal unit vector.

It will be shown in the following how an attempt can be made to solve (2.120) to obtain the one kinematically admissible velocity field which, by means of constitutive equations, produces a stress field that is in equilibrium and satisfies traction boundary conditions.

This means a solution for a given configuration and, at most, how the procedure is to be continued.

Because temperatures rarely reach a steady state and materials work harden, most metal forming problems are transient, and it is necessary to follow the history of deformation. All the Finite Element examples given in later chapters fall into this category. Therefore, such problems are solved by increments, i.e. by a sequence of solutions to equation (2.120), with an updating of the appropriate variables from configuration to configuration.

As has been previously mentioned, there are two schools for dealing with the problems: the elastic-plastic approach and the rigid-plastic, or flow, approach. Since our experience is confined to the latter approach, the following will describe only the general philosophy and the kind of problems dealt with by the elastic-plastic approach. All the Finite Element equations and solution detailing will be presented in conjunction with the flow approach.

2.6.3.1 Elastic-Plastic Formulations

These formulations seek a full solution to the metal forming
problem. In fact, even if the elastic strains are negligible with
respect to the plastic strains present in metal forming, only a
full solution will be able to predict the residual stresses at
the end of a process. Moreover, in many cases there are regions
of the body that deform elastically only. In order to solve
(2.120) it is necessary to use a constitutive relation between
stresses and strains. Substituting these in the equation, the
problem can be solved in kinematic terms first and, afterwards,
stresses calculated out of the same constitutive relations.

Plastic behavior is a flow type of behavior. This means stresses
are related not with strains, as in elasticity, but with strain
rates. As a result, elastic-plastic constitutive equations relate
stress rates, stresses, and strain rates. During deformation,
when large strains and large rotations are present, finite deformation theory is necessary for the appropriate measure of stress
rate to be used. In addition, strain measures as in (2.18) become totally inappropriate.

These characteristics of large deformation theory, used in a
step-by-step manner, are the backbone of the elastic-plastic
formulations used in metal forming, commonly known as updated
Lagrangean. At every time step, variables and configurations are
updated; the reference configuration for the large deformation
variables is the current configuration. Such an approach to
solving forming problems was pioneered by McMeeking and Rice
[2.25]. Since then, several variations and improvements have
appeared in the literature.

The rate of virtual work is obtained from (2.120) as:

$$\int_{v} \dot{\sigma}_{ij}\,\delta v_{i,j}\,dv - \int_{s_{f}} \dot{f}_{i}\,\delta v_{i}\,ds_{f} = 0 \qquad (2.122)$$

Equation (2.122), as it is written, refers to a current configuration using a Cartesian axis system, denoted by lower case
designations. It can be referred to some reference configuration
[2.26], still using a cartesian axis system, denoted by upper

case designations:

$$\int_V \dot{s}_{iJ} \, \delta v_{j,I} \, dV - \int_{S_F} \dot{F}_i \, \delta v_i \, dS_f = 0 \qquad (2.123)$$

where s_{iJ} is the first Peola-Kirchoff stress tensor (non-symmetric), that relates to the Cauchy stress through:

$$J \, \sigma_{ij} = \frac{\partial x_i}{\partial X_K} s_{jK} \qquad (2.124)$$

with:

$$J = \frac{dv}{dV} \qquad (2.125)$$

being the determinant of the deformation gradient tensor $\frac{\partial x_i}{\partial X_J}$.

In (2.123), $\dot{F}_i$ is the traction rate, measured as the rate of change of force per unit area in the reference state.

If the present configuration is used as the reference configuration, all stress measures are equal to the Cauchy stress, $dv_{i,J} = dv_{i,j}$, but the stress rate measures are different.

In order to choose a stress rate to be used in the constitutive relations, a material derivative should be used, as well as one invariant upon rigid body rotation. There are several possibilities which fall into this category [2.27], such as the Lie derivative of the Kirchoff stress tensor τ, the Lie derivative of the Cauchy stress tensor, the Truesdell rate of the Cauchy stress tensor, or the Jaumann rate of the Cauchy stress tensor.

It must be remembered that:

$$\tau_{ij} = J \, \sigma_{ij} \qquad (2.124)$$

It is commonly accepted that elastic-plastic processes are governed by hypo-elasticity constitutive equations. Under such assumptions, certain stress rates, such as the Lie derivative of Cauchy stress, or the Truesdell rate of Cauchy stress, are written in the form:

$$\overset{o}{\sigma}_{ij} = L_{ijkl}\ d^{e}_{kl} \tag{2.125}$$

where d^{e}_{kl} is the elastic part of the rate of deformation tensor, and L_{ijkl} is the elasticity tensor, derivable from a potential.

Either through kinematical considerations [2.28], or thermo-dynamic considerations [2.29], it is possible to show that the rate of deformation tensor:

$$d_{ij} = \frac{1}{2}\ (v_{i,j} + v_{j,i}) \tag{2.126}$$

can be decomposed into an elastic and a plastic part:

$$d_{ij} = d^{e}_{ij} + d^{p}_{ij}\ . \tag{2.127}$$

The same thermodynamic considerations lead to:

$$\overset{o}{\sigma}_{ij} = L_{ijkl}\ (d_{ij} - d^{p}_{ij}) \tag{2.128}$$

where $\overset{o}{\sigma}_{ij}$ is the Truesdell rate of Cauchy stress.

The classical infinitesimal elastic-plastic constitutive equations, using a von Mises yield criterion and a flow rule based on Drucker's postulate [2.1], produce in large strain analysis the relation:

$$\overset{*}{\sigma}_{ij} = L^{e-p}_{ijkl}\ d_{kl} \tag{2.129}$$

with:

$$L^{e-p}_{ijkl} = 2\ G\ [\delta_{ik}\ \delta_{jl} + \frac{\lambda}{2G}\ \delta_{ij}\ \delta_{kl} - \frac{3}{2}\ \frac{\sigma'_{ij}\ \sigma'_{kl}}{\bar{\sigma}^{2}}] \tag{2.130}$$

where:

$\overset{*}{\sigma}$ is a corotational rate, for example $\overset{\triangledown}{\sigma}_{ij}$

λ, G are elasticity constants

$\bar{\sigma} = \sqrt{\dfrac{3}{2}\, \sigma'_{ij}\, \sigma'_{ij}}$ is the effective stress

$\overset{\triangledown}{\sigma}_{ij} = \dot{\sigma}_{ij} - \omega_{ik}\,\sigma_{kj} - \sigma_{ik}\,\omega_{jk}$ is the Jaumann rate of Cauchy stress

$\sigma'_{ij} = \sigma_{ij} - \delta_{ij}\,\sigma_m$ is the deviatoric stress

$\omega_{ij} = \dfrac{1}{2}\,(v_{i,j} - v_{j,i})$ is the spin tensor

Relation (2.129) is not derivable from a potential, therefore it is not of the (2.128) type. Furthermore, it would lead to non-symmetrical Finite Element equations. Manipulations of such constitutive equations are necessary in order to refer them to other stress rate measures.

Using (2.124) and the relation:

$$\overset{o}{\sigma}_{ij} = \dot{\sigma}_{ij} - v_{i,k}\,\sigma_{kj} - \sigma_{ik}\,v_{j,k} + \sigma_{ij}\,v_{k,k} \qquad (2.131)$$

(2.123) can be transformed, with the current configuration as the reference configuration, into:

$$\int_V (\overset{o}{\sigma}_{ij}\,\delta v_{i,j} + \sigma_{ij}\,v_{k,i}\,\delta v_{k,j})\, dv -$$

$$- \int_{S_f} \dot{f}_i\,\delta v_i\, ds_f = 0 \qquad (2.132)$$

There are several ways in the literature [2.25, 2.27, 2.29, 2.30, 2.31] to manipulate equation (2.123) or (2.132), together with constitutive assumptions in the form (2.128) or (2.130) leading to equations of the type:

$$\int_V [\ell_{ijkl}\,d_{ij}\,\delta d_{kl} + f\,(\sigma_{ij},\, v_{ij},\, \delta v_{ij})]\, dv -$$

$$- \int_{s_f} \dot{f}_i \, \delta v_i \, ds_f = 0 \qquad\qquad (2.133)$$

with ℓ_{ijkl} an equivalent elastic-plastic tensor.

These forms of the rate of virtual power equation are used in the Finite Element formulations.

The next problem posed in an elastic-plastic formulation is how to integrate the equations in time.

Earlier procedures would solve the Finite Element equations obtained from (2.133) for the velocity field. Then they would update configurations and, finally, integrate the stress rate equations. This integration of stresses is done at the numerical integration point level and poses some problems.

First an appropriate transformation of the stress rate used in the constitutive equations has to be performed to produce a Cauchy stress rate. Next, there is the problem of the elastic-plastic tensor containing a hardening rate that changes along the time step. If it is considered to be constant and equal to the starting value - tangent modulus method - the algorithm is conditionally stable and, therefore, time steps have to be very small. To obviate that, implicit procedures may be used which consider the hardening rate at the middle of the time step (mean-normal method), or, at the end of the time step (radial return method), which are always stable.

However, if large time steps producing large amounts of rotations and/or deformations are to be used, equation (2.133) has to be written in an incremental form because changes in displacement directions occur during a time step. The first problem this poses is the fact that constitutive equations relate stress rates to the rate of deformation tensor. The latter does not relate directly to any strain measure increment. Secondly, appropriate measures of stress increments capable of coping with configuration changes have to be used.

Nagtegaal and Jong [2.29] integrate the constitutive relations over an increment of time, assuming constant strain rate during such increment even if the integral cannot be obtained in an exact closed form. Instead of writing the rate form, the virtual work equation itself is written both at the beginning and at the end of the time step. Both equations are written in Lagrangean form, i.e., both are referred to the reference (beginning) configuration. Then, an incremental form of the constitutive relations is obtained by approximately integrating their rate form. Upon subtracting the two equations, an incremental virtual work equation is obtained which is non-linear in displacement increments. Pinsky [2.27], making an extension of previous work done by Hughes [2.32], uses the concept of directional derivatives to deal conveniently with the aforementioned problems.

Several developments have been made recently [2.34-2.36] producing second order accuracy in the integration of the constitutive relations, and allowing for even larger time steps (of the order of a few percent strain) in cases where large rotation and strains are present.

As both these approaches produce proper incremental finite element equations that are non-linear in displacement increments, a linearization must be made which must include a proper treatment of various kinematic and stress related variables.

Such linearization leads to some kind of variation on the Newton-Raphson iterative solution of the equations. As something similar is done in the rigid-plastic approach, no further comment on it is needed.

In all the processes described in this section, there are iterations that lead to successive approximate values of stresses at the end of a time increment. In order to determine such values, a decision has to be made as to whether the next value falls in the elastic range or on the yield surface. Typically, the problem is first solved as if it were elastic. Then a check is made as to whether the stress exceeded the yield surface. If such is the case, plasticity terms are added to make the stress return to the yield surface. There are several algorithms possible for this [2.33]. The implementation is very much intertwined with the intricacies of the whole formulation procedures.

2.6.3.2 Rigid-Plastic Formulations

These formulations discard the elastic part of deformation. In most metal forming processes, strains reach large values of many percent - even a few hundred percent is quite common. When compared with less than one percent elastic strains, the approximation made is very reasonable.

The major drawback is that residual stresses are not obtained at the end. The solutions are obtained only when there is a load applied able to plastically deform a body. However, it is always possible to unload a body from the last deforming configuration, using an elastic (or infinitesimal elastic-plastic) Finite Element formulation. Experiments along this line have shown very good results when compared with elastic-plastic formulations [2.38].

Of course, there may be regions in a plastically deforming body that deform elastically only. In a rigid-plastic formulation, such regions are rigid, posing the problem that stresses are not uniquely defined there. It will be shown that there are ways of circumventing this problem, but the stress values obtained in a rigid zone are not correct.

Rigid-plastic formulations in non steady-state problems operate in a very similar fashion to updated Lagrangean formulations and have been available since the early seventies with the work of Kobayashi [2.39, 2.40]. A solution is obtained for a given configuration, the configuration is updated, and the process is repeated. However, constitutive equations are simpler now. Stresses relate directly to rate of deformation tensors or infinitesimal strain rates. Therefore, using symmetry of the Cauchy stress tensor, equation (2.120) can be written in the form:

$$\int_V \sigma_{ij}\, \delta\dot{\varepsilon}_{ij}\, dv - \int_{s_f} f_i\, \delta v_i\, ds_f = 0 \qquad (2.134)$$

with:
$$\dot{\varepsilon}_{ij} = d_{ij}$$

which is ready for use, without further developments.

Due to the character of the constitutive relations, there is no
need to update stresses because they are calculated from scratch
at every configuration. Updating the configurations is made by
multiplying the velocities by the time increment. If the classical plasticity theory is used, the only other parameter that need
be updated is a measure of work hardening. In isotropic work
hardening this can be done by means of the scalar effective, or
equivalent strain, which is obtained by adding successive increments of effective strain rates, multiplied by the time increment. This total integration is equivalent to the logarithmic
strain of a uniaxial test.

These characteristics show that problems can be solved using infinitesimal theory. The strain tensor can be updated in the same
way as the effective strain, but whenever there are large rotations, it will be meaningless and large deformation theory strain
measures should be used. Because everything is always dealt with
in a current configuration, a rigid-plastic approach can be called updated Eulerian.

It should be mentioned here that rigid-plastic formulations are
used very often to solve forming problems considered as steady
state ones [2.41, 2.42]. In fact, the material behavior is mathematically identical to a non-Newtonian fluid (hence flow formulation), and such problems are usually treated in a Eulerian form.
If there is no work hardening present, a single solution for the
given configuration produces the desired results. However, if
the material work hardens, it is necessary to integrate strains
along the flow lines [2.43] in order for material properties to
be known. This produces an iterative procedure between Finite
Element solutions with approximate material properties (and such
material properties), until a final result is obtained. The
method can even be extended to produce elastic-plastic solutions
by considering the elastic effects as a correction to be made to
a fully plastic analysis [2.44].

In addition, coupled thermal analysis, of which more mention
will be made later, produces different equations (and solution

schemes) in a steady state analysis, due to the introduction of convective terms.

Going back to equation (2.134), the constitutive relations are now required. Because of the simplicity of an infinitesimal approach, it was easy to include here strain rate effects [2.45], based on a summary of theoretical work done by Perzyna [2.46].

A static yield function, assumed regular and convex, is introduced:

$$F\left(\sigma_{ij},\, \varepsilon_{kl}^{p}\right) \;=\; \frac{f\left(\sigma_{ij},\, \varepsilon_{kl}^{p}\right)}{K} - 1 \qquad (2.135)$$

where $f\left(\sigma_{ij},\, \varepsilon_{kl}^{p}\right)$ is a function of the state of stress and plastic strain, and $K\left(w_{p}\right)$ is a strain-hardening parameter, function of the plastic work w_{p}. F is such that for F < 0 the material is rigid; for F = 0, it is on the verge of yielding with zero strain rate; and for F > 0, the material yields with a non-zero strain rate. The constitutive relations, using the yield function and its associated flow rule, are expressed as:

$$\dot{\varepsilon}_{ij} \;=\; \gamma < \phi\,(F) > \frac{\partial f}{\partial \sigma_{ij}} \qquad (2.136)$$

where $< \phi\,(F) >$ is zero for $F \leq 0$, and $\phi\,(F)$ for F > 0. The function $\phi\,(F)$ is a parameter used to fit material behavior obtained experimentally. γ is a viscosity constant.

Equation (2.136) states that the strain rate is a function of the excess of stress over the static yield stress.

Making the usual assumptions of isotropy, isotropic work hardening, incompressibility, and the von Mises yield criterion, the following equation is given:

$$f \;=\; \left(\tfrac{1}{2}\, \sigma_{ij}'\, \sigma_{ij}'\right)^{1/2} \qquad (2.137)$$

from where (2.136) becomes:

$$\dot{\varepsilon}_{ij} = \gamma < \phi (F) > \frac{\sigma'_{ij}}{2f} \tag{2.138}$$

and from which it is possible to conclude that:

$$\left(\frac{1}{2} \dot{\varepsilon}_{ij} \dot{\varepsilon}_{ij}\right)^{1/2} = \frac{\gamma}{2} < \phi (F) > \tag{2.139}$$

Substituting in (2.138):

$$\dot{\varepsilon}_{ij} = \frac{3}{2} \frac{\dot{\bar{\varepsilon}}}{\bar{\sigma}} \sigma'_{ij} \qquad \text{is obtained,} \tag{2.140}$$

with $\qquad \dot{\bar{\varepsilon}} = \sqrt{\dfrac{2}{3} \dot{\varepsilon}_{ij} \dot{\varepsilon}_{ij}} \qquad$ effective strain rate. $\tag{2.141}$

Constitutive relations (2.140) have exactly the same form as the ones obtained in infinitesimal rate-independent theory based on similar simplifying assumptions; the difference now is that $\bar{\sigma}$, besides being a function of strain, temperature, and possibly other internal parameters, is a function of strain rate as well.

It should be noted that (2.140) relates strain rates to deviatoric stresses, and implicitly states incompressibility:

$$\dot{\varepsilon}_{ii} = 0$$

For it to be possible to substitute directly into (2.134), the kinematically admissible velocity fields must be restricted to the ones that satisfy incompressibility. Such restrictions are better imposed mathematically by means of a Lagrangean multiplier or by means of a penalty term in equation (2.134). In the first case, the Lagrangean multiplier λ is the average stress and (2.134) becomes:

$$\int_V \sigma'_{ij} \, \delta\dot{\varepsilon}_{ij} \, dv - \int_{S_f} f_i \, \delta v_i \, dS_f + \int_V \lambda \delta\dot{\varepsilon}_{ii} +$$

$$+ \int_V \dot{\varepsilon}_{ii} \, \delta\lambda \, dv = 0 \tag{2.142}$$

Now equilibrium equations and traction boundary conditions are
satisfied for the kinematically admissible velocity field and
average stress field that for any arbitrary variation δv and
$\delta \lambda$, satisfy equation (2.142).

In the case of a penalty term, (2.134) becomes:

$$\int \sigma'_{ij} \, \delta \dot{\varepsilon}_{ij} \, dv - \int_{S_f} f_i \, \delta V_i \, ds_f + \int \xi \, \dot{\varepsilon}_{jj} \, \delta \dot{\varepsilon}_{ii} \, dv = 0 \qquad (2.143)$$

where ξ is an arbitrarily large constant, such that:

$$\xi \, \dot{\varepsilon}_{jj} = \sigma_{average} \, . \qquad (2.144)$$

Thus, equilibrium equations and traction boundary conditions are
satisfied for the kinematically admissible velocity field that
for any arbitrary variation δv, satisfies equation (2.143).

These two methods are used equally in practice. Obviously, the
penalty procedure requires less computer storage space and produces fewer equations in the Finite Element discretization. On the
other hand, the Lagrangean multiplier procedure produces faster
convergence in the solution of the non-linear Finite Element
equations obtained. There seems to be no clear advantage of one
method over the other. In this book, the Finite Element equations will be written based on equation (2.142).

2.6.4 Thermal Formulations

During any metal forming or thermomechanical treatment, there is
a certain amount of heat generated by deformation as well as heat
lost to the environment (or supplied by it). This produces two
effects: one, if material properties are very sensitive to temperature, deformation patterns can be modified substantially;
two, if temperature gradients are large, induced thermal stresses
and deformations can be quite severe. Hot-die forging is an
example of the first case, where the billet is at a much higher
temperature than the die, thus producing large temperature gradients at the surface. Quenching is an example of the second
case, where thermally induced stresses can warp and crack parts.

It is therefore important in many cases to perform a Finite
Element analysis of the thermal aspects of a problem in parallel
with the mechanical analysis. They are (weakly) coupled in the
sense that temperatures influence material properties and deformation generates heat.

Thorough reviews of thermodynamic theory of deformation using
internal parameters have been made by Perzyna [2.47, 2.48], an
adaptation of which was made for the rigid viscoplastic case in
[2.49].

On the other hand, thermodynamic considerations are always present when large deformation theory developments are made [2.27].
In the context of coupled Finite Element analysis, several applications have been made both in conjunction with the updated
Lagrangean approach [2.50, 2.51], and with the flow approach
[2.52, 2.53].

Because internal parameters are very difficult to quantify, a
thermal formulation is practically based solely on the energy
balance equation, and invariably looks like the following: considering that a hot metal-working process does not take long,
and that temperature changes are due mainly to plastic work and
losses to the environment, a simple Fourier law for heat flux is
used, with isotropic conductivity:

$$q_i = K_1 T,_i \qquad (2.145)$$

where q_i is the heat flux through a unit surface normal to the
direction i, and T is the temperature.

The first law of thermodynamics, stating balance of energy, can
be written locally as:

$$\sigma'_{ij} \dot{\varepsilon}_{ij} - R + K_1 T,_{ii} - \rho c \dot{T} = 0 \qquad (2.146)$$

where $\rho \cdot c$ is the volume specific heat of the material, $\sigma'_{ij} \dot{\varepsilon}_{ij}$
represents the work heat per unit volume due to plastic deforma-
tion and R is some kind of variation of free energy produced by
structural changes in the material. In practical applications,

equation (2.146) is simplified to:

$$\alpha \, \sigma'_{ij} \, \dot{\varepsilon}_{ij} + K_1 \, T,_{ii} - \rho c \, \dot{T} = 0 \qquad\qquad (2.147)$$

with α an arbitrary parameter to be determined experimentally. In fact, experiments have shown that, in general, only between 90 to 95 % of the deformation energy is dissipated into heat, the remainder being used, for instance, to create dislocations in the material. At the same time, this disparity between these two energies ensures, in the rigid-plastic case, the satisfaction of the Clausius-Duhem inequality that is a direct consequence of the second principle of thermodynamics.

The thermal problem is an initial value problem in which initial temperatures are specified, energy balance is satisfied through equation (2.147), and on the surface of normal n a heat flux is prescribed. This latter can be either a direct one:

$$K_1 \; \frac{\partial T}{\partial n} = q_n^d \qquad\qquad (2.148)$$

or a radiation heat flux:

$$K_1 \; \frac{\partial T}{\partial n} = q_n^r = \sigma\varepsilon \, (T_e^4 - T_s^4) \qquad\qquad (2.149)$$

where:

$$\sigma \quad - \quad \text{Stephan Boltzman constant}$$
$$\varepsilon \quad - \quad \text{emissivity of the surface}$$
$$T_e \quad - \quad \text{environment temperature}$$
$$T_s \quad - \quad \text{surface temperature}$$

or a convection heat flux

$$K_1 \; \frac{\partial T}{\partial n} = q_n^c = h \, (T_e - T_s) \qquad\qquad (2.150)$$

where h is the surface heat transfer coefficient.

To solve the initial value problem, a weak form of the heat balance equation is used. Multiplying (2.147) by an arbitrary variation of the temperature field, and integrating over the volume gives:

$$\int_V K_1 \, T,_{ii} \, \delta T \, dv - \int_V \rho c \, \dot{T} \, \delta T \, dv +$$

$$+ \int_V \alpha \sigma'_{ij} \, \dot{\varepsilon}_{ij} \, \delta T \, dv = 0 \qquad (2.151)$$

Use of the divergence theorem produced the following equation:

$$\int_V K_1 \, T,_i \, \delta T,_i \, dv + \int_V \rho c \dot{T} \, \delta T \, dv - \int_V \alpha \sigma'_{ij} \, \dot{\varepsilon}_{ij} \, \delta T \, dv -$$

$$- \int_S q_n \, \delta T \, ds = 0 \qquad (2.152)$$

The solution to the problem is the sequence of temperature fields over time which, for any arbitrary perturbation δT, satisfy (2.152).

Again, integration in time of the process is necessary, made in conjunction with the time evolution of the deformation part. It will be presented next, together with the Finite Element equations.

2.6.5 <u>Finite Element Equations</u>

The following pages give the Finite Element equations that are obtained when the rigid-viscoplastic and energy balance equations discussed previously are discretized in two dimensions.

Consider the body being studied as having been divided into m elements connected by n nodes. The velocity field in (2.142) and the temperature field in (2.152) are approximated by:

$$\underset{\sim}{u} = \underline{N}^T \underset{\sim}{v} \qquad (2.153)$$

$$\underset{\sim}{T} = \underset{\sim}{N}^T \underset{\sim}{T} \qquad (2.154)$$

where:

$\underset{\sim}{v}$ - vector of nodal point velocities
$\underset{\sim}{T}$ - vector of nodal point temperatures
$\underset{\sim}{N}$ - interpolation function vector
$\underline{N}$ - interpolation function matrix

The integrals of the previous equations are the sum of integrals over each element. The derivations for bilinear isoparametric four-node elements, which are the most commonly used in applications, will be made. Thus, the same interpolation functions apply for the coordinates

$$x = \underset{\sim}{N}^{T} (s, t) \underset{\sim}{x} \qquad\qquad (2.155)$$

$$y = \underset{\sim}{N}^{T} (s, t) \underset{\sim}{y}$$

Figure 2.24 shows a general element, s and t being the natural coordinates.

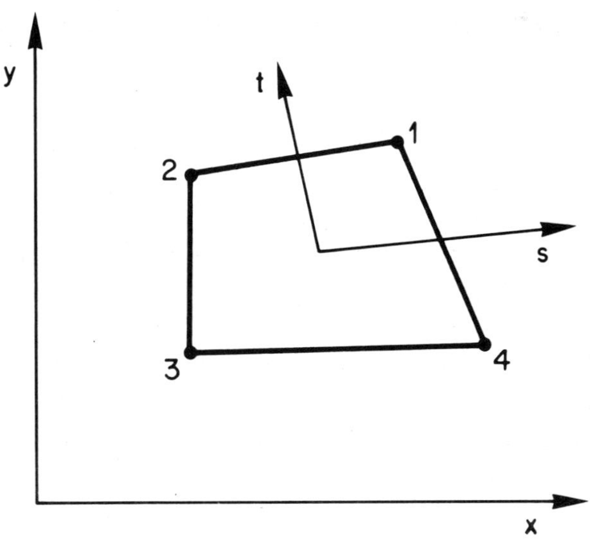

Figure 2.24: Quadrilateral element and natural coordinate system.

A single value of the Lagrangian multiplier λ is assigned to each element. This lower order interpolation is used in order not to overconstrain the system. It produces the same effect as the well-known technique of reduced integration when a penalty function constraint is imposed.

2.6.5.1 Velocity Equations

The strain rate tensor can be written in a vector form (with all of its components):

$$\dot{\underset{\sim}{\varepsilon}} = \underline{B}\,\underset{\sim}{v} \tag{2.156}$$

matrix $\underline{B}$ containing appropriate spatial derivatives of interpolation functions. And the effective strain rate is expressed in the form:

$$\dot{\bar{\varepsilon}} = (\dot{\underset{\sim}{\varepsilon}}^T \underline{D}\, \dot{\underset{\sim}{\varepsilon}})^{1/2} = (\underset{\sim}{v}^T \underline{B}^T \underline{D}\,\underline{B}\,\underset{\sim}{v})^{1/2} = (\underset{\sim}{v}^T \underline{P}\,\underset{\sim}{v})^{1/2} \tag{2.157}$$

Substitution in (2.143) yields:

$$\sum_{j=1}^{m} \left\{ \int_{v_j} \frac{\bar{\sigma}}{\dot{\bar{\varepsilon}}} \delta\underset{\sim}{v}^T \underline{P}\,\underset{\sim}{v}dv_j - \int_{s_{f_j}} \delta\underset{\sim}{v}^T \underline{N}\,\underset{\sim}{f}\,ds_{f_j} \right.$$

$$\left. + \int_{v_j} \lambda\,\delta\underset{\sim}{v}^T\underline{B}^T \underset{\sim}{c}dv_j + \int_{v_j} \underset{\sim}{v}^T\underline{B}^T \underset{\sim}{c}\,\delta\lambda\,dv_j \right\} = 0 \tag{2.158}$$

where:

$$\underset{\sim}{c}^T\dot{\underset{\sim}{\varepsilon}} = \dot{\varepsilon}_{ii} \tag{2.159}$$

Due to the arbitrariness in $\underset{\sim}{v}$ and λ, the following system of non-linear equations in $\underset{\sim}{v}$ and linear in λ is obtained:

$$\sum_{j=1}^{m} \left\{ \int_{v_j} \bar{\sigma}\,\frac{\underline{P}\,\underset{\sim}{v}}{\dot{\bar{\varepsilon}}}\,dv_j - \int_{s_{f_j}} \underline{N}\,\underset{\sim}{f}\,ds_{f_j} + \right.$$

$$\left. + \lambda_j \int_{v_j} \underline{B}^T \underset{\sim}{c}\,dv \right\} = 0 \tag{2.160}$$

$$\int_{V_j} \underset{\sim}{v}^T \underline{B}^T \underset{\sim}{c} \, dv_j \ = \ 0$$

These equations have the form:

$$\underset{\sim}{f} \, (\underset{\sim}{v} \cdot \lambda) \ = \ \underset{\sim}{0} \qquad\qquad (2.161)$$

and are solved by the Newton-Raphson procedure. Making a series expansion and keeping only the first two terms:

$$\underset{\sim}{f} \, (\underset{\sim}{v}, \, \lambda) \ \underset{\sim}{\approx} \ \underset{\sim}{f} \, (\underset{\sim}{v}_n, \, \lambda) + \frac{\partial \underset{\sim}{f}}{\partial \underset{\sim}{v}} \, (\underset{\sim}{v}_n, \, \lambda) \, \Delta \underset{\sim}{v}_n \ = \ 0 \qquad (2.162)$$

The result is the following system of equations linear in λ and $\Delta \underset{\sim}{v}$:

$$
\left[
\begin{array}{c|c}
\sum\limits_{j=1}^{m} \{ \int_{V_j} \bar{\sigma} \, \dfrac{P}{\dot{\bar{\varepsilon}}} \, dv_j + \int_{V_j} \underline{R} \, dv_j \} & \int_{V_j} \underline{B}^T \underset{\sim}{c} \, dv_j \\[6pt]
\hline
\int_{V_j} \underline{B}^T \underset{\sim}{c} \, dv_u & 0
\end{array}
\right]
\cdot
\left\{
\begin{array}{c}
\Delta \underset{\sim}{v} \\[6pt]
\hline
\lambda
\end{array}
\right\}
$$

$$\qquad\qquad\qquad\qquad\qquad\qquad\qquad\qquad\qquad\qquad\qquad\qquad (2.163)$$

$$
= \left\{
\begin{array}{c}
\sum\limits_{j=1}^{m} [\int_{S_{f_j}} \underset{\sim}{N} \underset{\sim}{f} \, ds_{f_j} - \int_{V_j} \bar{\sigma} \, \dfrac{P}{\dot{\bar{\varepsilon}}} \, \underset{\sim}{v} \, dv_j] \\[6pt]
\hline
\int_{V_j} \underset{\sim}{v}^T \underline{B}^T \underset{\sim}{c} \, dv_j
\end{array}
\right\}
$$

where:

$$\underline{R} \ = \ (-\bar{\sigma} \, \dot{\bar{\varepsilon}}^{-3} + \frac{\partial \bar{\sigma}}{\partial \dot{\bar{\varepsilon}}} \, \dot{\bar{\varepsilon}}^{-2}) \, \underset{\sim}{v}^T \underline{P}^T \underline{P} \, \underset{\sim}{v} \qquad\qquad (2.164)$$

Numerical quadrature using 2 x 2 points is used in the calculation of the integrals. After each iteration, the values of v are updated as:

$$\underset{\sim}{v}_{n+1} = \underset{\sim}{v}_n + r\Delta\underset{\sim}{v}_n \tag{2.165}$$

where r is a deceleration coefficient, $0 < r < 1$. The values of r are chosen according to the convergence character of the system of equations.

A drawback in the Newton-Raphson method is that it requires an initial guess for the velocity field. This initial guess, when not easy to find, can be calculated as follows. Assuming any reasonable value for $\bar{\varepsilon}$, then equation (2.160) becomes linear in $\underset{\sim}{v}$ and λ, for which they are solved. Then $\bar{\varepsilon}$ is updated, and the procedure is repeated a couple of times.

2.6.5.2 Temperature Equations

Making:

$$M_{ij} = N_{i,j} \tag{2.166}$$

and substituting with (2.154) into (2.152):

$$\sum_{j=1}^{m} \left[\int_{v_j} K_1 \, \delta\underset{\sim}{T}^T \, \underline{M} \, \underline{M}^T \underset{\sim}{T} \, dv_j + \int_{v_j} \rho c \, \delta\underset{\sim}{T}^T \, \underset{\sim}{N} \, \underset{\sim}{N}^T \, \dot{\underset{\sim}{T}} \, dv_j - \right.$$

$$\left. - \int_{v_j} \alpha\sigma'_{ij}\dot{\varepsilon}_{ij} \, \delta\underset{\sim}{T}^T\underset{\sim}{N} \, dv_j - \int_{s_{q_j}} q_n \, \delta\underset{\sim}{T}^T\underset{\sim}{N} \, ds_{q_j} \right] = 0. \tag{2.167}$$

Due to arbitrariness in $\underset{\sim}{T}$ the following system of equations is obtained:

$$\sum_{j=1}^{m} \left[\int_{v_j} K_1 \, \underline{M} \, \underline{M}^T dv_j \underset{\sim}{T} + \int_{v_j} \rho c \, \underset{\sim}{N} \, \underset{\sim}{N}^T \, dv_j \dot{\underset{\sim}{T}} - \right.$$

$$\left. - \int_{v_j} \alpha\sigma'_{ij}\dot{\varepsilon}_{ij} \, \underset{\sim}{N} dv_j - \int_{s_{q_j}} q_n \underset{\sim}{N} \, ds_{q_j} \right] = 0 \tag{2.168}$$

or:

$$\underline{C}\,\underline{\dot{T}} + \underline{K}\,\underline{T} - \underline{Q} - \underline{Q}_n = \underline{0} \qquad (2.169)$$

where:

$$\underline{Q} = \int_V \alpha\sigma'_{ij}\,\dot{\varepsilon}_{ij}\,\underline{N}dv \qquad (2.170)$$

$$\underline{Q}_n = \int_{s_r} \sigma\varepsilon\,(T_e^{\,4} - T_s^{\,4})\,\underline{N}\,ds_r + \int_{s_c} h\,(T_e - T_s)\,\underline{N}\,ds_c +$$

$$+ \int_{s_l} h_1\,(T_d - T_w)\,\underline{N}\,ds_l + \int_{s_f} q_f\,\underline{N}\,ds_f \ . \qquad (2.171)$$

In (2.171) the first component is the contribution of the heat radiated into the body and the second, the contribution of the heat transferred by convection. The third term is identical to the second and corresponds to the heat transferred between a workpiece and a die, described in a convection-like manner. The fourth term is the heat generated by friction at an interface, where q_f is half of the friction force multiplied by a relative velocity.

2.6.5.3 <u>Time Dimension</u>

In order to describe a deformation process in time, integration of equation and (2.169) has to be done.

The method of updating the configurations has already been discussed.

To integrate equation (2.169) a one-step method is used. Otherwise, the storage requirements would escalate too much. Convergence of a scheme requires consistency and stability [31]. Consistency is satisfied by an approximation of the type:

$$\underline{T}_{t+\Delta t} = \underline{T}_t + \Delta t \cdot [(1 - \beta)\,\underline{\dot{T}}_t + \beta\underline{\dot{T}}_{t+\Delta t}] \qquad (2.172)$$

where β is a parameter varying between 0 and 1. Unconditional stability is obtained if $\beta \geq 0.5$. This is important, because it is desirable to take time steps as large as the deformation formulation allows, since this is the most expensive part of the process.

Equations (2.163) and (2.169) are coupled, making a simultaneous solution of their Finite Element counterparts necessary.

An implementation of such an algorithm, considering $\underset{\sim}{T}_{t + \Delta t}$ as a primary dependent variable, has been elaborated in order to include the coupling of the solution of stress equilibrium equations with heat balance.

The sequence of calculations to be performed in one time step Δt is as follows:

a) assume that the initial temperature field $\underset{\sim}{T}_o$ is known;
b) calculate initial velocity field;
c) calculate initial temperature rates $\underset{\sim}{\dot{T}}_o$ (equation 2.169);
d) calculate:

$$\hat{\dot{T}} = -\frac{1}{\beta \Delta t} \cdot \underset{\sim}{T}_o - \frac{(1-\beta)}{\beta} \cdot \underset{\sim}{\dot{T}}_o$$

e) update displacements and strains;
f) use the old velocity field for first approximate temperature calculations:

$$\underset{\sim}{R} = \underset{\sim}{Q}^{(1)}_{\Delta t} - \underline{C}\,\underset{\sim}{\hat{\dot{T}}}$$

   equation (2.169) written at the end of the interval Δt, together with (2.172) produces:

$$(\underline{K} + \frac{1}{\beta \Delta t}\,\underline{C}) \cdot \underset{\sim}{T}^{(1)}_{\Delta t} = \underset{\sim}{R}$$

   solved for $\underset{\sim \Delta t}{T^{(1)}}$

g) calculate new velocity field;
h) calculate second values for temperatures:

$$\underset{\sim}{R} = \underset{\sim}{Q}^{(2)}_{\Delta t} - \underline{C}\,\underset{\sim}{\hat{\dot{T}}}$$

   solve:

$$(\underline{K} + \frac{1}{\beta \Delta t} \underline{C}) \cdot \underline{T}_{\Delta t}^{(2)} = \underline{R} \quad \text{for} \quad \underline{T}_{\Delta t}^{(2)}$$

i) iterate until convergence;

j) calculate new temperature rate:

$$\dot{\underline{T}}_{\Delta t} = \dot{\hat{\underline{T}}} + \frac{1}{\beta \Delta t} \underline{T}_{\Delta t} \; .$$

The workpiece and the dies are treated separately. In this way the number of equations to be solved simultaneously is greatly reduced, thereby reducing computing costs. Because of the heat transfer at the interface, die and workpiece equations contain overlapping temperatures, and the solution is found iteratively. This brings no increase in cost because iterations with respect to velocities have to be done anyway. Realistic heat transfer coefficients at the interface do not produce instabilities in the iteration process.

2.6.6 Forming Specifics

Once the formulation and the Finite Element equations are written, it would be expected that problems could be solved immediately. In classical structural analysis type of problems, this might be so because real problems are not far from the typical boundary value problem discussed already in abstract terms. Consider again the schematics of it, as shown in Figure 2.25.

There is given a volume in which equilibrium equations, compatibility equations and constitutive equations, together with heat balance equations have to be satisfied. There is a surface on part of which velocities are imposed and on the rest of which forces are imposed, together with convection and radiation heat transfer boundary conditions.

Figure 2.26 takes a closer look at one of the simplest conceivable forging problems - ring compression.

There is the ring in which are sought variables such as displacements, velocities, strains, strain rates, stresses, temperatures, and deformation by dies that have a certain speed. The dies deform also (hopefully elastically) with heat conduction through them. The boundaries are made of ring-free boundaries, contact boundaries, and die-free boundaries. Along the contact boundaries, there are prescribed velocities in the vertical direction but prescribed forces (friction) in the horizontal direction. Heat transfer includes radiation and convection along the free boundaries and convection-like terms through the interface in the contact areas.

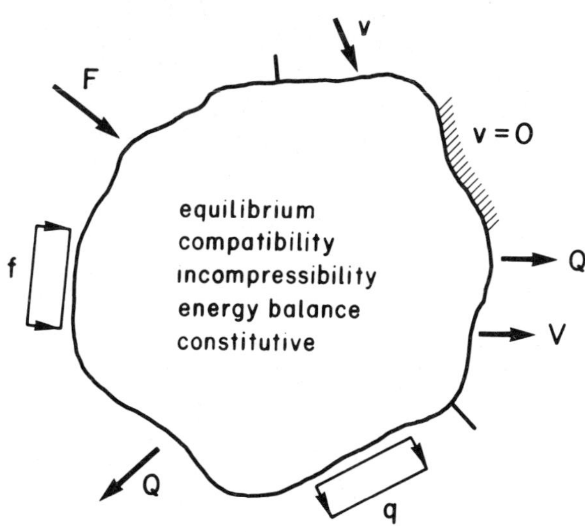

Figure 2.25: General form of a solid mechanics problem.

Current programs consider the dies as rigid, i.e., the elastic deformations of such dies are too small to interfere with the deformation of the workpiece and with its dimensional accuracy. However, simultaneous solution of heat transfer equations in workpiece and dies is necessary in warm forging, hot-die forging and isothermal forging because heating of dies not only interferes with the heat transfer but may affect its structural integrity.

In a problem like ring compression, when reasonably high friction is considered, a neutral zone develops which brings up two important features.

90

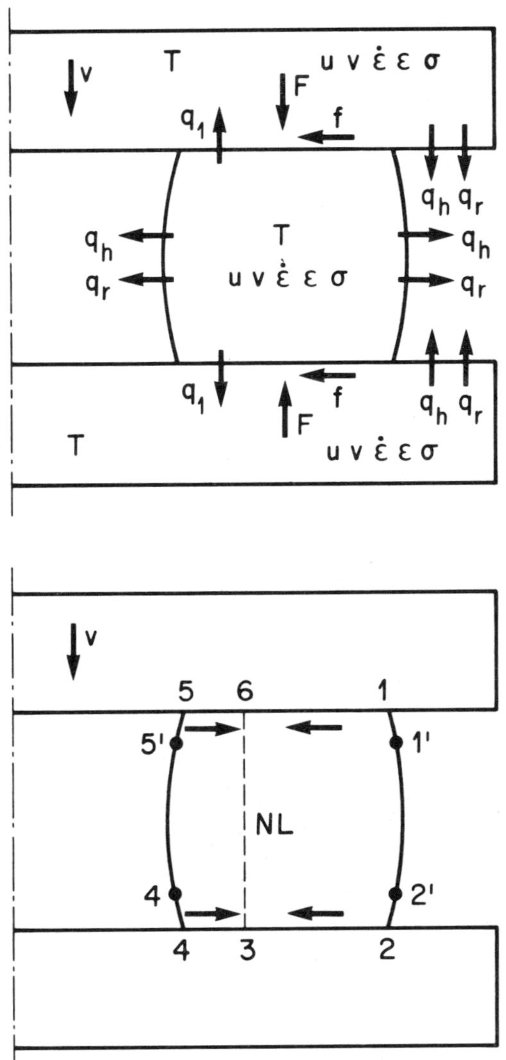

Figure 2.26: Typical forging problem.

First, a rigid zone within the deforming body develops. In rigid-plastic analysis, stresses are undetermined for zero strain rate. Whenever some regions in a non-uniform deformation process become rigid or almost non-deforming, the system of equations (2.163) becomes very ill-conditioned. In fact, $\dot{\varepsilon}$, being in the denominator, produces some very large values in the matrices. To obviate that, material behavior is modified, and a strain rate offset, $\dot{\varepsilon}_o$, below which stress drops linearly to zero, is imposed [2.56], as shown in Figure 2.27. This offset value is chosen for numerical reasons only, and should not be related to real material behavior. Values of 2 orders of magnitude smaller than an average strain rate in the deformation have given successful results.

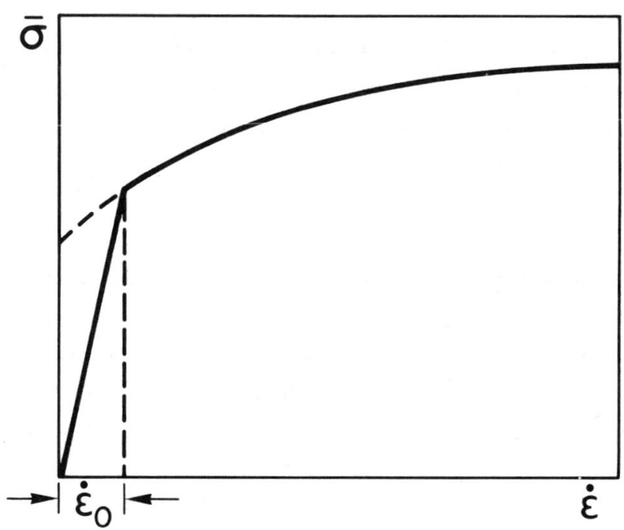

Figure 2.27: Schematic stress-strain curve for rigid-viscoplastic materials and modification with offset strain rate.

In the elements in which $\bar{\dot{\varepsilon}} < \dot{\varepsilon}_o$, equations (2.163) and (2.164) are conveniently modified.

Boundary tractions imposed in metal forming problems are mainly frictional forces. The modelling of these forces is traditionally made in two ways: by means of a friction coefficient and by means of a friction factor. A directly applied friction coefficient, the ratio between shear friction force and normal force, leads to non-symmetric matrices and is not commonly used in bulk deformation metal forming. A friction factor, defined by the equation:

$$\tau = mk \tag{2.173}$$

where:

τ -   shear friction stress
k   -   shear yield stress of the material
m   -   friction factor

was used in the examples presented in this book. Although friction modelling is still empirical and somewhat far from reality, as has been pointed out in a survey paper by Wilson [2.56], we believe that until further research clarifying the correct mechanisms is done, consistent ways of modelling should be used so that different analyses may be compared and, more importantly, experimental data can be quantified.

Secondly, a major problem is faced when a point of reverse relative velocity between workpiece and die exists and its position is not known a priori, like the neutral line. The orientation of the friction force changes at that point, so an abrupt jump in its value shows up, rendering equations (2.163) very ill-conditioned. Chen and Kobayashi [2.55] have made the problem amenable by substituting such step functions for an arc tangent function, as close to the step function as desired (Figure 2.28). The final equation for the friction force is given by:

$$ f = - mk \left\{ \left(\frac{2}{\pi} \right) \tan^{-1} \left(\frac{v_r}{a} \right) \right\} \underaccent{\tilde}{t} \tag{2.174}$$

where:

m   -   friction factor ($0 \leq m \leq 1$);
k   -   shear yield stress;
v_r -   magnitude of relative velocity between die and workpiece;
a   -   constant several orders of magnitude less than the die velocity.

The introduction of this frictional force requires linearization in the corresponding surface integrals of equation (2.160), modification of the contents of equation (2.163) but not its form.

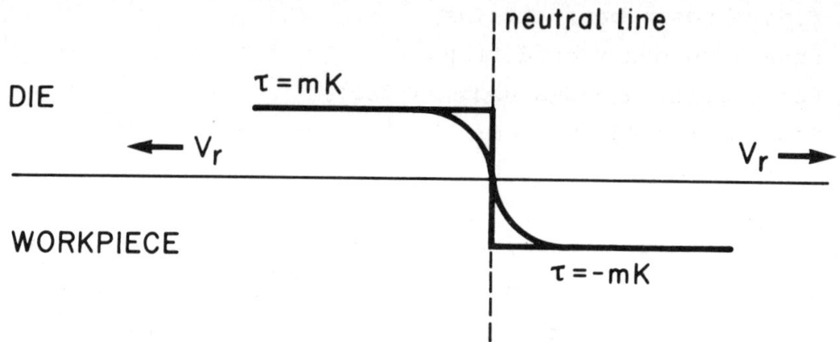

<u>Figure 2.28</u>:   Modelling of friction forces.

The next specific problem encountered is the fact that, with time, the geometry changes. Up to reasonable amounts of deformation, the bulk changes do not present a problem; however, the boundary conditions do change. Part of the surface that was free comes in contact with the dies, becoming restricted in the normal direction with frictional forces applied in the tangential direction. The type of heat transfer changes too, of course. The possibility of defining one or more dies, moving or not, with arbitrary shape, together with all the geometric verifications and adaptations, requires some programming, but does not really follow any particular theory. Experience has been obtained with a scheme, which, geometry-wise [2.57], defines dies by a sequence of points which have to be entered in a certain sequence as defined in Figure 2.29. Every two consecutive points define a segment to which an inclination corresponds. If there is a node touching a given segment, the inclination of the segment defines the direction of the friction force; the normal of the segment, the direction of the velocity restriction. In terms of deformation, the sequence of activities that the scheme executes is:

1)   Start with the preform in contact with one or more dies;
2)   Approach the other dies until first point is touched;
3)   Initialize the problem;
4)   Calculate velocity field;
5)   Determine minimum time for another point to touch a die;
6)   Choose time increment;

7) Update positions, strains;

8) Update boundary conditions

9) Force sliding nodes onto surface;

10) Repeat from 4) on.

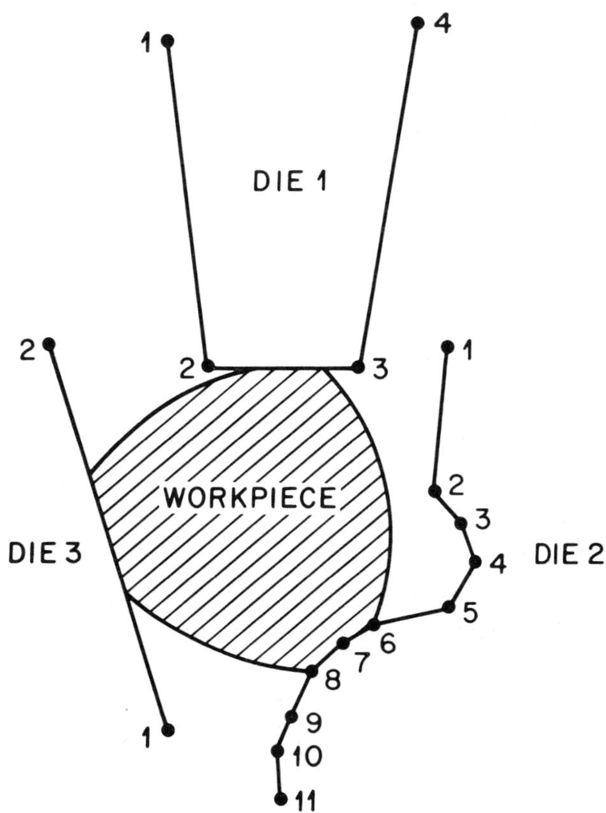

Figure 2.29: Definition of arbitrarily shaped dies.

After very large amounts of deformation, the initial mesh may distort and some elements may even degenerate, after which not only will the calculations become unreliable, but it may also be impossible to obtain any results. In this situation it is necessary to start again with a new mesh, to which the relevant variables will have to be transposed. As far as the deformation part is concerned, the strains need to be carried from step to step

because of work hardening, so these must be interpolated from
the element values of the old mesh into the element values of a
new mesh. With respect to heat transfer analysis, it is the temperatures that must be interpolated from nodal values of the
old mesh into nodal values of a new mesh. These procedures cannot
be done by hand and intermediate programs have to be written for
them. The software developed in conjunction with graphical pre-
and post-processors that deals with this problem in a semi-auto-
matic way will be described. Figure 2.30 shows the connections
between such software.

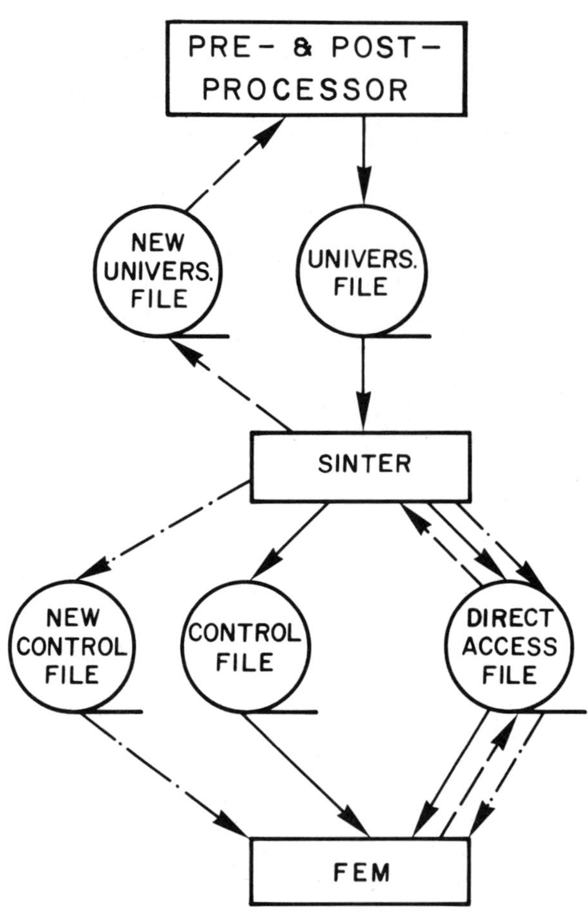

Figure 2.30: Software for forming simulation.

A commercially available Finite-Element pre- and post-processor
is taken and an original mesh is created, as well as the geometry
of the dies by means of a series of linear beams. Information on
nodal restrictions is also fed-in at this level. This program
writes all the information received into a file in a format
called "Universal".

Then, an intermediate, completely interactive program named
"Sinter" reads the "Universal"-file and asks for all other necessary information for running the Finite Element program. This
information includes the boundary nodes, precisions on die interface nodes, operating conditions, and so on. It writes a binary
file with most of the information and a very small formatted file
which contains control information.

The Finite Element code accesses both the control and the binary
file and executes the number of steps required. It appends the
results of pre-selected steps to the binary file, and prints
whatever information was indicated in the control parameters.

With the intermediate program the binary results of the stored
steps can again be accessed and written in another "Universal"-
file or a new control file that will make the Finite Element
code restart the analysis at any intermediate point can then be
created.

The pre- and post-processor can access the new "Universal"-file
and display the results in terms of distorted meshes, isoline
plots of all result variables, nodal forces or plots of variables
along selected lines.

If, upon examining the results, the analyst decides that a remesh
is necessary, he can again use the intermediate program to create
a new type of "Universal"-file. This one should contain only the
dies and the boundary points. With the pre-processor he can start
with the data of such a file and create a new mesh, which in turn
will be re-written on a start-like "Universal"-file. Finally, the
intermediate program will access both the old and the new mesh,
automatically interpolating the necessary variables and interactively asking for the operating parameters. As a result, a new

control file is written, and a new binary file is started. The Finite Element code interprets these as in a starting process.

It should be emphasized again that during a complete analysis no files are created by hand. Geometry and meshes are interactively created on a graphics screen, and the intermediate code "Sinter" operates uniquely by questions and answers with some suggested values for less obvious parameters.

A few comments should be made on the interpolations done during remeshing, in the light of the fact that several ways of doing this are reported in the literature [2.58, 2.59]. Our own procedures have evolved since starting this work.

In the analysis referred to in this text, an average value of strain is calculated at each element which can be seen as the value at the center. The interpolation procedure for strains goes as follows: networks of element centers, old and new, are set aside as shown in Figure 2.31; then, for each new center, the three closest old centers are sought for; finally, the strain values are linearly interpolated (or extrapolated).

The temperatures are nodal values. In order to interpolate them, it is necessary to find into which old element a new node falls. This element is then divided into triangles and the temperature is linearly interpolated within them, Fig. 2.32. At the boundaries, a new node may fall outside the old mesh, in which case its value is extrapolated from the nearest old triangle.

2.6.7   <u>Summary and Future Developments</u>

This chapter has described what is being done today in Finite Element process modelling. With technological advances being made at a tremendous speed, it is difficult to predict how things will evolve. However, there are a few developments on the horizon that it will not take long to implement.

First, Finite Element analysis needs to be extended to three-dimensional forming processes. Actually, very elementary shapes have already been done [2.60]. There are no conceptual difficul-

ties in making this extension and only computing times stand in
the way of making feasible analysis.

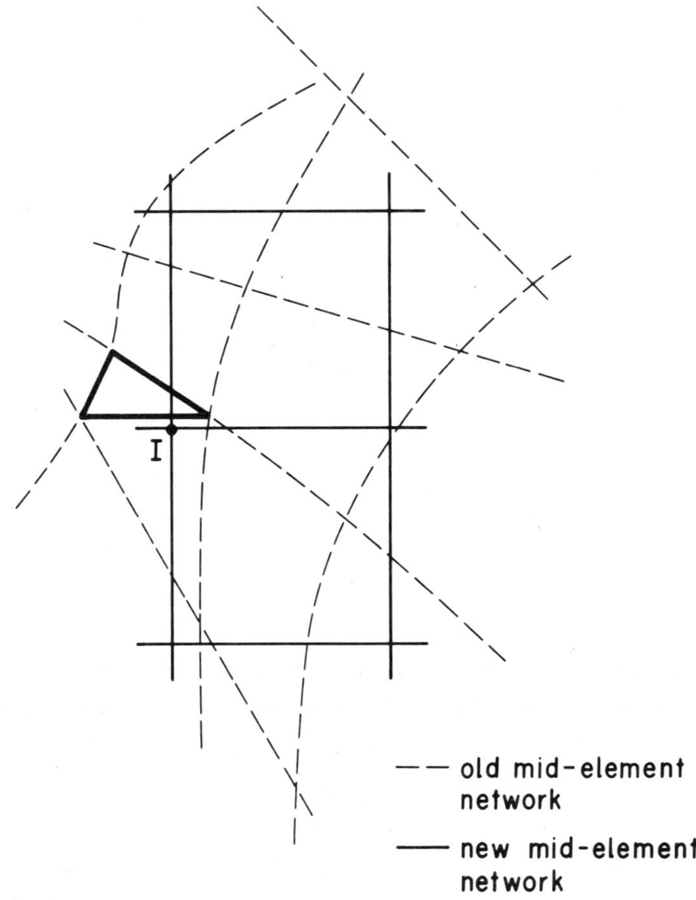

Figure 2.31: Strain interpolation scheme during remeshing -
interpolating values of element I.

The developments of section 2.6.6 are the first steps of a full
integration of Finite Element analysis with CAD/CAM systems. If,
from the design stages, CAD databases contain all the geometrical
data necessary for modelling, it should not be necessary to re-
enter the data into the pre-processor. This same data, in fact,
is already used for fabrication purposes.

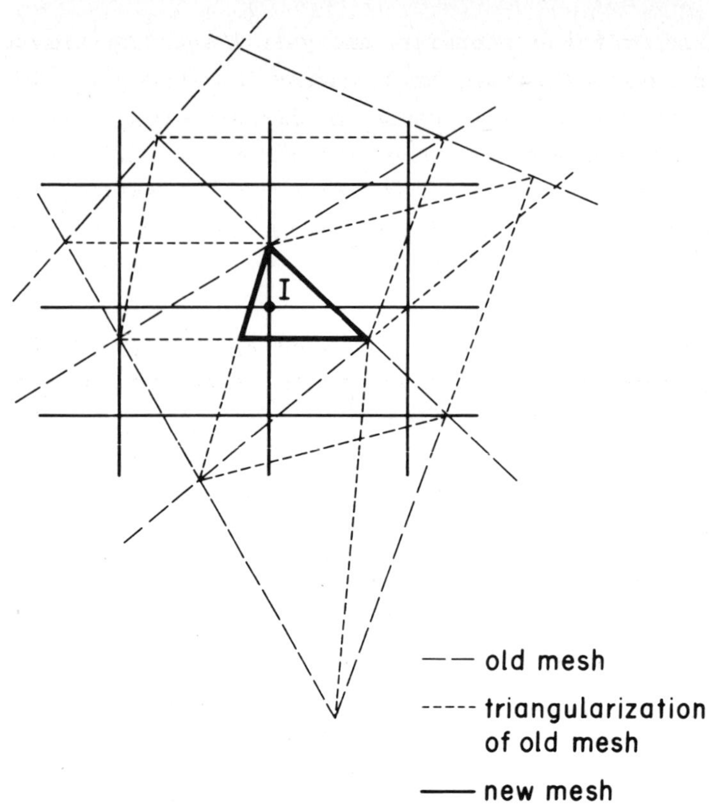

<u>Figure 2.32</u>: Temperature interpolation scheme during remeshing -
interpolating values of node I.

By using standard protocols of graphical data exchange, the pre-
processor can access the geometrical definition of the dies and
the preforms. The same applies to databases of material properties.

In fact, the full designing process will be totally computer-
integrated. Information stored in databases about materials,
shapes, parting lines, preforms and blocker designs will help
the designer do his job. Expert systems using artificial intelligence techniques will have a significant role in the decision-
making process of less analytical tasks.

Improvements in modelling friction can be made in order to include the knowledge collected on tribology during processes.

In addition, Finite Element analyses lend themselves to the calculation of parameters that change locally with deformation, such as, the parameters used in damage rules, which allow a prediction to be made of when the material will fail upon excessive deformation; or, metallurgical parameters that will allow the determination of mechanical properties or structural parameters.

Another interesting possibility, recently introduced [2.61] is to try to reverse the deformation process from end to beginning, in order to optimize preform design.

References Chapter 2

[2.1]   Naghdi P.M., Stress-strain Relations in Plasticity and Thermoplasticity, in Plasticity, Proceedings 2nd Symposium on Naval Structural Mechanics, eds. E.H. Lee, P.S. Symonds, Pergamon (1960).

[2.2]   Hill R., The Mathematical Theory of Plasticity, Oxford Clarendon Press (1950).

[2.3]   Boër C.R., Schröder G., Process Modelling of Hot-Die Forging-Application to a Nickel Base Alloy, Annals CIRP, 31/1 (1982) 137-140.

[2.4]   Boër C.R., Rydstad H., Schröder G., Choosing Optimal Forging Conditions in Isothermal and Hot-Die Forging, J. Applied Metalworking, 3, No. 4 (January 1985) 421-431.

[2.5]   Shah S.N., Monti G.J., Forging Simulation and Properties Achievable for Large Waspaloy Forgings for Land-based Gas Turbines, ASME Paper Nr. 81-GT-84 for Meet. Mar. 9-12 (1981) 1-8.

[2.6]   Kanetake N., Tozawa Y., Kato T., Ishikawa T., Computer aided Experimental Techniques in a Forming Laboratory, Annals of the CIRP, 32/1 (1983) 219-222.

[2.7]   Knight W.A., Poli C., Product Design for the Economical Use of Forging, Annals of the CIRP, 30/1 (1981) 337-342.

[2.8]   Lahoti G.D., Subramanian T.L., Altan T., Computer-aided Prediction of Metal Flow, Temperatures and Forming Load in Selected Metal-Forming Processes, ASME AMD, 28, Appl. of Numer. Methods to Form. Processes, ASME Winter Annu. Meet., San Francisco, Calif., Dec. 10-15 (1978) 183-195.

[2.9]   Altan T., Lahoti G.D., Nagpal V., Application of Process Modelling in Massive Forming Processes, Process Modelling Tools ASM Materials & Process Congress Proceedings (1980) 77-99.

[2.10]   Shahaf M., Bercovier M., Guez D., Blades I., Interactive
         Simulation of a Forging Process for Blades, Numerical
         Methods in Industrial Forming Processes (1982) 343-350.

[2.11]   Nagpal V., General Kinematically Admissible Velocity
         Fields for some Axisymmetric and Metal Forming Problems,
         Transactions of the ASME, Journal of Engineering for
         Industry (November 1974) 1197-1201.

[2.12]   Bishop J.F.W., An Approximate Method for Determining the
         Temperatures Reached in Steady-State Motion Problems of
         Plane Plastic Strain, Quarterly J. Mechanics and Appl.
         Math., 9 (1956) 236-246.

[2.13]   Kobayashi S., Metalworking Process Modelling and the
         Finite Element Method, Proceedings NAMRC (1981) 16-21.

[2.14]   Mouton J.-P., Combined numerical and Algebraic Computer
         Processing Applied to Plasticity Problems, Annals of the
         CIRP, 28/1 (1979) 131-134.

[2.15]   Oudin J., Ravalard Y., A General Method for Computing
         Plane Strain Plastic Flows, Proceedings 20th Interna-
         tional Machine Tool Design and Research Conference (1979)
         211-216.

[2.16]   Kiuchi M., Murata Y., Simulation of Contact Pressure
         Distribution on Tool Surface by UBET, Proceedings 21st
         International Machine Tool Design and Research Conference
         (1980) 13-20.

[2.17]   Venter R.D., Hewitt R.L., Johnson W., An Engineering
         Approach to the Matrix Operator Technique for Slip Line
         Field Construction, Proceedings NAMRC (1978) 111-118.

[2.18]   Boër C.R., Avitzur B., Schneider W.R., Eliasson B., An
         Upper Bound Approach for the Direct Drawing of Square
         Rod from Round Bar, Proceedings 20th MTDR Conference
         (1979) 149-156.

[2.19]   Clough R.W., The Finite Element in Plane Stress Analysis,
         Proc. 2nd A.S.C.E. Conf. on Electronic Computation, Pitts-
         burg, PA. (1960).

[2.20]   Lee C.H., Kobayashi S., Elastoplastic Analysis of Plane
         Strain and Axisymmetric Flat Punch Indentation by the
         Finite Element Method, Int. J. Mech. Sciences, 12, 4
         (1970) 349-370.

[2.21]   Alexander J.M., Gunasekera J.S., On the Geometrically
         Similar Expansion of a Hole in a Thin Infinite Plate,
         Proc. R. Soc. A, 326, 1566 (1972) 361-373.

[2.22]   Zienkiewicz O.C., The Finite Element Method, McGraw-Hill,
         U.K. (1977).

[2.23]   Bathe K.J., Finite Element Procedures in Engineering
         Analysis, Prentice-Hall (1982).

[2.24]   Irons B., Ahmad S., Techniques of Finite Elements, Ellis
         Horwood (1980).

[2.25]   McMeeking R.M. and Rice J.R., Finite Element Formulations
         for Problems of Large Elastic-Plastic Deformation, Int.
         J. Solids and Structures, 11 (1975) 601-616.

[2.26]   Hill R., Some Basic Principles in the Mechanics of Solids
         without a Natural Time, J. of the Mechanics and Physics
         of Solids, 7 (1959) 209.

[2.27]   Pinsky P.M., A Numerical Formulation for the Finite De-
         formation Problem of Solids with Rate-Independent Con-
         stitutive Equations, Report No. UCB/SESM-81/07, Univer-
         sity of California, Berkeley (1981).

[2.28]   Lee E.H., Some Comments on Elastic-Plastic Analysis, Int.
         J. Solids and Structures, 17 (1981) 859-872.

[2.29] Nagtegaal J.C., deJong E.J., Some Computational Aspects
        of Elastic-Plastic Large Strains Analysis, Int. J. Num.
        Meth. Eng., 17 (1981) 15-41.

[2.30] Lubarda V.A., Lee E.H., A Correct Definition of Elastic
        and Plastic Deformation and Its Computational Signifi-
        cance, J. Appl. Mechanics, 48 (1981) 35-40.

[2.31] Key S.W., Krieg R.D., Bathe K.J., On Application of
        the Finite Element Method to Metal Forming Processes -
        Part I, Comp. Meth. Appl. Mech. Eng., 17/18 (1979)
        597-608.

[2.32] Hughes T.J.R., Pister K.S., Consistent Linearization in
        Mechanics of Solids and Structures, Computers and Struc-
        tures, 8 (1978) 391-397.

[2.33] Krieg R.D., Krieg D.B., Accuracies of Numerical Solution
        Methods for the Elastic Perfectly-Plastic Model, J. of
        Pressure Vessel Tech., 99 (1977) 510-515.

[2.34] Nagtegaal J.C., On the Implementation of Inelastic Con-
        stitutive Equations with Special Reference to Large
        Deformation Problems, Comp. Meth. Appl. Mech. Eng., 33
        (1982) 469-484.

[2.35] Nagtegaal J.C., Veldpaus F.E., On the Implementation of
        Finite Strain Plasticity Equations in a Numerical Model,
        in Numerical Analysis of Forming Processes, John Wiley
        & Sons (1984).

[2.36] Simo J.C., Ortiz M., A Unified Approach to Finite Defor-
        mation Plasticity based on the Use of Hyperelastic Con-
        stitutive Equations, Report 10, UCB/SESM-84/13, Univer-
        sity of California, Berkeley (1984).

[2.37] Simo J.C., On the Computational Significance of the In-
        termediate Configuration and Hyperelastic Stress Rela-
        tions in Finite Deformation Elastoplasticity, Workshop
        on Inelastic Deformation and Failure Modes, Northwestern
        University (1984).

[2.38]   Oh S.I., Private communication.

[2.39]   Lee C.H., Kobayashi S., New Solutions to Rigid-plastic
         Deformation Problems Using a Matrix Method, J. Eng.
         Ind., 95 (1973) 865.

[2.40]   Kobayashi S., Rigid-Plastic Finite Element Analysis of
         Axisymmetric Metal Forming Processes, Numerical Modelling
         of Manufacturing Processes, ASME, PVP-PB-025 (1977) 49-
         68.

[2.41]   Zienkiewicz O.C., Godbole P.N., Flow of Plastic and
         Viscoplastic Solids with Special Reference to Extrusion
         and Forming Processes, Int. J. Num. Meth. Engr., 8
         (1974) 3.

[2.42]   Zienkiewicz O.C., Jain P.C.; Oñate E.; Flow of Solids
         During Forming and Extrusion: Some Aspects of Numerical
         Solutions, Int. J. Solids and Structures, 14 (1978) 15.

[2.43]   Chen C.C., Oh S.I., Kobayashi S., Ductile Fracture in
         Axisymmetric Extrusion and Drawing - Part I, J. of Eng.
         for Industry, 101 (1979) 23.

[2.44]   Shimazeki Y., Thompson E.G., Elasto-Viscoplastic Flow
         with Special Attention to Boundary Conditions, Int. J.
         Num. Meth. Engrg., 17 (1981) 97.

[2.45]   Oh S.I., Rebelo N., Kobayashi S., Finite Element Formu-
         lation for the Analysis of Plastic Deformation of Rate-
         Sensitive Materials in Metal Forming, IUTAM Symposium on
         Metal Forming Plasticity (1978) 273.

[2.46]   Perzyna P., The Study of the Dynamic Behaviour of Rate-
         sensitive Plastic Materials, Arch. Mech. Stos., 15 (1963)
         113.

[2.47]   Perzyna P., Thermodynamic Theory of Viscoplasticity, Adv.
         Appl. Mech., 11 (1971) 313.

[2.48]   Perzyna P., Sawczuk A., Problems of Thermoplasticity, Nucl. Eng. Design, 24 (1973) 1.

[2.49]   Rebelo N., Kobayashi S., A Coupled Analysis of Visco-Plastic Deformation and Heat Transfer I - Theoretical Considerations, Int. J. Mech. Sci., 22 (1980) 699-705.

[2.50]   Argyris J.H.; St. Doltsinis J., On the Natural Formulation and Analysis of Large Deformation Coupled Thermo-mechanical Problems, Comp. Meth. Appl. Mech. Eng., 25 (1981) 195.

[2.51]   Wertheimer T.B., Thermal Mechanically Coupled Analysis in Metal Forming Processes, Numerical Methods in Industrial Forming Processes (1982) 425-434.

[2.52]   Zienkiewicz O.C, Oñate E., Heinrich J.C., Plastic Flow in Metal Forming - I. Coupled Thermal Behaviour in Extrusion - II. Thin Sheet Forming, Appl. Num. Meth. Form. Proc. ASME AMD 28 (1978) 107-120.

[2.53]   Rebelo N., Kobayashi S., A Coupled Analysis of Visco-Plastic Deformation and Heat Transfer II - Applications, Int. J. Mech. Sci., 22 (1980) 707-718.

[2.54]   Rahlston A., A First Course in Numerical Analysis, McGraw-Hill, New York (1965).

[2.55]   Chen C.C., Kobayashi S., Rigid-Plastic Finite Element Analysis of Ring Compression, Appl. Num. Meth. Form. Proc., ASME AMD 28 (1978) 163-174.

[2.56]   Wilson W.R.D., Friction and Lubrication in Bulk Metal Forming Processes, J. of Appl. Metalworking, 1 (1979) 7.

[2.57]   Rebelo N., Rydstad H., Schröder G., Simulation of Material Flow in Closed-Die Forging by Model Techniques and Rigid-Plastic FEM, Numerical Methods in Industrial

[2.58] Gelten C.J.M, Konter A.W.A., Application of Mesh-Rezoning
        in the Updated Lagrangian Method to Metal Forming Ana-
        lysis, Numerical Methods in Industrial Forming Processes
        (1982) 511-521.

[2.59] Badawy A., Oh S.I., Altan T., A Remeshing Technique for
        the FEM Simulation of Metal Forming Processes, Proc. 1983
        ASME Int. Computer Engr. Conf., Chicago (1983).

[2.60] Park J.J., Kobayashi S., Three-Dimensional Finite Element
        Analysis of Block Compressions, Int. J. Mech. Sic., 26
        (1984) 165.

[2.61] Park J.J., Rebelo N., Kobayashi S., A New Approach to
        Preform Design in Metal Forming with the Finite Element
        Method, Int. J. Mach. Tool Des. Res., 23 (1983) 71-79.
        Forming Processes (1982) 237.

3 Physical Modelling

3.1 Similarity Theory

Modelling laws permit the transfer of information obtained from model tests to the full-scale component, which may differ from the model in terms of size, material, physical characteristics or boundary conditions. For a long time, model tests were a decisive aid to the engineer when designing complex structures and machines, particularly for aircraft and ships. The development of modern computer technology and computer simulation has reduced the importance of the model technique.

In many respects, forming technology is informative as an example of the present state of modelling theory. For simple geometries, it can be seen that computer simulation has clearly replaced model testing, whereas for complex geometries (e.g. closed die forging) and for complex materials laws (crack formation, instabilities), the modelling technique still has advantages not only in terms of cost.

3.1.1 Basic Principles

The basic requirement for the use of modelling theory is that the model (M) and full-scale component (F) are geometrically similar. Corresponding lengths (ℓ) must be in a fixed ratio to one another:

$$\lambda \;=\; \ell_M / \ell_F \qquad\qquad \text{Geometrical similarity.}$$

If geometrically similar objects are subjected to physical processes or effects, then the procedure is referred to as physically similar if the values of the physical characteristics at the appropriate positions of the model and full-scale component have the same relationship. Two procedures are completely similar if similarity also applies to the boundary conditions.

In model tests, it is generally only possible to attain incomplete similarity of the boundary conditions and transfer of the physical characteristics (extended similarity). This has led to the development of model techniques which provide special models with limited predictive capabilities for special requirements and questions [3.1, 3.2, 3.3].

For the basic quantities length ℓ, time t, force F and temperature T there is geometrical, temporal, dynamic or thermal similarity between the full-scale component (F) and the model (M) provided that the scale factors are complied with:

$$\lambda = \ell_M/\ell_F$$
$$\tau = t_M/t_F$$
$$\chi = F_M/F_F \qquad \text{Scale factors}$$
$$\vartheta = T_M/T_F$$

The ratio values λ, τ, χ, ϑ are also called basic scale factors. Derived quantities can be obtained from the basic quantities e.g. for the velocity:

$$V_M/V_F = \lambda/\tau$$

It is advantageous to determine the scale factor for any physical quantity from the units. If a physical quantity

$$B = F^{n1} \ell^{n2} t^{n3} T^{n4}$$

has the unites $N^{n1} m^{n2} s^{n3} K^{n4}$ (with the SI units kg, m, s, K), than the transfer scale factor is obtained directly from the units as:

$$B_M/B_F = \chi^{n1} \cdot \lambda^{n2} \cdot \tau^{n3} \cdot \vartheta^{n4}$$

Through the use of non-dimensional characteristic factors, the number of variables is reduced and the appropriate equations/ differential equations which describe a process can be expressed as a function of the non-dimensional characteristic factors [3.1]. The description of the test results with non-dimensional characteristic factors permits a simple concentrated illustration using only a few diagrams. Similarity laws are also generally defined using non-dimensional characteristic factors.

3.1.2 Similarity Laws in Metal Forming

Forming technology and heat treatment and combinations of the
two are dealt with using a limited number of differential equations, materials properties and boundary conditions. Accordingly, a number of non-dimensional characteristic factors suffice
for the formulation of modelling laws and for the description
of the effective physical processes [3.4, 3.5, 3.6]. The ten
well-known basic equations of plasticity, Chapter 2, provide
the starting point for the actual forming process.

According to [3.4, 3.6], the following non-dimensional characteristic factors can be derived:

$$K_1 \; = \; \frac{F}{\sigma_f \cdot L^2} \; = \; \frac{F}{A \cdot \sigma_f} \qquad \text{Force characteristic of static plasticity.}$$

with σ_f as mean flow stress.

The force characteristic refers the stress characteristic (F/A)
to the mean flow stress:

$$K_2 \; = \; \frac{F}{G \cdot A} \quad \text{or} \quad K_3 \; = \; \frac{F}{E \cdot A} \qquad \text{Force characteristic of static elasticity.}$$

with E as modulus of elasticity and G as shear modulus.

$$K_4 \; = \; \frac{F}{\rho \cdot A \cdot v^2}$$

The characteristic factor K_4, also known as the Newton-Bertrand
similarity law, is important at high velocity or acceleration
at which inertial forces appear. This is only applicable for
high velocity forming e.g. explosive forming.

$$K_5 \; = \; \frac{A \cdot C_p \cdot \rho}{t \cdot \lambda}$$

with λ: thermal conductivity; C_p: Specific heat; ρ: density.

K_5 describes the thermal conduction and is known as the Fourier similarity law. Two thermal conduction processes are similar if the Fourier characteristic parameters (K_5) agree.

$$K_6 = \frac{F}{A \cdot \Delta T \cdot C_p \cdot \rho}$$

with the temperature increase ΔT.

K_6 is the similarity law for adiabatic heating as a result of plastic deformation. In this way, the temperature change in the model caused by the forming process can be compared with that in the full-scale component.

Friction

In forming technology, frictional processes such as sliding, adherence and shearing occur in the gap between the tool and workpiece. These boundary conditions of the forming processes can be described approximately using laws of friction:

$$\mu = \frac{\tau}{\bar{p}} \qquad\qquad \text{Coulomb law of friction.}$$

$$m = \frac{\tau}{k}$$

with
τ: frictional shear stress

$\bar{p}$: mean normal stress

k: shear flow stress.

The dimensional analysis of μ and m shows immediately that these are already non-dimensional characteristic factors. This means that the similarity law for the friction requires identical frictional characteristic factors (μ, m) in the model and in the full-scale component. Often this represents a major difficulty in the modeling of forming processes with model materials.

Materials Laws, Flow Stress

The most important materials property in forming technology is the flow stress, σ_f:

$$\sigma_f = f(\varepsilon, \dot{\varepsilon}, T)$$

The following simplification is usual for the selection of model materials:

<u>Cold forming</u>: $\sigma_f = f(\varepsilon)$
with the representation $\sigma_f = C_1 \cdot \varepsilon^n$

<u>Hot forming</u>: $\sigma_f = f(\dot{\varepsilon}, T)$
with the representation $\sigma_f(T) = C_2 \cdot \dot{\varepsilon}^m$

The following requirement is then obtained for the flow curves of model and full-scale component in cold forming:

from $\qquad K_1 = \dfrac{F}{A \cdot \sigma_f} \qquad$ follows

$$\sigma_{f/M} = \frac{\chi}{\lambda^2} \cdot \sigma_{f/F} \qquad \text{or}$$

$$\frac{\sigma_{f/M}}{\sigma_{f/F}} = \frac{\chi}{\lambda^2} = \frac{C_{1/M} \cdot \varepsilon^{n_M}}{C_{1/F} \cdot \varepsilon^{n_F}} = \text{const.}$$

This shows that the work hardening exponent n must be the same for the model and the full-scale component, whereas the constants C_1 for the model material can be chosen freely. This is the basis for the use of soft model materials such as wax for the modelling of forming processes.

The conditions are complex for hot forming, since even with the simplification $\sigma_f = f(\dot{\varepsilon}, T)$ a function of two variables is present. For different materials in the model and full-scale component, it is practically impossible to achieve agreement with the similarity laws.

The following simplification is valid for constant (but not identical) temperatures in the model and full-scale component:

$$\frac{\sigma_{f/M}}{\sigma_{f/F}} = \frac{\chi}{\lambda^2} = \frac{C_{2/M} \cdot \dot{\varepsilon}^{m_M}}{C_{2/F} \cdot \dot{\varepsilon}^{m_F}} = \text{const.}$$

The similarity law requires agreement of the strain rate exponent m of the material of the model and of the workpiece (full-scale component), whereas the constants C_2 can differ. These considerations show that model tests are only of very limited use for the modelling of non steady-state temperature distributions such as in forging. However, (approximately), isothermal processes (isothermal forging, hot die forging, extrusion, cold forming) can be well simulated.

3.2 Application and Basic Techniques

Physical modelling of metal forming is based on the idea that metal flow can be simulated by using a soft model material using cheap dies and simple machinery, thereby making it easy and relatively cheap to change die and preform geometries, as well as forming parameters such as reduction per step etc. The idea of studying metal flow by observing flow of non-metals is not new; it dates back to the last century when Tresca and Obermeier carried out studies using plasticine and clay [3.7, 3.8]. T. Wanheim, V. Maegaard and J. Danckert have made extensive references to the literature in this field [3.9].

The model material technique is well developed for different applications in metal forming, mainly

-   die filling, strain rate and strain distribution in the deformation zone during deformation

-   hardness and strength distribution of the parts after forming

-   load distribution on the dies.

Table 3.1 summarizes the cases where model techniques could be used and those cases where it can be said that the state-of-the-art is such that they can be used for routine investigations. It has also been demonstrated [3.11] that the temperature increase

Part of process to be analyzed	state-of-the-art	Where in system
- filling of tool	a	
- strain increments	a	
- strain distribution	a	Deformation
- strain rate distribution	a	Zone
- temperature distribution	b	
- stress distribution	b	
- obtainable deformation	d	
- fibre structure	a	
- hardness distribution	a	
- strength	a	Part after
- ductility	d	Deformation
- porosity	d	
- residual stresses	d	
- elastic deformation during unloading	d	
- geometrical surface properties	d	
- mechanical surface properties	d	
- tangential stress	b	
- normal stress	b	
- sliding length	a	Interface
- wear	b	
- surface deformation	b	
- strain distribution	b	
- stress distribution	b	
- temperature distribution	d	Tool
- elastic deformation	b	
- yield/fracture load	b	
- post-failure condition	c	
- geometrical, mechanical and chemical surface properties due to surface reactions	d	Free Surface
- magnitude and direction of forces	a	
- load-displacement curve	a	Machine
- mechanical deformation	c	
- thermal deformation	d	

<u>TABLE 3.1</u>

State-of-the-art in physical modelling (after [3.10]).

a: Techniques have been applied to different processes and are
    ready for daily use.

b: Techniques have been tested on a single or a few processes
    and seem promising for other processes.

c: Have not been tested but are considered as probably possible.

d: Probably not possible.

in the model material plasticine can be used as a measurement of the deformation. The case that was studied was the upsetting of a billet. The temperature increases were very small though, ($\sim$ 0.2°C) and the method can only be applied for processes where large deformations take place, for instance, in extrusion. In another interesting case, an attempt was made to study corner cracking in ingot rolling [3.12]. For the creation of corner cracks, the temperature gradient in the ingot is of prime importance. It was not possible to create a realistic gradient when scaling down the ingot to sizes suitable for laboratory tests using metallic materials. However, by using plasticine at room temperature where the surfaces were cooled with the help of cold steel plates at -18°C, a temperature gradient was created. The plasticine billet was then rolled with the gradient and the corner-cracking behavior studied. The method is based on the fact that plasticine has a very low thermal conductivity (1.12 (W/m°C)) compared with metals (Aluminum 220 (W/m°C), Iron 75 (W/m°C)), which made it possible to scale the temperature distribution.

A few attempts have been made to deal with porosity using model materials. The spring-back of model wax turned out to be a problem when simulating ingot rolling [3.13]. For a different reason, namely, to be able to estimate hydrostatic pressure, porous sintered copper was used in forging simulation [3.14].

Since model wax has low ductility in tension, cracks will appear where the material experiences tensile stresses. This effect, which in itself is not desirable, can be used to show where there is a risk of cracking when deforming brittle materials.

A wide range of materials are used in the modelling of metal forming. One group consists of metals which have low flow stresses and which are easy to deform. For instance, lead [3.15, 3.16, 3.17] or copper, but also aluminum, are used to perform low-temperature simulation of steel forming. Lead has also been used to simulate superplastic flow in isothermal forging of Ti6Al4V [3.18]. The other group is the non-metals such as waxes [3.19, 3.20], sodium [3.15], slipline wax [3.21] and plasticine [3.22], whereby plasticine is the preferred material. The production of

billets suitable for model experiments is explained in detail in an excellent handbook on model material techniques by J. Danckert [3.6].

There are two different ways of preparing plasticine specimens so that the deformation can be registered. One route is to paint a grid on the specimen with silk-screen techniques. The other method is to build up the specimen of strips or cubes of plasticine in different materials. The first method is preferable, but cannot be used for three-dimensional studies. Plasticine experiments are most suitable when the problem can be idealized as being a plane-strain one. In this case, it is reduced to a two-dimensional problem and a cross-section of the billet can be covered by a glass plate and the whole deformation can be registered using one specimen. In the three-dimensional case, the billet has to consist of different colors and a stop must be made at different stages during the deformation, the specimen cut up and the deformation registered. Reference [3.6] describes the experimental technique in detail. Other references describing it are [3.23, 3.24, 3.25 and 3.26].

A material used to help construct upper-bound solutions of the deformation is the slip-line wax [3.21]. The wax is black to start with, but when it is deformed, the zones where shearing occurs change color. Other methods that can be used to determine the slipline field are the photo-elasticity and photo-plasticity methods [3.10].

The die stresses can also be modelled. The methods used include using brittle coatings on the dies as well as making the dies out of plaster to see where the cracking occurs. Other methods include measuring local pressure in the die with pressure gauges [3.23] and one method where the tools are made from transparent polyurethane rubber printed with a moiré grating in the symmetry plane. When the rubber die is under load, i.e. during deformation, the moiré grating can be photographed and analyzed [3.6, 3.27]. The pictures of the deformed moiré gratings were superimposed with a line grating and the moiré patterns corresponding to the horizontal and vertical displacements were obtained by spatial filtering out of all but the first diffraction order in

an optical filtering system. The fringe orders in the moiré patterns at the node points of the grid were used as input data for a computer program which calculated the strains, the equivalent strains and the shear stresses at each node point.

Other methods utilize the change in refractive index that arises when a stress is applied to some transparent material [3.28]. Two different approaches using this phenomenon can be identified. In one, a sensitive layer (Epoxiresin) is placed in a die model made out of insensitive material (Acrylresin). When the load is applied, the stresses are measured using reflective or transparent lighting techniques. Problems with this technique include stress-free glueing of the layers and the fact that only one plane can be studied at a time. The other approach uses the so-called "freeze" technique, where a model is loaded above a certain temperature, the load kept constant and the temperature slowly reduced to room temperature. During this temperature change, the Young's modulus changes several orders of magnitude as also do the optical properties. The deformations introduced above the critical temperature are then fixed or "frozen" and remain even after removing the load. The model can then be cut into slices and investigated. One problem is the fairly large deformations that arise due to the low Young's modulus when the model is loaded.

The model material technique is suitable for die design. For example, the die design when extruding copper has been experimentally determined using model wax and a kind of build-up technique for the dies [3.29]. The final shape of the copper profile was upset to different deformations until an almost round shape was achieved. This was then assumed to be the starting billet size. A wooden die was built up using the different geometries that arose during the upsetting. In this built-up wooden die, which then consisted of different layers, the angles were smoothed out and a wax billet extruded.

In the literature, the main emphasis for model material simulations are in rolling and forging. In rolling, the technique has been used to predict force and torque [3.30], strip crown [3.16] and flow in profile rolling [3.31]. Crop loss is one important

118

aspect which is suitable for model material techniques, since
mathematical techniques, especially in rolling plate in different
directions, are difficult to use (fish tailing, three-dimensional
crop losses) [3.32 - 3.35].

The material losses when piercing material have been studied
using model wax to support slip-line field solutions [3.36].

3.3 Material Flow in Forging – Examples

The material flow in a circular symmetrical forging was investigated. A plane cross-section was used to simulate the process in dies which were wire-eroded and made from aluminum. The model material was FIMO[*] from which rectangular sections were made and glued together. These striped billets were then cut to size and forged. An even better impression of how the material is deformed was achieved by having different billets with strips in different directions. The study was undertaken to do an initial study of the deformation pattern and distribution when isothermally forging a nickel-base alloy. The way the model material flows can be seen in Figures 3.1a - 3.1d.

Figure 3.1: Upsetting of plasticine billet.

---

*) Registered trade name

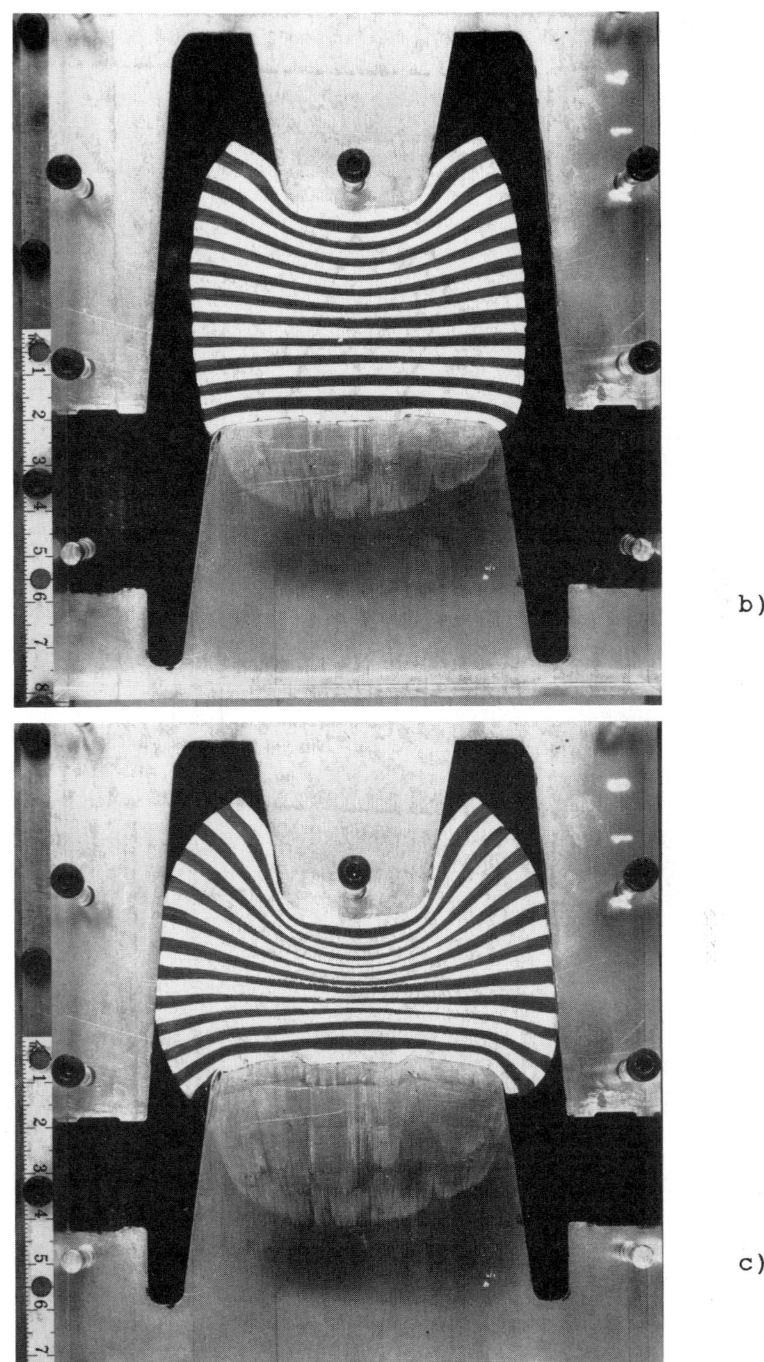

Figur 3.1:   Upsetting of plasticine billet.

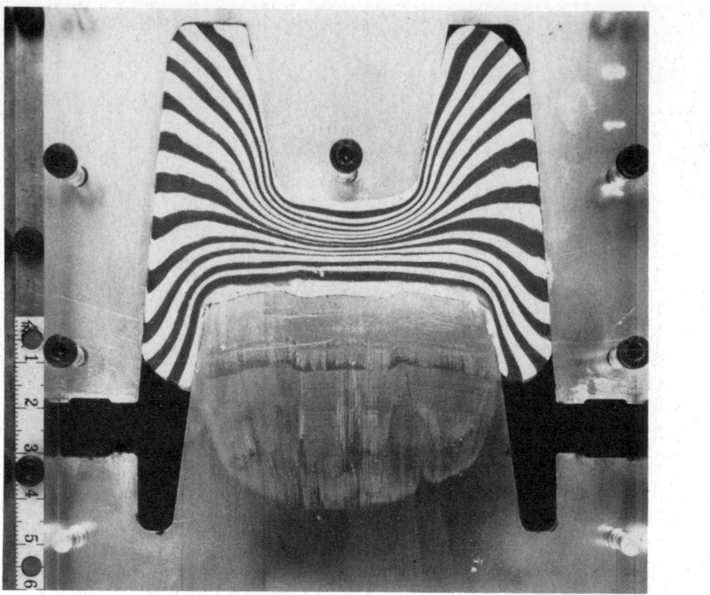

Figure 3.1:   Upsetting of plasticine billet.

The thick rib was almost completely filled before the material started to flow into the thin rib. Figure 3.2 shows the deformation pattern when the dies were completely filled.

By having the strips vertical or horizontal, the deformation pattern can be seen in different ways. In Figure 3.3, the flow of the material along the inside of the thin rib as well as flow in the middle can be seen. A feeling for the amount of deformation given to the material can be obtained from Figures 3.2 and 3.3. When studying the two pictures, an area in the outside corner in the thicker rib can be found where very little deformation takes place. This is in good agreement with the results from FEM-calculations and real forgings as presented in Chapter 4.4.

The technique used is only suitable for simulating the flow. A quantitative analysis is not possible. Another problem is that the differently-colored strips of plasticine tend to separate from each other during deformation, thereby changing the simulated flow.

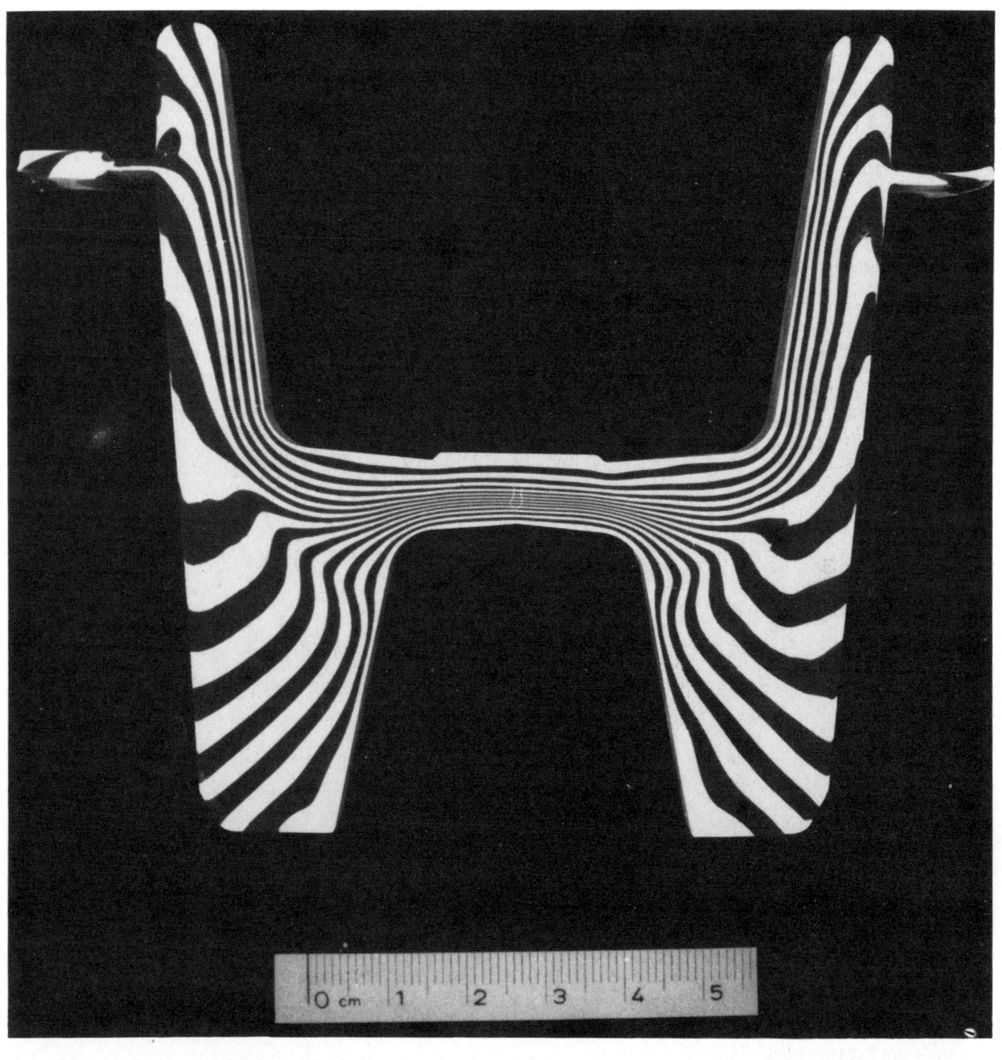

<u>Figure 3.2</u>:   Final deformation pattern. Original billet size
55 x 73 mm^2.

A cross-section of a three-dimensional simulation done in an axisymmetric die is shown in Figure 3.4. The wax was aged in an oven
after deformation so that it could be cut and surface ground.
Once again, a small area where very little deformation was found
can be seen. These findings led to the introduction of an upsetting step to introduce a minimum amount of deformation in the
billet before the shaping operation.

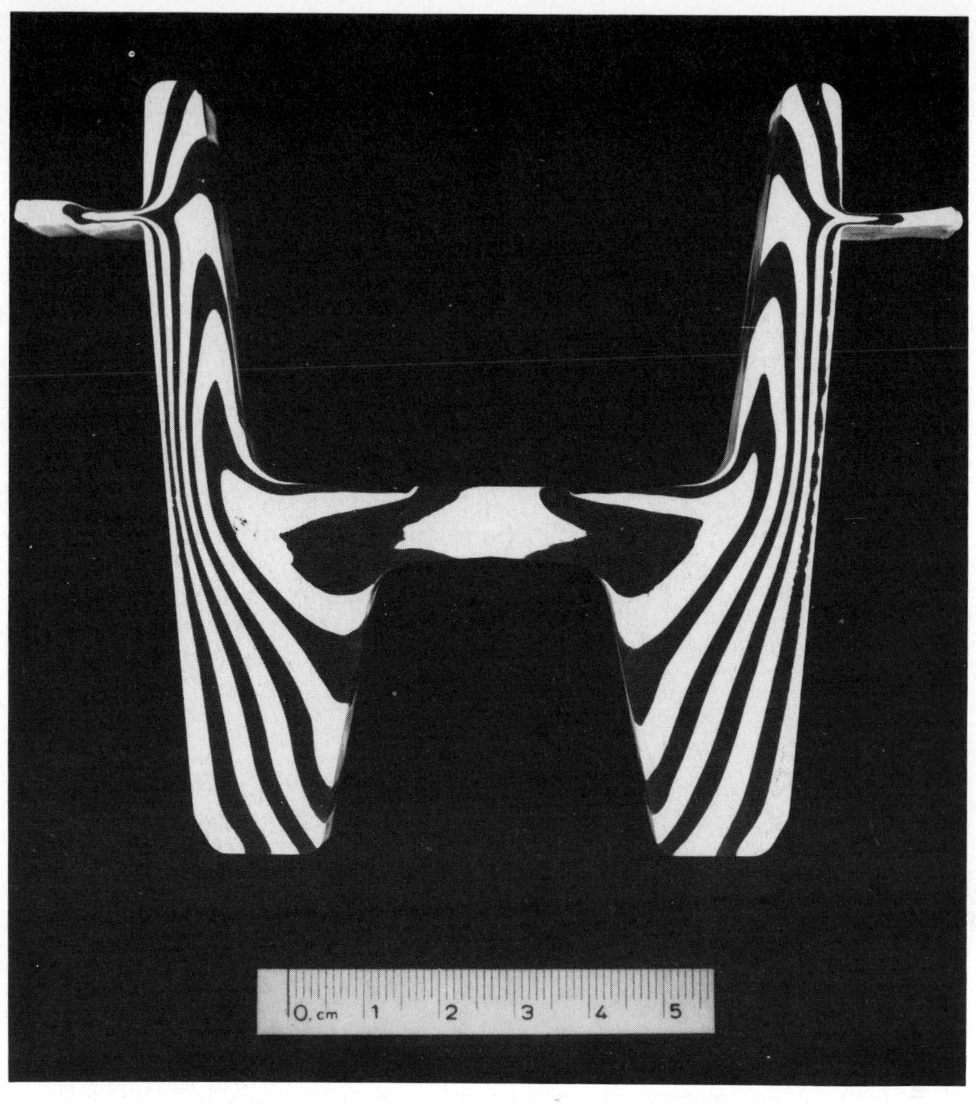

Figure 3.3:   Final deformation pattern using billet with vertical
striping. Original billet size 70 x 57 mm^2.

Figure 3.5 shows another example of material flow where a first
approximation of the deformations was obtained. In the forging,
in this case a turbine blade, different cross-sections were
studied, Figures 3.6 and 3.7. It gave a first visualization of
the material flow and the preform geometry could then be optimized.

124

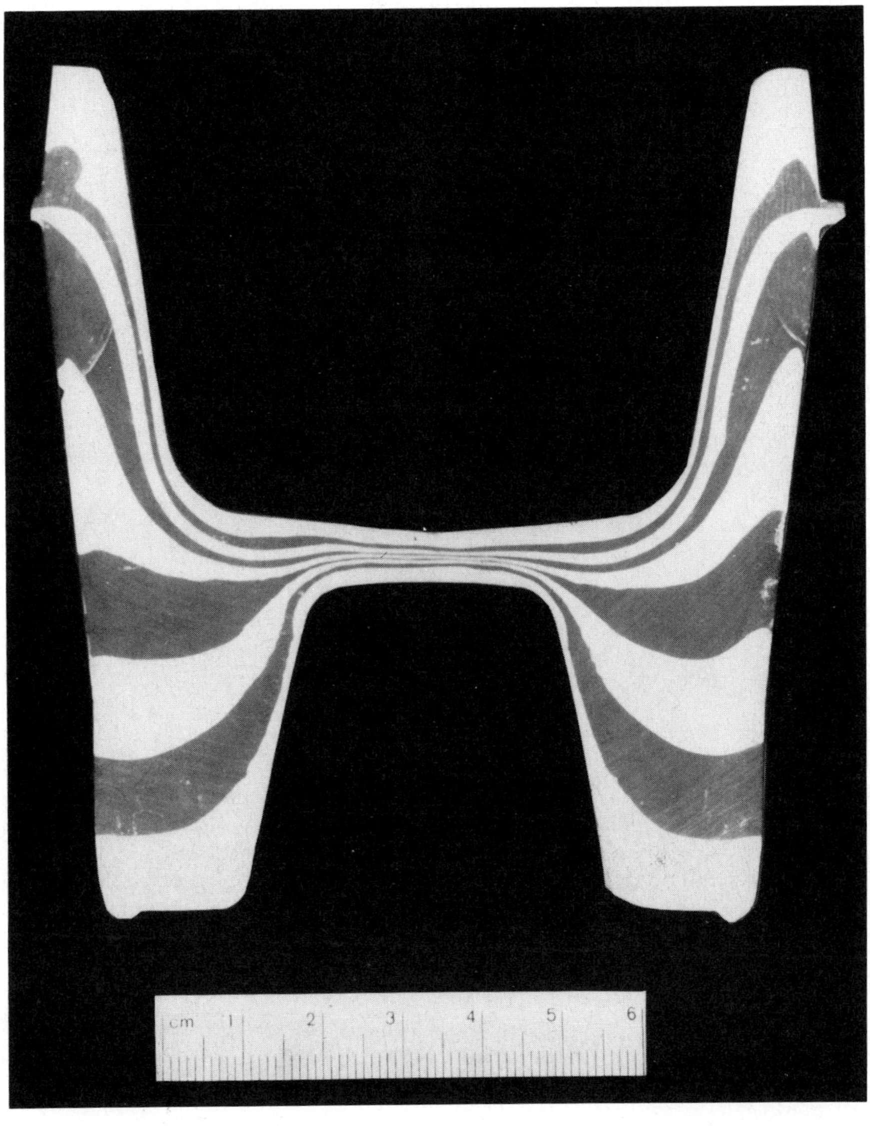

<u>Figure 3.4</u>: Cross-section of a three-dimensional simulation in an axisymmetric die.

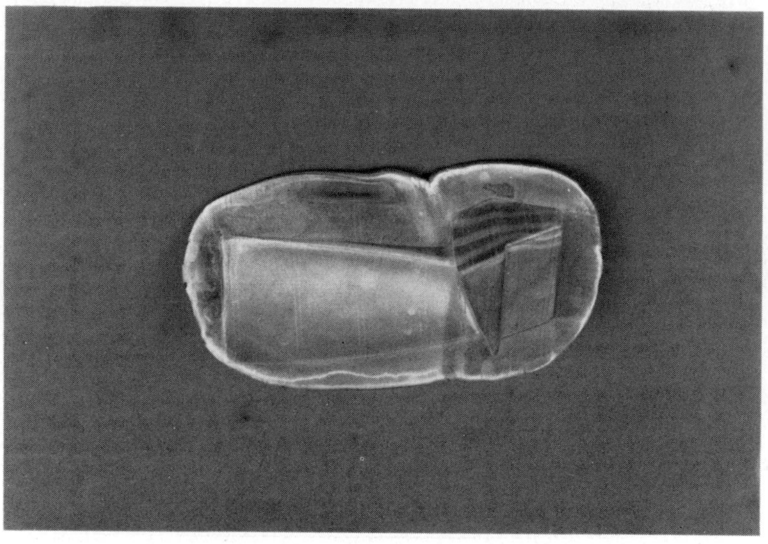

Figure 3.5: Model material simulation of turbine blade forging using a very simple preform.

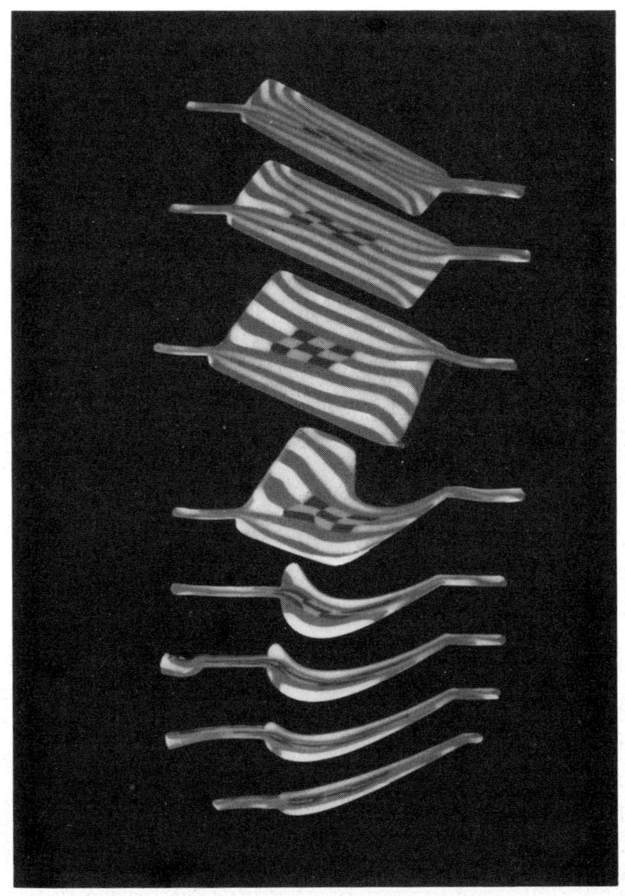

Figure 3.6: Cross-section taken from a model material simulation of a turbine blade forging.

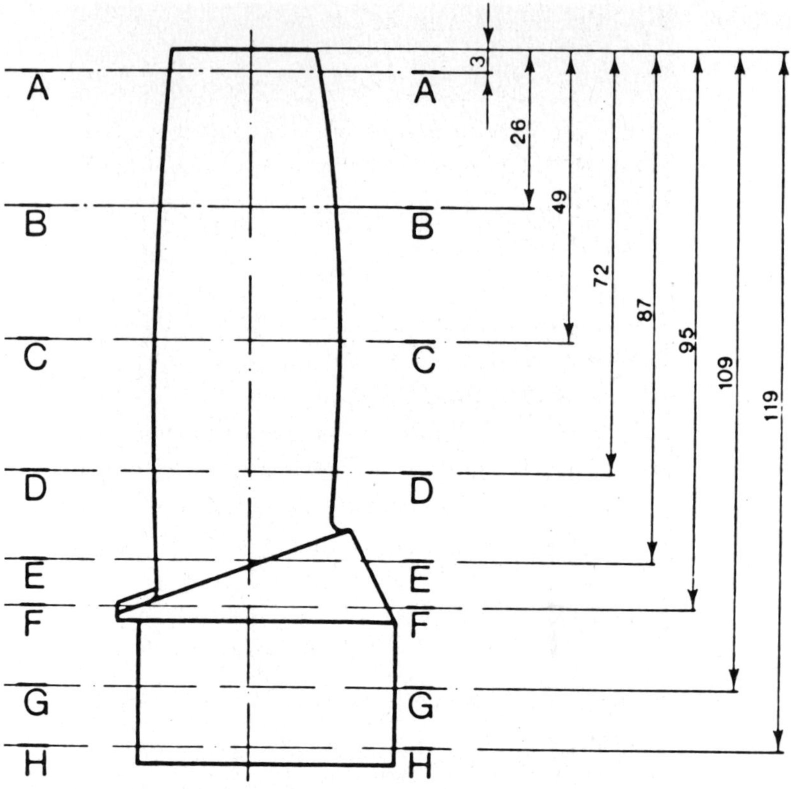

Figure 3.7: Position of the cross-sections shown in Figure 3.6.

3.4 Simulation of Material Deformation in Turbine Blade Forging

In this chapter, the simulation of material flow during isothermal forging of a turbine blade is described using plasticine
in the physical experiments. The experiments were compared
with FEM-calculations [3.37] using the rigid-plastic formulation
as described in Chapter 2.

In the model material study, plasticine with a rigid-plastic
behavior with no deformation hardening was used. The forging
simulation was done under plane strain conditions with a friction coefficient similar to the one expected in forging.
A square grid was painted on the wax billets so that the material
flow could be detected in the different incremental deformation
steps. When the grid pattern became difficult to resolve due to
deformation, the grid was renewed.

The problem simulated was the forging of a turbine blade, at a
given cross-section. Here it was assumed as an approximation that
plane strain conditions occur. The final desired profile was
divided into two, defining an upper and a lower die. A circular
preform was used. These three elements define the geometry of
the problem. The model material was plasticine of 20 mm thickness
that was deformed between two wooden dies. The initial diameter
of the billet was 70 mm, on which a 4 x 4 mm grid was painted,
Figure 3.8. During compression, the convex die was fixed and the
concave die was moving. The calculated mesh and the mesh from the
experiments are directly comparable up to a deformation of about
40 %. After this, the mesh on the model material was renewed.
These dies were unlubricated to simulate the friction conditions
in actual forging. Ring tests with the model material on wooden
dies showed that the m-value corresponded to 0.56.

The material properties used in the FEM calculations were identical to the properties of the model material. The flow stress
of the model material was experimentally determined, and numerically represented as in Figure 3.9. 145 nodes and 120 elements
defined the FEM-mesh.

Three different friction conditions were simulated with the FEM-

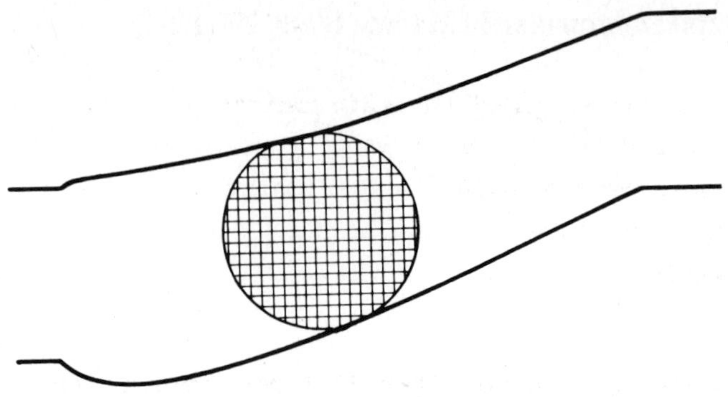

Figure 3.8: Model material preform with painted grid.

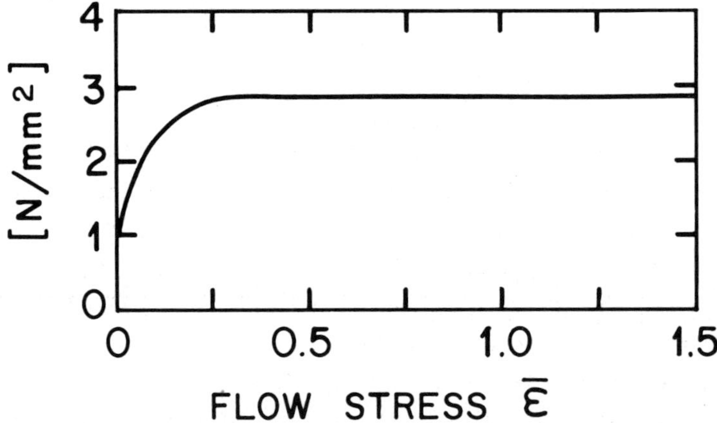

Figure 3.9: Flow stress curve of the model material used.

program. The influence of the friction conditions on the material flow are described in Chapter 4.3.4.4.

Figure 3.10 shows the model material after 31.5 mm compression. This should be compared with the deformed FEM-grid after the same compression, Figure 3.11. Isostrain curves for the same condition were plotted in Figure 3.12. The isolines show the effective strains of 0.1, 0.4, 0.8 and 1.1.

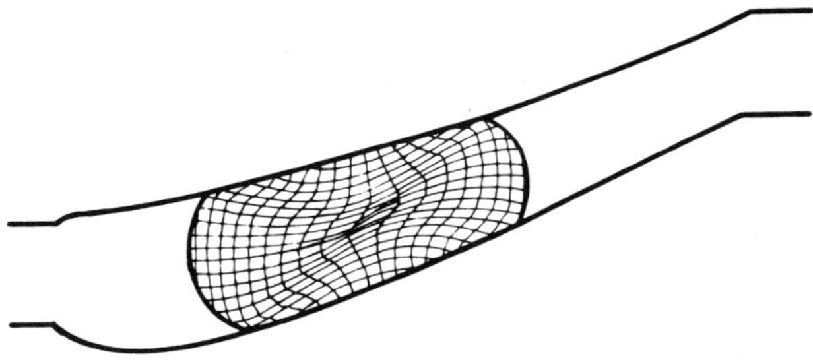

Figure 3.10: Model material grid after 31.5 mm compression.

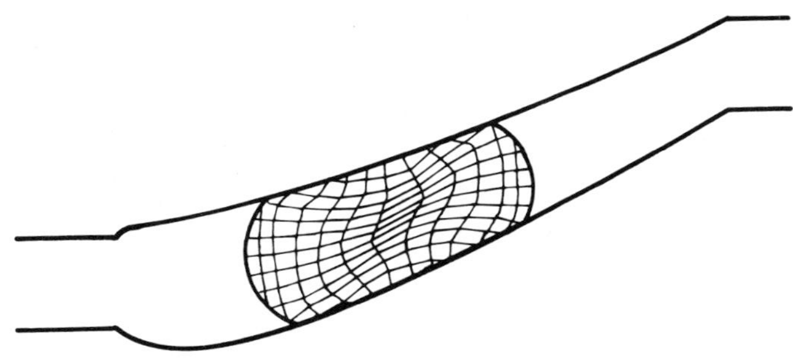

Figure 3.11: Calculated grid distortions using FEM after 31.5 mm
compression. m = 0.56.

A computer-aided strain analysis was also performed using the
information obtainable directly from the deformed mesh painted
on the plasticine billet. The billet to be deformed was painted
with a net of 4 x 4 mm squares. When this was placed between the
dies, a picture was taken. This picture was later used to define
the position of the nodes in the grid. After each deformation
step, a new photo was taken. In each step, the strain increment
was calculated and added to the total strain. The method used was
the so-called coefficient method. A description of the method
and of the computer program used for evaluating the experiments
is found in [3.38, 3.39].

The deformed model material grid after two deformation steps, but
without updating as interpreted by the computer after digitizing

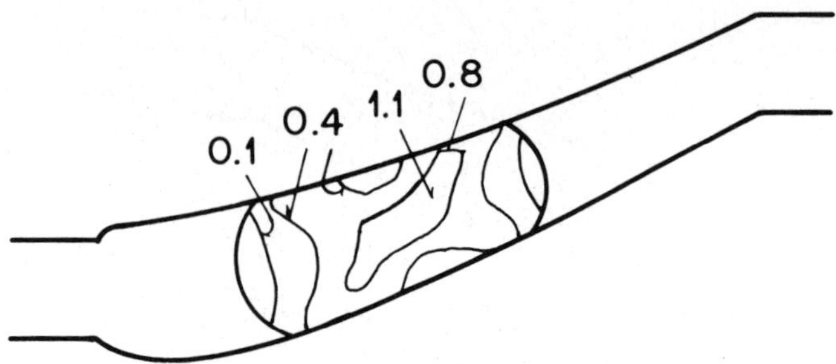

Figure 3.12: Calculated effective strain distribution after 31.5 mm compression. m = 0.56.

the deformed grid, is shown in Figure 3.13 a). The maximum shear stress directions are shown in Figure 3.13 b) and the equivalent strains in Figure 3.13 c). The grid after the first updating is shown in Figure 3.13 d). It is apparent that by using this method, the information on the flow behavior in the boundary zones is very limited.

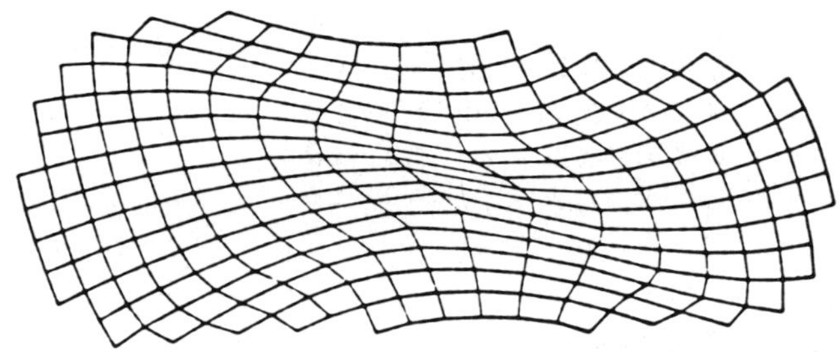

Figure 3.13 a): Deformed grid after two steps as interpreted by the computer after digitizing the deformed grid.

Figure 3.14 shows a model material experiment with well-lubricated dies, giving a coefficient of friction very close to 0 and the grid from a FEM-simulation using m = 0. The compression in this case was 24.5 mm. The material flow is well predicted, including the large translation of the billet. The disagreement is due to the slight differences in compression and to friction.

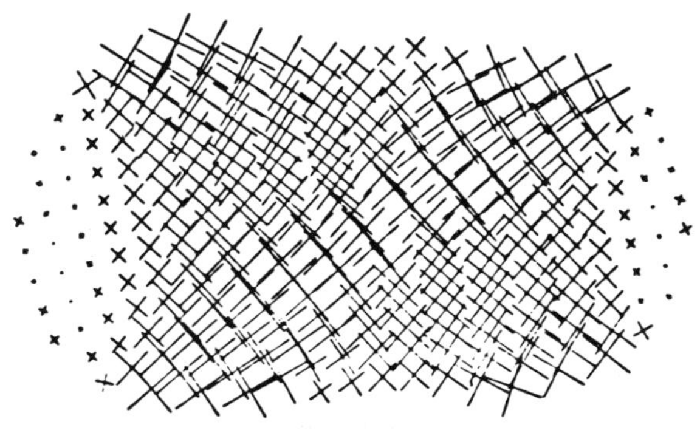

Figure 3.13 b): Maximum shear stress directions in the deformed grid shown in Figure 3.13 a).

A	0.0 - 0.1	M	1.0 - 1.1
B	0.1 - 0.2	N	1.1 - 1.2
C	0.2 - 0.3	P	1.2 - 1.3
D	0.3 - 0.4	R	1.3 - 1.4
E	0.4 - 0.5	S	1.4 - 1.5
F	0.5 - 0.6	T	1.5 - 1.6
G	0.6 - 0.7	U	1.6 - 1.7
H	0.7 - 0.8	V	1.7 - 1.8
K	0.8 - 0.9	X	1.8 - 1.9
L	0.9 - 1.0	Z	1.9 - 2.0

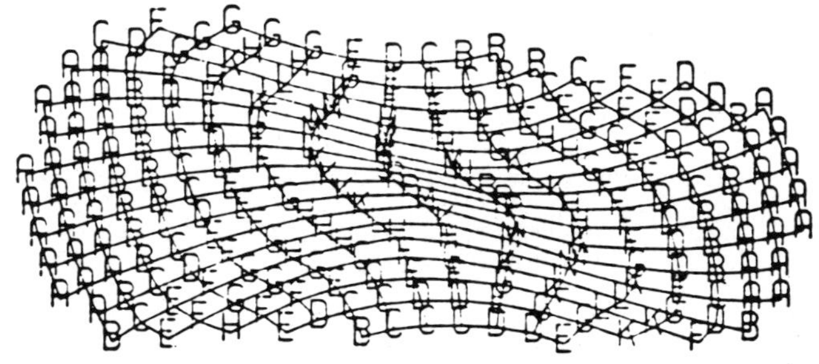

Figure 3.13 c): Equivalent strain distribution in the billet after two deformation steps.

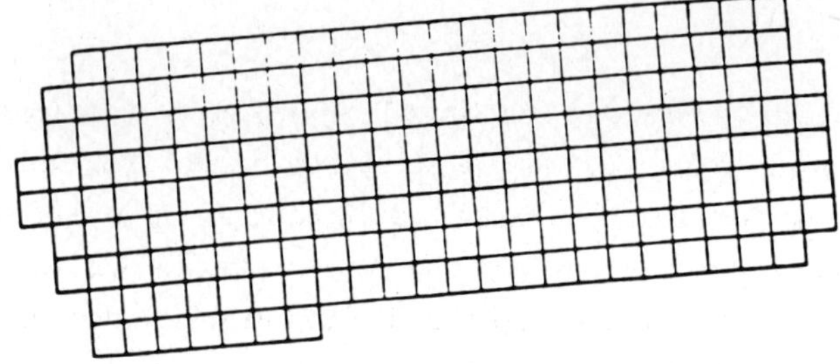

<u>Figure 3.13 d)</u>: Updated grid after two deformation steps.

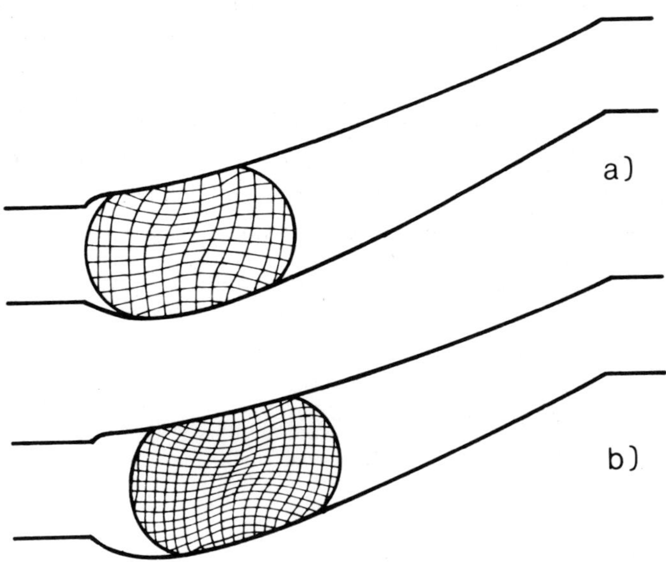

<u>Figure 3.14</u>: 24.5 mm Compression: a) Calculated grid distortion
m = 0.0; b) Model material grid distortion using
vaseline as lubricant.

3.5 Conclusions

As has been clearly shown, physical modelling reached maturity
a number of years ago. It is well established for certain purposes such as understanding and visualizing the material flow
in a deformation process, studying the filling of dies and for
use in education to explain metal deformation. Its main advantages lie in the facts that the capital costs are low and changes
in tooling or billet geometries can be made quickly and cheaply.

However, for more detailed simulation, the different mathematical
modelling techniques are superior. The mathematical modelling
techniques also fit better into the modern concept of CAD/CAM
where the designer can simulate the process, design the tools
and send the raw data directly to the workshop for NC-machining.

The main drawback of the FEM-simulation, namely the long computational times, are becoming less important as faster and cheaper
computers are introduced. Other drawbacks with FEM such as lack
of user-friendliness are also being overcome.

References Chapter 3

[3.1]   Weber M., Das allgemeine Ähnlichkeitsprinzip in der Physik und sein Zusammenhang mit der Dimensionslehre und der Modelwissenschaft. Jahrb. Schiffbautechn. Ges. (1930) 274-388.

[3.2]   Görtlar H., Dimensionsanalyse, Springer, Berlin (1975).

[3.3]   Katanek S., Grögner R., Bode C., Ähnlichkeitstheorie, VEB Deutscher Verlag für Grundstoffindustrie, Leipzig (1967).

[3.4]   Pawelski O., Beitrag zur Ähnlichkeitstheorie der Umformtechnik, Arch. f. d. Eisenhüttenwesen 35 (1964) 27-36.

[3.5]   Kast D., Modellgesetzmässigkeiten beim Rückwärtzsfliesspressen geometrisch ähnlicher Näpfe. Ber. a. d. Inst. f. Umformtechnik, Univ. Stuttgart, W. Girardet, Essen, Nr. 13 (1969).

[3.6]   Danckert J., Modelmaterialeteknik, AMT - Publikation 78.13 A, Danmarks tekniske Höjskole, Lyngby, Denmark (1977).

[3.7]   Tresca H., Sur l'écoulement des corps solides soumis á de fortes pressions. C.R. Acad. Sci. Paris 59 (1864 II) 754.

[3.8]   V. Obermayer A., Versuche über den Ausfluss plastischen Tons. S.-B. Akad. Wiss. Wien 58 (1868) 737.

[3.9]   Wanheim T., Maegaard V., Danckert J., The Physical Modelling of Plastic Working Processes, Proceedings of the First International Conference on Technology of Plasticity, Tokyo, Vol. II (1984) 984-997.

[3.10] Wanheim T., Schreiber M.P., Grönbaek J., Danckert J., Physical Modelling of Metal Forming Processes, Materials and Processing Congresses 1978-1979, American Society for Metals, Metals Park, Ohio (1980) 145-166.

[3.11]   Engelbert T., Side Extrusion of Tube, Ph.D. Thesis, Royal Institute of Technology, Department of Metal Working, Stockholm, Sweden (1980).

[3.12]   Andrén G., Götvalsningens betydelse för uppkomst av hörnsprickor, Ph.D. Thesis, Royal Institut of Technology, Department of Metal Working, Stockholm, Sweden (1980).

[3.13]   Wallerö A., SIMON-Simulering med modellmaterialier i Norden, Copenhagen, Denmark, Nordforsk (Sept. 1983) 34-35.

[3.14]   Ono S., Tanaka M., Tsukada H., Iwadate T., Analysis of Distribution of Deformation during Hot Forging, Proceedings of the First International Conference on Technology of Plasticity, Tokyo, Vol. II (1984) 997-1003.

[3.15]   Altan T., Henning H.J., Sabroff A.M., Use of Model Materials in Predicting Forming Loads in Metal Working, J. Eng. Ind. - Trans. ASME (May 1970) 444-452.

[3.16]   Shida S., Awazuhara H., Yasada K., Tsu-mura S., Simulation of Hot Rolling of Steel Using Lead, Trans. Iron Steel Inst. Japan, 19 (1979) 700-705.

[3.17]   Tsukamoto H., Taura Y., Ibushi J., Simulation of Hot Steel in Plastic Working with Plasticine and Lead, Proceedings of the First International Conference on Technology of Plasticity, Tokyo, Vol. II (1984) 1003-1009.

[3.18]   Corti C.W., Gessinger G.H., Shabaik A.H., Superplastic Isothermal Forging: A Model Metal Flow Study, J. Mech. Working Techn., 1 (1977) 35-51.

[3.19]   Bodsworth C, Halling J., Barton J.W., The Use of Paraffin Wax as a Model Material to Simulate the Plastic Deformation of Metals, Part I - A Preliminary Investigation into the Mechanical Properties of Paraffin Wax, J. of the Iron and Steel Institute (March 1957) 375-383.

[3.20] Lee R.S., Blazynski T.Z., Mechanical Properties of a Composite Wax Model Material Simulating Plastic Flow of Metals, Journal of Mechanical Working Technology, 9 (1984) 301-312.

[3.21] Danckert J., Wanheim T., Slipline Wax, Experimental Mechanics (August 1976) 318-320.

[3.22] Green A.P., The Use of Plasticine Models to Simulate the Plastic Flow of Metals, Philosophical Magazine, 42, No. 327 (1951) 365-373.

[3.23] Wanheim T., Danckert J., Johansen E., Anwendung der Modelltechnik bei Massivumformvorgängen, Industrie-Anzeiger Nr. 70 (1977).

[3.24] Wanheim T., Schreiber M.P., Grönbaek J., Danckert J., Physical Modelling of Metal Forming Processes, J. Appl. Metalwork., Vol. 1, Nr. 3-5, (1980) 5-14.

[3.25] SIMON-Simulering med modellmaterialier i Norden, Copenhagen, Denmark, Nordforsk (Sept. 1983).

[3.26] Wanheim T., Simulation with Model Materials in the Nordic Countries, Symposium "Fundamentals of Metal Forming Technique", Oct. 1983, Institut für Umformtechnik, University of Stuttgart (1983).

[3.27] Danckert J., and Wanheim T., The Friction Shear Stress Distribution at the Material-Tool Interface, when Upsetting a Flat Circular Cylinder between Flat Parallel Plates, 1979 ASM Materials and Processing Congress, Chicago (1979) 429-447.

[3.28] Kuske A., Schmidt O., Entwicklung eines spannungsoptischen Modellverfahrens zur Festigkeitsermittlung von Umformwerkzeugen (z.B. Strangpressmatrizen) under Verwendung eines plastischen, spannungsoptisch aktiven Modellumformgutes, Fortschr.-Ber. VDI-Z., Reihe 2, Nr. 43 (1980).

[3.29]   Jounio S., Kivivuori S., Nieminen M., Pihalainen H., SIMON-Simulering med modellmaterialier i Norden, Copenhagen, Denmark, Nordforsk (Sept. 1983) 27-29.

[3.30]   Tsukamoto H., Egawa T., Ibushi J., Oomori S., Yagishita K., Simulative Model Test on Metal Forming using Plasticine as a Model Material, SME Technical Paper, Dearborn, Mich. (1974).

[3.31]   Barlow K.N., Lancaster P.R., Investigation of Internal Flow Occurring During a Square-Diamond Rolling Pass, Using Plasticine as a Model Material, Met. Technol., 11 (1976) 503-509.

[3.32]   Ikushima H., Hirasawa T., Nakauchi I., Settai Y., Yamagishi, Y., Plasticine Model Tests to Determine Effects of Rolling and Ingot Geometry Variables on Bottom Crop Losses of Slab Ingots, Ironmaking Steelmaking, 3 (1977) 176-180.

[3.33]   Isuda O., A Comprehensive Approach to Minimizing Crop Loss in Slabbing, International Conference on Steel Rolling, 29 Sept. - 4 Oct., 1980, Vol. I, Science and Technology of Flat Rolled Products, Tokyo, Japan (1980) 169-180.

[3.34]   Ståhlberg U., Söderberg J.-O., Wallerö A., Overlap at the Back and Front End in Slab Ingot Rolling, Int. J. Mech. Sci., Vol. 23 (1981) 243-252.

[3.35]   Private communication, BHP Research Laboratories, Melbourne, Australia.

[3.36]   Nilsson T., Ståhlberg U., Reduction of the Discard Formed in Piercing, Scand. J. Metall., 9 (1980) 41-45.

[3.37]   Rebelo N., Rydstad H., Schröder G., Simulation of Material Flow in Closed-Die Forging by Model Techniques and Rigid-Plastic FEM, in Numerical Methods in Industrial Forming Processes, edited by J.F.T. Pittman, R.D. Wood,

J.M. Alexander and O.C. Zienkiewicz, Pineridge Press, Swansea, U.K. (1982) 237-246.

[3.38] Danckert J., Wanheim T., The Use of a Square Grid as an Alternative to a Circular Grid in the Determination of Strains, Journal of Mechanical Working Technology, 3 (1979) 5-15.

[3.39] Bredendick F., Methoden der Deformationsmittlung an verzerrten Gittern, Wissenschaftliche Zeitschrift der Technischen Universität Dresden (1969) Heft 2.

4 Modelling of Forging

<u>List of symbols</u>

(other symbols are defined in the text)

c_p specific heat

D_D die diameter

F force

h_o initial billet height

h_f final billet height

H billet height

H_D die height

t_2 lubricant or thermal barrier thickness

T_b billet temperature

T_D die temperature

T_{DO} initial die temperature

T_{BO} initial billet temperature

v ram velocity

ε strain

$\dot{\varepsilon}$ strain rate

θ time

λ thermal conductivity

ρ density

σ_{fb} flow stress of an axisymmetric billet

$\sigma_{0.2}$ 0.2 % yield stress

4.1 Introduction

The forging operation can be seen as a system with a large number of interacting variables. Furthermore, these variables can have a rather large field of values which they can span when the die and workpiece temperatures and resulting contact times for conventional forging, hot-die forging and isothermal forging are rather different. The selection of the optimum conditions during isothermal forging is not simple. In conventional and hot-die forging, the problem is even more complex due to the heat transfer phenomena between billet and die.

This chapter will be concentrated into three main parts: the elementary analysis (or system approach), the Finite Element analysis (or detailed mechanics approach) and a discussion on the integration of CAD/CAM and process modelling for forging simulation and die design.

The elementary analysis or "slab method" is a very useful tool to be used for a preliminary study of forging processes. The rather simple implementation in computer programs, together with the good results obtained in terms of stresses and forces, makes the elementary analysis particularly suitable for a system approach to forging [4.1] as described in 4.2.

More sophisticated implementation of the method allows the study of rather complex geometric forms in plane strain situations as shown in 4.3 and 4.4.

The method is very well suited for the optimization of forging parameters as described in 4.2.4, because of the relatively large number of results obtained in a rather short computer time.

The elementary analysis can also be integrated in a CAD/CAM system (see Chapter 4.3.2) and become a particularly useful tool for the forging industry for the design of forging dies.

4.2 System Modelling in Hot Die Upsetting

Forging is a deformation process which involves several variables interconnected by more or less complex functions. Figure 4.1 illustrates the forging as a system of several variables. The system consists mainly of three physical components: the billet, the lubricant (or thermal barrier) and the die. Each component can be represented by geometrical, physical and mechanical properties.

The billet is defined by geometry (height and diameter), material (or physical properties such as the thermal conductivity, specific heat and density) and mechanical properties (such as the flow stress).

The die is defined (it is supposed that its dimensions are much larger than those of the billet) by physical as well as mechanical properties (i.e. allowable stress).

The lubricant or thermal barrier is defined by its thickness and physical properties and friction coefficient. For the purpose of evaluation of different forging techniques, the major interactions between the variables of the system must be known. The more important variables are the force developed during the process, the ram velocity, the friction and heat transfer at the interface between die and billet, the initial temperature in the die and in the billet and the temperature distribution during deformation in the die and in the billet.

The method discussed here, used in the hot-die forging model to determine the stresses and the load, is the well known "slab method" or elementary analysis. Several authors have been working on this method, which can calculate the stress distribution for almost any practical plane strain and axisymmetric shape. The model for cylindrical upsetting (with or without flash) as described in [4.2] was used in the present model.

The following information is used when implementing the slab analysis:

- the geometry of the die including flash height and width when required. The model will calculate whether the material flows by sliding against the die or by shearing internally.

- the flow stress of the billet material which is a function
  of the temperature, strain and strain rate.

  There is a direct and indirect influence of the rate of deformation on the flow stress. In general, the flow stress increases with the strain rate and this effect can be directly
  calculated. The strain rate is a function of the ram speed
  which in turn determines the contact time between the die and
  the billet. The temperature in the billet and in the die is
  therefore a function of the ram speed. This indirect effect
  of the deformation rate on the temperature must be combined
  with the gain in temperature as a result of deformation and
  friction (see Figure 4.1).

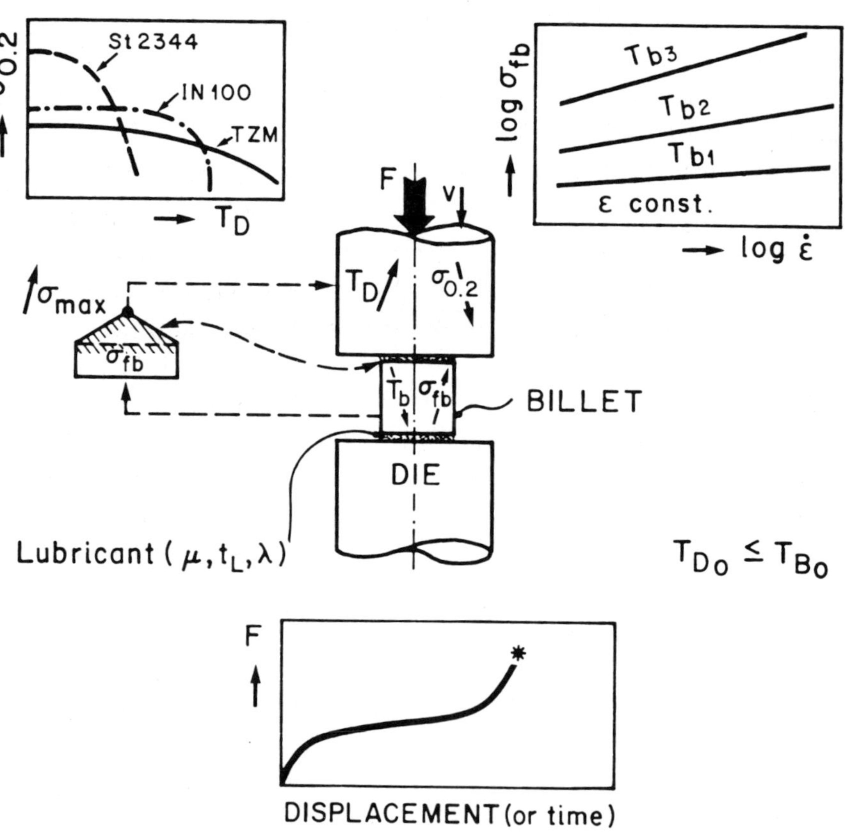

Figure 4.1: The forging system.

The flow stress relationship is implemented in the model as a
subroutine. Different material behavior can therefore be easily
represented. In general the form is:

$$\sigma_{fb} = C \, \dot{\varepsilon}^{m} \qquad\qquad (4.1)$$

where: σ_{fb} is the flow stress of the billet,
    $\dot{\varepsilon}$ is the strain rate,
    $C = C(\varepsilon, T_b)$,
    ε is strain,
    T_b is the billet temperature.

The functions C and m are generally expressed in the form of
tables and the appropriate values are linearly interpolated.

From the above consideration, it is evident that the temperature
distribution in the system as a function of time is quite important for an evaluation of the overall performance of the dies.
The assumption of an average temperature with time is therefore
unacceptable and hence it is mandatory to calculate the heat
transfer more precisely. It is also important to be able to simplify the model in order to reduce the calculation time.

The heat transfer problem during hot upsetting in heated dies was
solved by the Finite Element method [4.3]. To simplify the approach, a non-dimensional analysis was attempted in order to
reduce the number of variables to a reasonable number of non-
dimensional parameters [4.4]. The result is a set of non-dimensional curves which can be easily implemented in a computer program. The experimental validation of the heat transfer model has
also been demonstrated [4.5]. The effect of the initial die temperature and billet temperature can be determined as well as the
influence of the thermal barrier thickness.

4.2.1 Heat Transfer Analysis

In conventional and hot-die forging, the die material reaches
temperatures that are critical or near critical.

The reasons are many:

- increase of temperature in the die with consequent reduction of the allowable die stress,

- fast transfer of heat from billet to the surface layer of the die with developing of high thermal stresses,

- rapid chilling of the billet with increasing flow stresses, and high contact stresses on the die,

- high contact pressures between die and billet with breaking of the lubricant layer and higher heat transfer by conduction.

There are many variables involved in the heat transfer, both geometrical (dimensions of the billet, but neglecting the dies by assuming they are considerably larger than the billet), physical (specific heat, conductivity, density of the material of the die, billet and lubricant), environmental (insulation of the dies, heat supplied by the furnace, convection and radiation from the billet and the die) and temporal.

Little information is available in the literature on the subject of heat transfer for this specific problem. For some references see [4.3].

The present study is a simulation of the heat flow during hot-die forging. The temperature distribution within the billet and the die is predicted taking into consideration the heat transfer between the billet and the environment. The determination of the temperature is important for the following reasons:

1. the stress distribution in the workpiece and in the die as well as the metal flow are considerably affected by temperature changes,

2. in practical forging operations, the die wear, due to erosion, is associated with thermal softening of the die surface, and the contact stresses between workpiece and die,

3. in the surface layer of the forging dies, temperature gra-
    dients are present which in turn can cause thermal fatigue
    failure of the die surface [4.6, 4.18].

With reference to Figure 4.2, it is possible to distinguish the
following phases in the hot forming operation:

I. phase: both the billet and the die are heated to the re-
               quired temperature; the billet is heated in a sepa-
               rate furnace with controlled atmosphere,

II. phase: when the temperature distribution is stabilized, the
               dies are opened to allow the billet to be set into
               position; the billet is transferred from the furnace
               to the dies and its temperature drops,

III. phase: the billet is compressed and heat flows from the
               billet to the die.

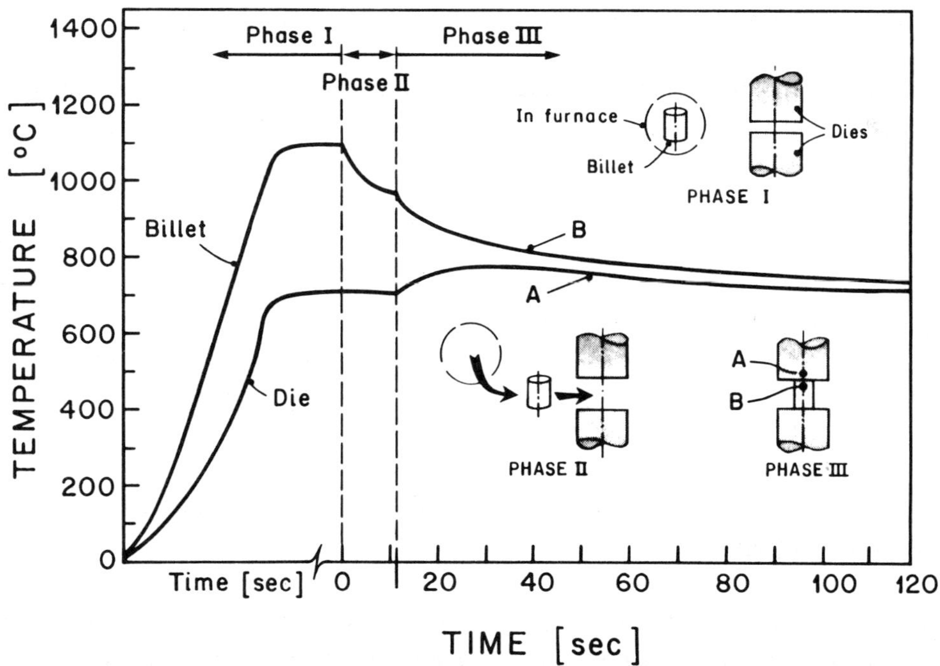

Figure 4.2: Schematic of the hot upsetting heating phases.

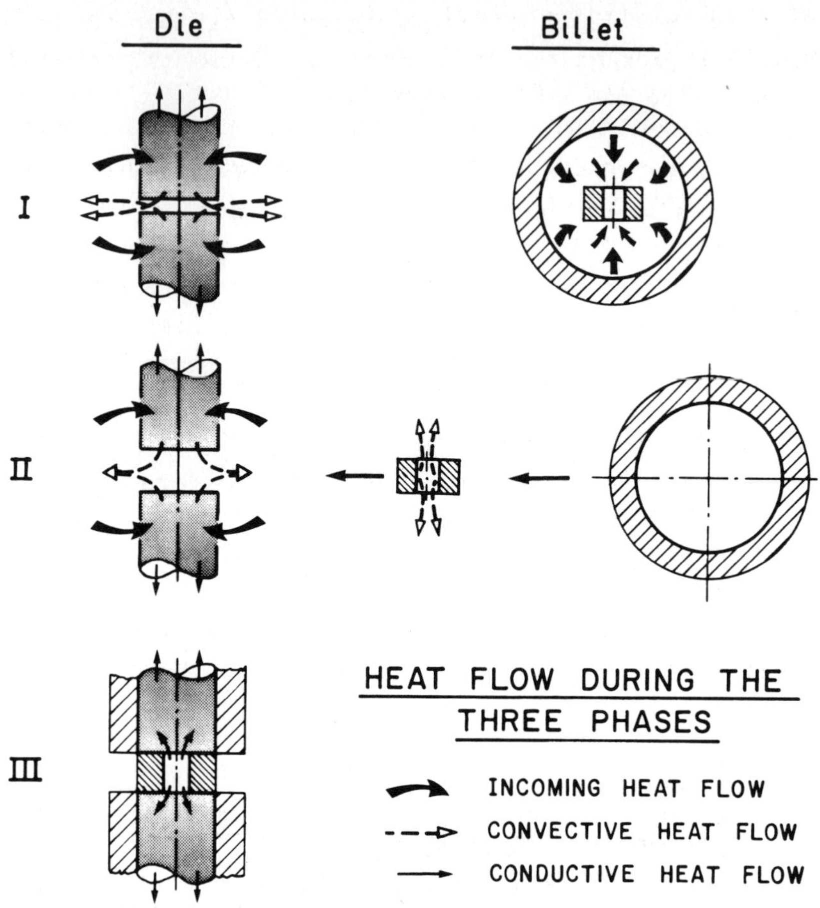

Figure 4.3:   Heat flow during the three phases.

The heat flow is shown in Figure 4.3. The radiating furnace provides heat to the dies. Heat is lost through the upper and lower
supports of the dies and by convection and radiation to the air
gap between the dies. The billet is insulated around its circum-
ferential area and it loses heat by convection and radiation
through the upper and lower faces during the transfer from the
furnace to the dies. During the pressing phase, the system die-
lubricant-billet can be considered an insulated system except for
some loss through the upper and lower supports of the dies. In
this system, heat flows from the billet to the die by conduction
in a non steady-state manner.

The heat transfer between the billet, lubrication layer, the die and the surrounding environment was studied with the help of the Finite Element method (FEM). This type of analytical-numerical analysis was chosen due to its adaptability to complex geometrical shapes. In the present study, an assessment of the accuracy, computational effort and costs was obtained by comparison with the experimental results.

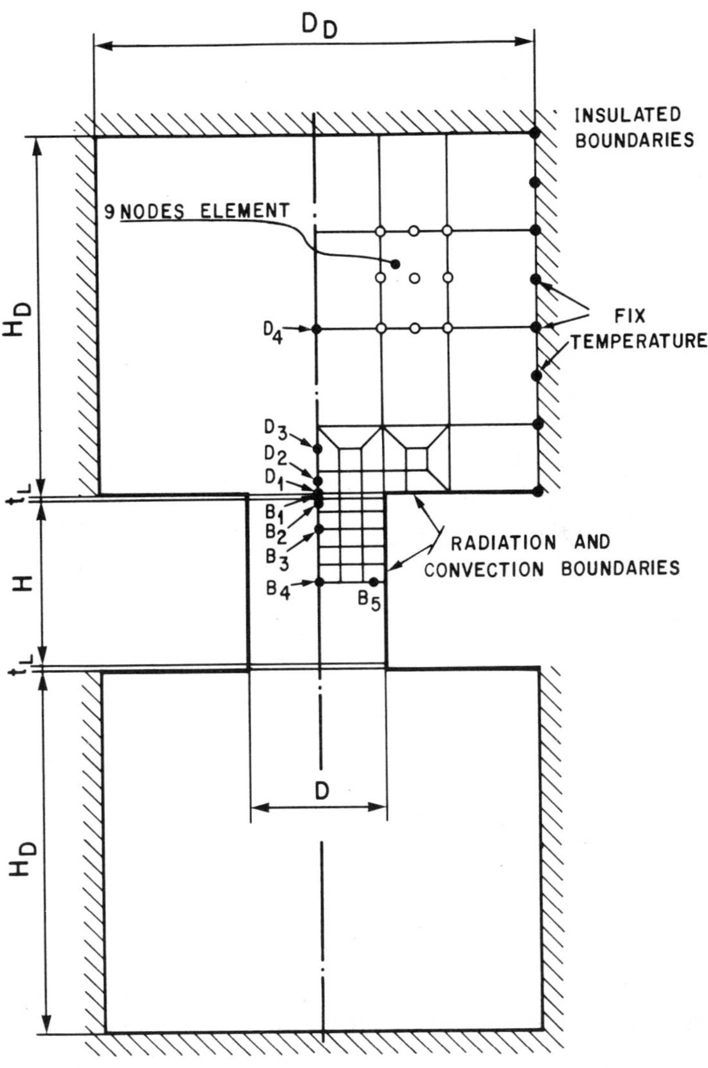

Figure 4.4: Finite Element discretization and location of points for graphs 5 - 8.

The FE discretization is shown in Figure 4.4. The element chosen
is a 9-node axisymmetric quadrilateral. For reasons of symmetry,
only one quarter of the total system was considered. The lubricant layer was also simulated to study the effect of different
lubricants. Only phase 1 and phase 3 were simulated at the present stage and the following assumptions were made:

Phase 1

1. the upper (lower) part of the upper (lower) die is insulated,

2. the heat transfer coefficient, the conductivity and the spe-
    cific heat are constant with temperature,

3. the radiation of the die horizontal surfaces is not conside-
    red due to the opposite effects of the upper and lower dies.

Phase 3

1. the total system die-lubricant-billet is insulated,

2. as Phase 1.2 above,

3. the billet has an initial uniform temperature.

Calculations were performed with the same experimental conditions
to check the accuracy of the solution and to assess the validity
of the assumptions. Then, other calculations were made with different initial temperature conditions.

A first set of experiments was conducted with a "radiating type"
furnace with a control on the temperature. The billets were
heated in a separate furnace and insulated around the lateral
area throughout the experiment. The temperatures in the upper and
lower dies ($\sim$ 5 mm below the surface) and in the billet ($\sim$ 1.5 mm
below the surface) were recorded by means of thermocouples.

In Figure 4.5 the FEM calculations are compared with the experimental results with lubricant glass (the values with no lubricant
being only slightly above or below the results obtained with

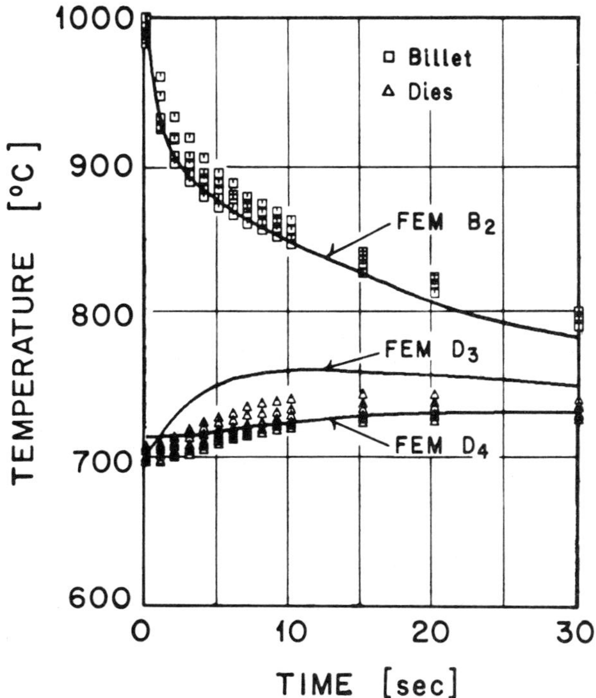

Figure 4.5:   Temperature versus time. Experiments versus FEM.

glass). The capital letters with subscripts refer to points in the die and the billet (see Figure 4.4). The upper line is the cooling of the billet in the span of 30 seconds, the lower lines are due to temperature increase in the die. B2 is the point corresponding to the position of the thermocouple in the billet; between D3 and D4 there is the thermocouple for the control of the die temperature. The experiments and the FEM calculations agree fairly well. In Figure 4.6 the results of 50 % compression tests are given for the two lubricating conditions. The FEM calculations for the same initial temperature conditions, but without compression simulation, are also given for reference purposes only. Examining closely the experimental results with compression, it is possible to draw the following conclusions:

1. during the deformation of the billet, internal heat is developed with a slowing down of the cooling of the billet it-

self. This phenomenon happens just at the beginning of the
compression operation due to the relatively high ram velocity
(4 mm/sec) (see A in Figure 4.6).

2. once the billet has been reduced to 50 % of the original
    height, a larger contact surface between the billet and the
    die brings a faster cooling of the billet as reflected in the
    curve after ∿ 20 seconds.

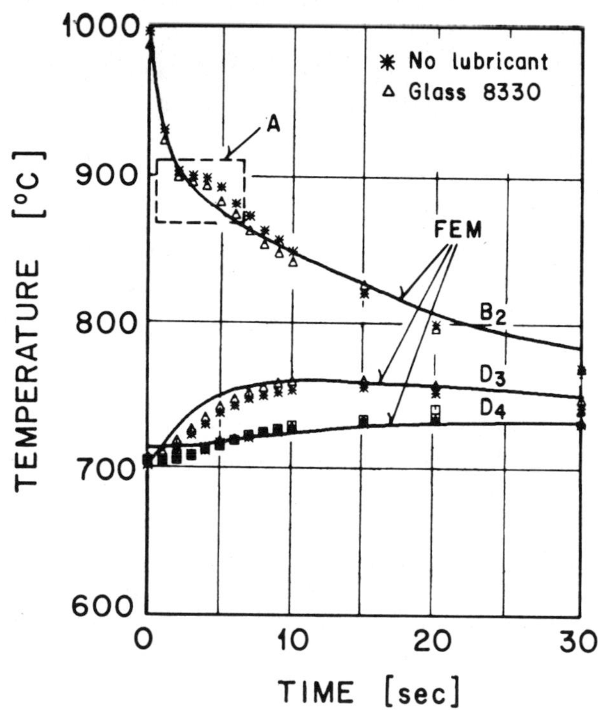

Figure 4.6: Temperature versus time for 50 % upsetting tests.

In Figure 4.7 the results of the FEM calculations for different
points in the die and in the billet are given. The surface of the
dies reaches temperatures 50 - 80°C higher than the temperature
recorded by the thermocouple which is located ∿ 6 mm away from
the surface. This has to be closely controlled to avoid a quick
deterioration of the die surface. The billet temperature distribution during cooling is also important to determine the flow

152

stress.

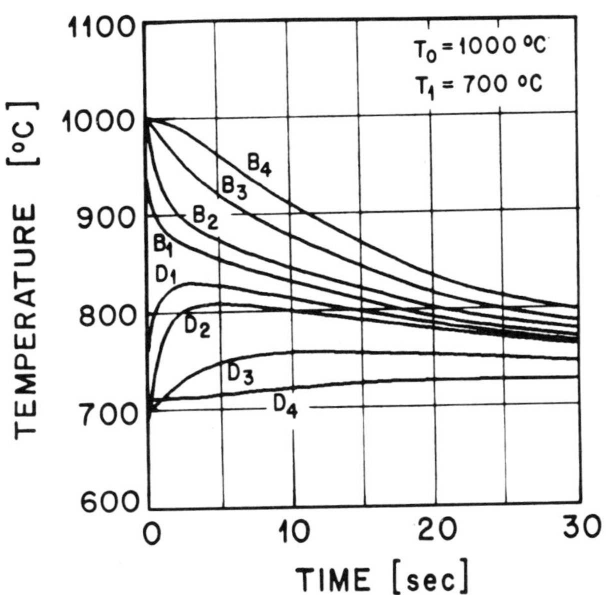

<u>Figure 4.7</u>: Die and billet temperatures versus time.

In the second series of experiments, the radiating furnace was
substituted with an induction heating coil. The die temperature
was recorded and when the required temperature was reached, the
current was shut off, the induction coil was moved away from the
dies, the dies were opened and the billet, preheated in a furnace, was placed between the dies. The billet and the die are now
without insulation. The temperature in the billet was recorded at
two different points: on the axis of the specimen near the surface and near the lateral surface at a height equal to H/2.
These two points correspond approximately with the nodes B_2 and
B_5 of the FE mesh (see Figure 4.4).

In this case, therefore, the die boundary conditions are quite
different from those mentioned above. The dies are no longer
insulated, heat is not supplied by the induction coil and there
is a greater heat loss than mentioned above.

The same Finite Element discretization was used to assess the second set of experimental results. The boundary conditions of the model were changed in order to simulate the heat losses of the billet by radiation and convection. The FE discretization is shown in Figure 4.4. The element chosen is a 9-node axisymmetric quadrilateral. The lubricant was also simulated by a thin layer of elements (0.1 mm thick for this first calculation). This dimension was arbitrarily chosen due to the impossibility of knowing the real thickness of the lubricant trapped between die and billet. The conductivity of the lubricant was such that the best fit to experimental data was obtained. Therefore the value of conductivity of the lubricant is an expression of the heat transfer coefficient between the die and the lubricant.

In general, a good agreement is found between the experiments and the FEM calculations. With the Finite Element model, it is possible to predict the temperature at every point of the physical system (as long as a node of the mesh can be located).

4.2.2 Non-Dimensional Analysis

The parameters which are involved in the heat transfer between die and billet are of a considerable number. To try to simplify the analysis, a non-dimensional approach was attempted in order to bring together certain variables in non-dimensional parameters. Geometry, material properties, temperatures and time all influence the problem. If it is assumed that the heat flux due to radiation and convection are negligible with respect to the heat transfer due to conduction, as was proved by our FEM analysis, then the variables of the problem are given in Table 4.1. The thermal conductivity, the specific heat and the density may be included in one variable called thermal diffusivity:

$$\alpha = \frac{\lambda}{\rho c_p} \qquad \text{with dimension } [L^2/\theta] \qquad (4.2)$$

According to the Buckingam theorem or π theorem:

$$\pi = H^a \ D^b \ \alpha^c \ \theta^d \ \Delta T_i^e \ \Delta T^f \qquad (4.3)$$

TABLE 4.1

Variable	Symbol	Dimensional Equation
Height	H	[L]
Diameter	D	[L]
Thermal Conductivity	λ	$[ML/\theta^3 T]$
Specific Heat	c_p	$[L^2/\theta^2 T]$
Density	ρ	$[M/L^3]$
Time	θ	$[\theta]$
Intial Die-Billet temperature difference	ΔT_i	[T]
Temperature difference ΔT in billet ΔT_B (or die ΔT_D) between time = 0 and time = θ	$\Delta T_B (\Delta T_D)$	[T]

where   L = length
          M = mass
          θ = time
          T = temperature

Substituting the dimensional equations:

$$\pi = [L]^a [L]^b [L^2/\theta]^c [\theta]^d [T]^e [T]^f \qquad (4.4)$$

The exponents a, b, ... f can be solved by the following system of three equations with six unknowns:

$$
\begin{array}{llr}
a + b + 2c = 0 & \text{for L} & (4.5\ a) \\
-c + d = 0 & \text{for } \theta & (4.5\ b) \\
e + f = 0 & \text{for T} & (4.5\ c)
\end{array}
$$

Three non-dimensional parameters are obtained:

$$\pi_1 = \frac{H}{D}$$
(4.6 a)

$$\pi_2 = \alpha \frac{\theta}{H^2}$$
(4.6 b)

$$\pi_3 = \frac{\Delta T}{\Delta T_i}$$
(4.6 c)

A graph can therefore be made when on the x-axis is plotted the non-dimensional time (π_2) and on the y-axis the non-dimensional temperature (π_3).

The third parameter (π_1) is plotted as lines of constant H/D ratios.

These non-dimensional parameters assume a precise physical meaning and they allow the change in temperature in the billet for determined external boundaries conditions to be found. These are:

    lubricant material
    lubricant thickness
    die material
    external heat flux on the die
    convection and radiation on the billet

The FE model was modified in order to obtain the change in temperature when the billet is deforming. The discretization of the billet was changed to simulate billets of H/D ratios 1.6, 1.0, 0.5, 0.2 with initial dimensions H = 16 and D = 10 (H = height, D = diameter). Volume constancy was considered. According to the non-dimensional analysis, the graph of Figure 4.8 presents the results in non-dimensional form. The non-dimensional time is $\alpha \frac{\theta}{H^2}$ and the non-dimensional temperature is $\Delta T_B / \Delta T_i$ where $\Delta T_B = T_{BO} - T_B$ or the difference between the initial and current temperature in the billet and $\Delta T_i = T_{BO} - T_{DO}$ or the difference between the initial temperature of the billet and of the die. Figure 4.9 is the graph of temperature variation for the die. The abscissa is the same as in Figure 4.8 but the non-dimensional temperature is $\Delta T_D / \Delta T_i$ where $\Delta T_D = T_D - T_{DO}$ or the difference between the current and the initial die temperature.

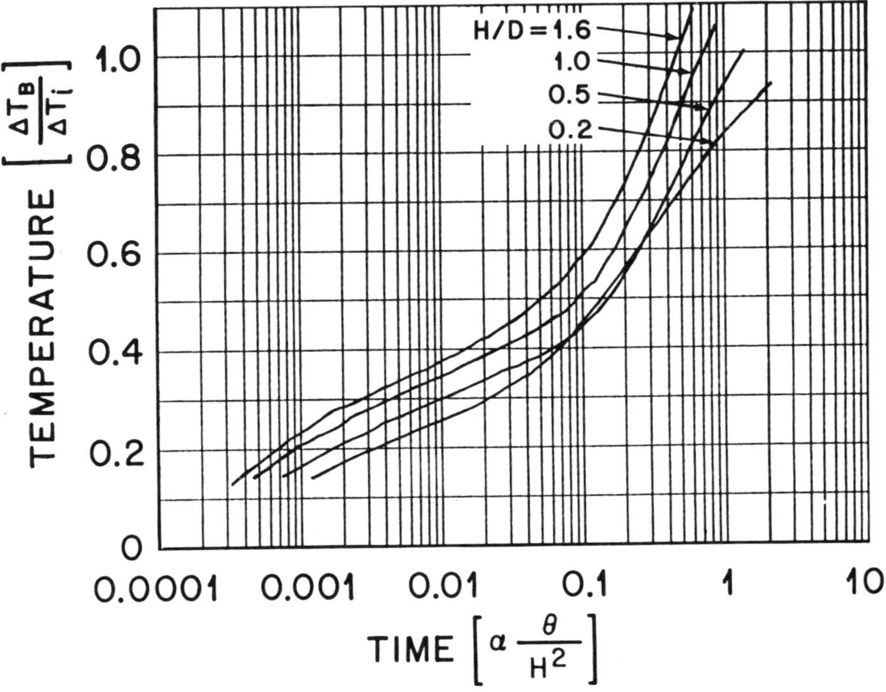

Figure 4.8: Billet temperature (Point B1) versus time. Non-dimensional graph.

The temperature T_B is for the point B_1 (see Figure 4.4), therefore for a point on the surface of the billet in contact with the lubricant. Figure 4.8 is the graph of temperature variation for the die (point D_1 in Figure 4.4). This set of curves allows determination of the temperature of the billet or of the die for billet dimension with H/D ratios in the range 1.6 to 0.2 and any initial billet-die temperature difference (ΔT_i). Interpolation has to be done for values between the lines of the graphs. These curves are therefore very important for the determination of the cooling rate of the billet and heating of the die.

The results are in good agreement with the experimental work [4.5].

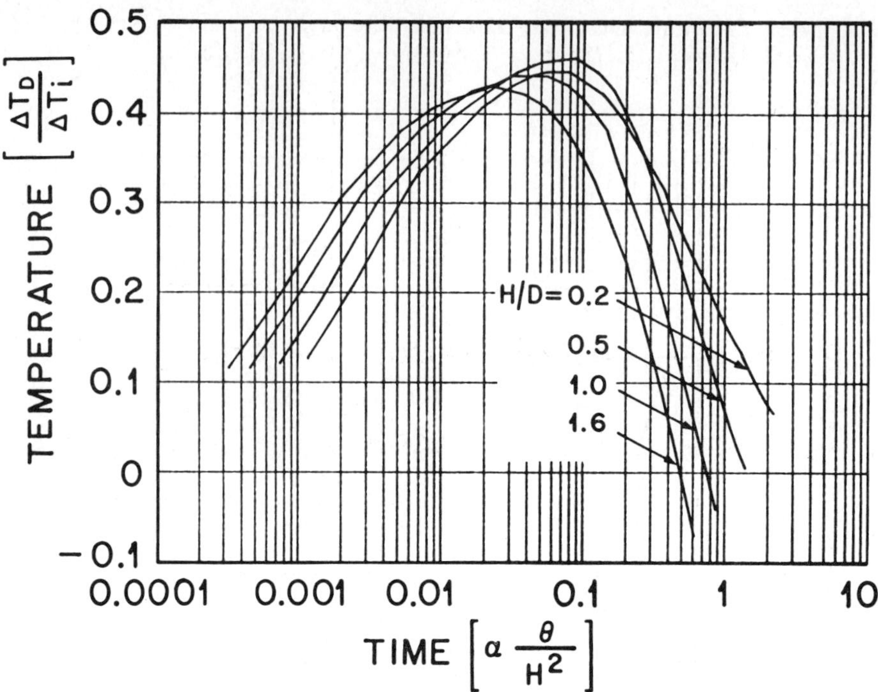

Figure 4.9: Die temperature (Point D1) versus time. Non-dimensional graph.

Using the non-dimensional plots, it is possible to analyze the influence of various parameters. For example, the effect of initial die and billet temperature is compared in Figures 4.10 and 4.11. The temperature profile for the centerline of the system billet-die is shown at different time steps. In Figure 4.10 the initial billet temperature is T_{BO} = 1000°C and the initial die temperature is T_{DO} = 800°C ("hot die"). In Figure 4.11 T_{BO} = 1100°C and T_{DO} = 200°C ("conventional forging"). The temperature profile is practically the same for the two cases but with a different temperature scale. For example, at 0.16 sec from the initial contact between the billet and the die, the billet surface temperature experiences a $\Delta T \cong$ 350°C and 75°C for conventional and hot-die respectively. The billet surface temperature

158

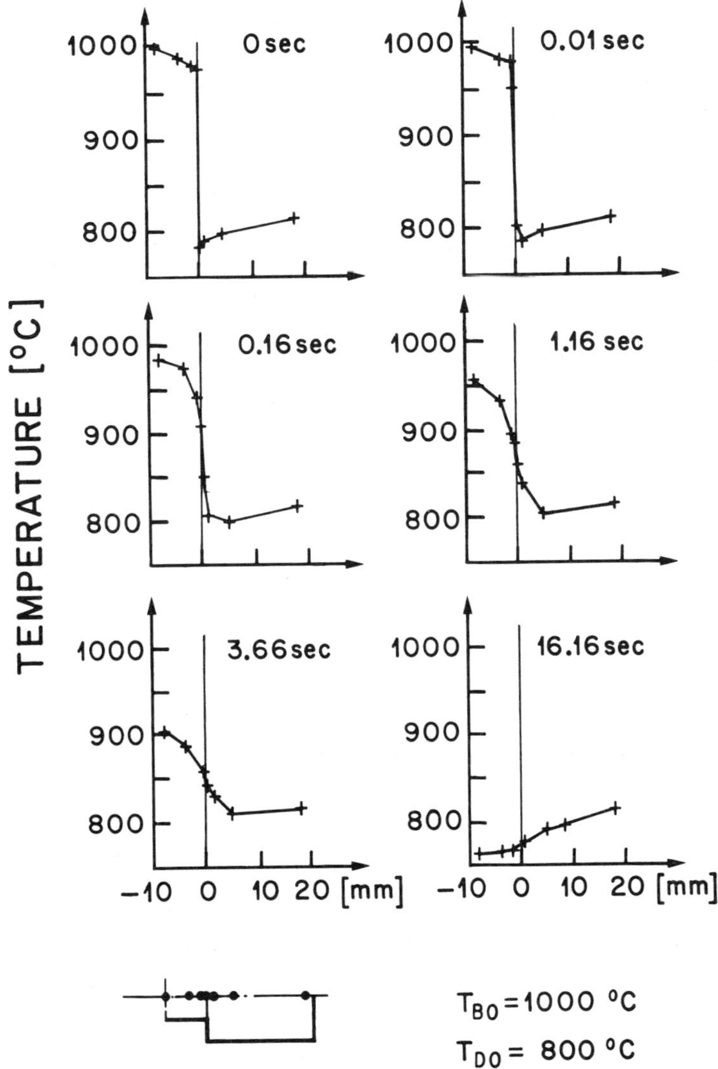

Figure 4.10: Temperature distribution for hot-die upsetting.

is at 750°C for conventional forging but is still around 900°C
for hot-die. The surface die temperature reaches a maximum of
∿ 600°C in conventional forging, but in hot die the maximum
temperature is ∿ 875°C. These results are important in order to
evaluate the change in the metallurgical properties of the billet

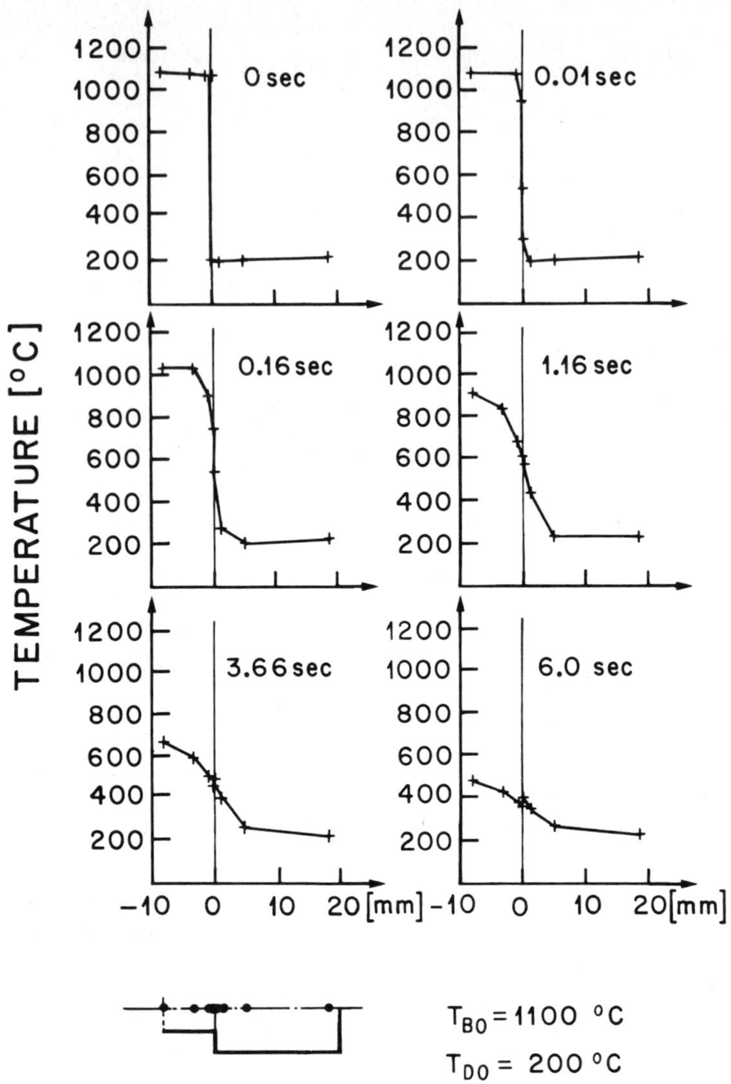

Figure 4.11:   Temperature distribution for conventional upsetting.

and the stresses which are induced in the die by thermal shock,
thermal fatigue and the lowering of the allowable die stress by
high temperatures. Both in hot-die and conventional upsetting,
the die experiences a temperature increase up to a depth of
around ∿ 5 mm. This has to be taken into account when making experiments to record the die temperature [4.5].

All the previous results assume a thickness of the lubricant between die and billet equal to 0.1 mm. By comparison with the experimental results, this seems to be a reasonable assumption.

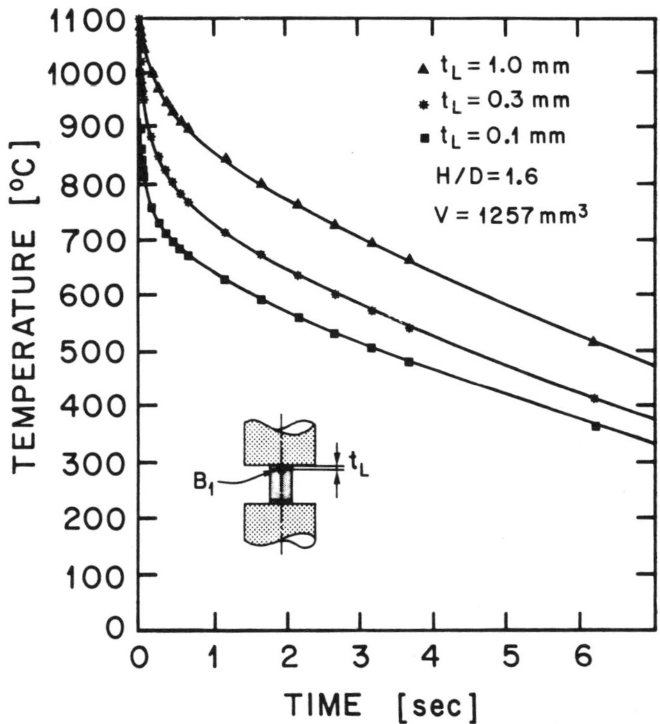

Figure 4.12:   Influence of the thermal barrier thickness on the
               billet temperature.

In Figures 4.12 and 4.13, the results for different lubricant
thickness (or better definition, in this case, thermal barrier)
are presented (0.1, 0.3 and 1 mm). The influence of the coating
is quite noticeable. For example, in Figure 4.12, a drop of
$\sim$ 400°C takes 0.4, 1.3, 3 sec for 0.1, 0.3, 1.0 mm thickness
respectively. The maximum temperatures in the die, Figure 4.13,
are 595, 500, 390 for 0.1, 0.3, 1.0 mm thickness respectively.
Furthermore, it takes a longer time to reach these maximum temperatures: 0.6, 1.2 and 1.8 seconds for 0.1, 0.3, 1.0 mm thickness respectively. The thermal barrier conductivity is equal to
2 W/(m°C).

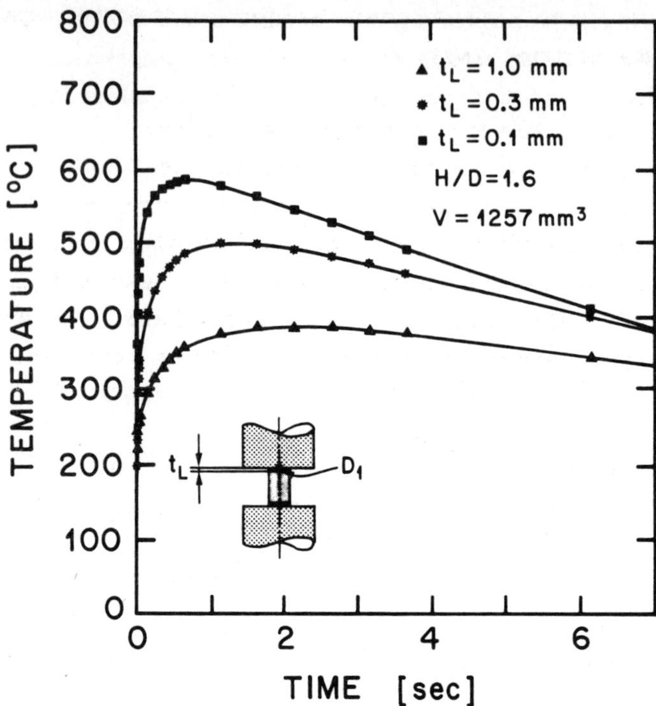

Figure 4.13: Influence of the thermal barrier thickness on the
die temperature.

The non-dimensional graphs also allow a study of the influence
of the other variables, such as the billet volume and the height-
to-diameter ratio.

In Figure 4.14 is reported the time that it takes the surface
of a billet of a certain volume to reach the die temperature T_B =
600°C for different H/D ratios. The larger the volume, the longer
the cooling time, as expected, but the graph provides a way to
quantify this time. It should be noticed that the height and the
diameter of the billet are uniquely determined if a volume and a
height-to-diameter ratio are selected.

The results of Figures 4.14 can be summarized as follows:

-   higher volumes and higher height-to-diameter ratios allow
    more time for deformation

-   a billet with low H/D ratio cools faster:
    a typical example is the material in the flash

-   if hot-die forging is to be a successful process, it seems
    necessary to reduce to a minimum the material flowing into
    the flash. Therefore the preforms should be studied very
    carefully in order to obtain a very high precision forming.

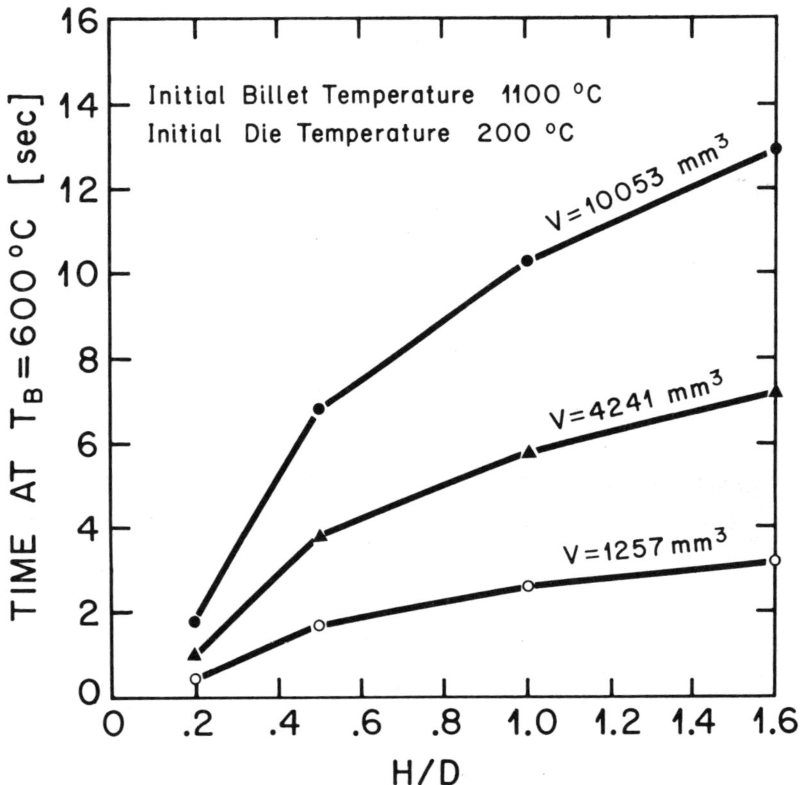

<u>Figure 4.14</u>:   Influence of volume and height-to-diameter ratio
              on the cooling rate of the surface of the billet.

The effect of different initial temperatures in the die and in
the billet was analyzed with the FEM in order to allow predictions of changes in the flow stress. The temperature distribu-

tion in the dies and in the billet can be calculated to predict regions of high temperature gradients. This is important in the design of dies in order to avoid thermal softening of the surface layer and in turn mechanical and thermal failure and short die-life.

This work has proved the validity of the FEM approach and non-dimensional analysis. The same approach could be used to study more complicated shapes such as axisymmetric billets with flash, plane strain shapes with flash and full three-dimensional analysis.

4.2.3 Hot-Die Upsetting System Modelling

The stresses in the upsetting of cylindrical specimens are calculated following the elementary analysis or slab method. During upsetting up to the formation of the flash, two different cases have to be considered as represented in Figure 4.15.

In case A, we have a situation of axisymmetric upsetting between flat dies and the solution is given in Chapter 2.4.3.1.

In case B, the lateral flow is constrained by the dies and shearing can occur as shown by the dashed lines. The lateral flow is therefore a combination between partially inclined dies and the solution is given in Chapter 2.4.2.2.

The heat transfer and stress analysis are coupled to model the complete system die-billet. The following procedure and equations are used.

4.2.3.1 Temperature Calculation

The temperatures during the process are calculated separately within the die cavity and in the flash.

Die cavity

The temperature distribution for a system die-lubricant-billet was previously calculated with the aid of Finite Element calcu-

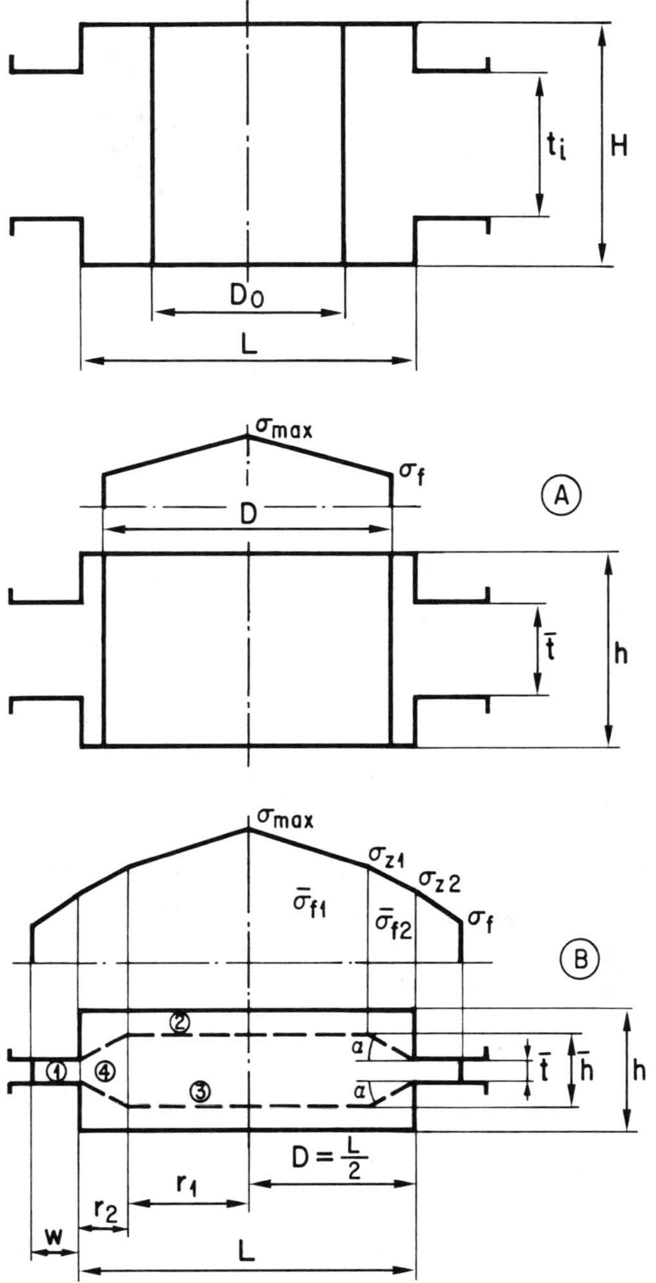

Figure 4.15: Billet and die geometry: initial configuration, upsetting (A), with flash (B).

lations (see Chapter 4.2.1.1). The results were summarized in non-dimensional form as:

$$FS_{Die} = \frac{T_D - T_{DO}}{T_{BO} - T_{DO}} \qquad (4.7)$$

and:

$$FS_{Billet} = \frac{T_{BO} - T_B}{T_{BO} - T_{DO}} \qquad (4.8)$$

where:

T_{DO} = initial die temperature

T_{BO} = initial billet temperature

T_D = current die temperature

T_B = current billet temperature

The FS_{Die} and FS_{Billet} are two non-dimensional functions representing the change in temperature in different locations within the die and the billet. For our program these locations were on the axis of symmetry and on the surface of the billet (FS_{Billet}) and on the contact surface on the die (FS_{Die}).

The two functions are plotted in Figure 4.8 and 4.9 for different H/D ratios.

The change in temperature with time can be calculated as:

$$T_D = T_{DO} + FS_{Die} (T_{BO} - T_{DO}) \qquad (4.9)$$

and:

$$T_B = T_{BO} - FS_{Billet} (T_{BO} - T_{DO}) \qquad (4.10)$$

The time is transformed in a non-dimensional parameter N_θ:

$$N_\theta = \alpha \frac{\theta}{H^2} \qquad (4.11)$$

where:

θ = time in seconds,

$\alpha = \dfrac{\lambda}{\rho c_p}$,

λ = thermal conductivity of the billet material,

ρ = density of the billet material,

c_p = specific heat of the billet material.

In this way, any material can be chosen for the billet. An average value of the thermal conductivity and specific density is assumed.

The program interpolates linearly for different H/D ratios.

During deformation, a very large portion of the deformation energy is transformed into heat. Thus, temperature of the deforming material may increase appreciably during the operation, particularly in zones of high strain as in the flash. The change in temperature is given by:

$$\Delta T = \frac{\int \bar{\sigma}_f \, d\varepsilon}{c_p \, \rho} \qquad (4.12)$$

where:

$d\varepsilon$ = strain increment

$\bar{\sigma}_f$ = flow stress (average)

c_p = specific heat

ρ = density

The flow stress $\bar{\sigma}_f$ of equation (4.12) is estimated on the basis of the temperature calculated by equation (4.15).

The heat developed during deformation is partially transferred to the die and therefore the temperature due to deformation is:

$$T_{Bd} = \Delta T \cdot \exp \left(- \frac{\alpha_T\, t}{c_p\, \rho\, h}\right) \qquad (4.13)$$

where:

T_{Bd} = Temperature rise of the billet due to deformation at time t

ΔT = Increase of temperature due to deformation

α_T = heat-transfer coefficient between dies and billet

ρ = density of the material

t = deformation time

c_p = specific heat of the material

h = height of the billet at time t

The final temperature of the billet is therefore given by:

$$T_{Bf} = T_B + T_{Bd} \qquad (4.14)$$

where:

T_B = temperature of the billet after heat transfer with the environment,

Flash land

Due to the relatively small thickness of the flash compared with the die cavity, a different approach was used to calculate the temperature with time.

The temperature gradient is neglected and the flash is considered as being a thin plate, with an average uniform temperature, cooled symmetrically from both sides. Thus, the average flash temperature during cooling is given by:

$$T = T_{DC} + (T_{BC} - T_{DC}) \exp \left(- \frac{\alpha_T (t - t_{cont})}{c_p \, \rho \, \bar{t}} \right) \qquad (4.15)$$

where:

t_{cont} = time of contact or start of flash formation

t = current time

$\bar{t}$ = height or thickness of flash

α_T = heat-transfer coefficient between dies and flash

ρ = density of the material

c_p = specific heat of the material

T_{DC} = die temperature at t_{cont}

T_{BC} = billet temperature at t_{cont}

At the present stage the T_{DC} and T_{BC} are calculated according to equations 4.9 and 4.10 but with a time shift equal to t_{cont}.

The final temperature in the flash is therefore given by:

$$T_F = T + \Delta T \qquad (4.16)$$

where:

T is calculated according to equation 4.17,

ΔT is calculated according to equation 4.18.

4.2.3.2 <u>Stress and Strain Calculation</u>

The strains and the stresses are calculated in the die cavity and in the flash. The "slab method" of analysis is used in order to calculate the stress distribution over the die-billet contact area.

<u>Strain</u>

o   Following Figure 4.15, the strain is, in case (A), before flash formation:

$$\varepsilon \; = \; \ln \left(\frac{H}{h}\right) \tag{4.17}$$

where:

    H = original height of the specimen
    h = current height of the specimen

o   During flash formation, case (B), the strain in the flash (at 1) is calculated according to:

$$\varepsilon_1 \; = \; \varepsilon_{tc} + \ln \left(\frac{\bar{t}_c}{\bar{t}}\right) \tag{4.18}$$

where:

    ε_{tc} = strain according to equation (4.17) at the beginning of flash formation

    $\bar{t}_c$ = thickness of the flash at the beginning of flash formation

    $\bar{t}$ = current flash thickness

At the end of forging, the material is assumed to flow into the flash by shearing along the surface indicated in Figure 4.15 (B) by a dashed line.

Therefore, in the zone between the shearing line and the die, the strain ε_2 is equal to the strain ε_{tc} and the strain ε_3 in the shearing zone is given by:

$$\varepsilon_3 = \varepsilon_{tc} + \ln\left(\frac{\bar{h}_c}{\bar{h}}\right) \qquad\qquad (4.19)$$

where:

$\bar{h}_c$ = average height of shearing zone at the beginning of flash formation

$\bar{h}$ = current average height of the shearing zone.

The average height of the shearing zone is calculated according to (see paragraph 4.2.3.3):

$$\bar{h} = 0.8\,\bar{t}\left(\frac{L}{2\,\bar{t}}\right)^{0.92} \qquad\qquad (4.20)$$

where: L = die diameter.

Note that:

If $\bar{h} \geq h$ then $\bar{h} = h$,
if $\bar{h} < h$ then $\bar{h}$ is calculated by equation (4.20).

In the convergence region 4 the height $\bar{h}_1$ is calculated as:

$$\bar{h}_1 = \frac{\bar{h} + \bar{t}}{2} \qquad\qquad (4.21)$$

and the strain:

$$\varepsilon_4 = \varepsilon_{tc} + \ln\left(\frac{\bar{h}_c}{\bar{h}_1}\right) \qquad\qquad (4.22)$$

Strain rates

The strain rates are calculated according to the following:

Case (A) $\qquad\qquad \dot{\varepsilon} = \dfrac{v}{h} \qquad\qquad (4.23)$

where:

$$v \quad = \quad \text{ram speed,}$$

Case (B):
$$\dot{\varepsilon}_1 \quad = \quad \frac{v}{\bar{t}} \tag{4.24}$$

$$\dot{\varepsilon}_3 \quad = \quad \frac{v}{\bar{h}} \tag{4.25}$$

$$\dot{\varepsilon}_4 \quad = \quad \frac{v}{\bar{h}_1} \tag{4.26}$$

4.2.3.3 <u>Determination of the Shearing Zone in Closed Die Forging</u>

The flow for axisymmetric forging is shown in Figure 4.15 (B).

The vertical stress at plane B is:

$$\sigma_{yB} \quad = \quad \frac{K_2}{K_1} \ln \frac{\bar{t}}{\bar{h}} + \sigma_{yC} \tag{4.27}$$

where:

$$K_2 \quad = \quad - \bar{\sigma}_f \cdot K_1 + 2\tau (1 + \tan^2 \alpha) \tag{4.28}$$

$$K_1 \quad = \quad - 2 \tan \alpha \tag{4.29}$$

$$\sigma_{yC} \quad = \quad \text{stress } \sigma_y \text{ acting at plane C known from the analysis in the flash}$$

The vertical stress at plane A is:

$$\sigma_{yA} \quad = \quad \sigma_{yB} + \frac{2\tau}{\bar{h}} \cdot r_1 \tag{4.30}$$

The dashed line in Figure 4.15 (B) is a shear surface, therefore equation (4.30) becomes [4.7]:

$$\sigma_{yA} = \frac{K_2}{K_1} \ln \frac{\bar{t}}{\bar{h}} + \sigma_{yC} + \frac{2\,\tau}{\bar{h}} \cdot r_1$$

and:

$$\sigma_{yA} = 2\bar{\sigma}_f \left(\frac{1 + \tan^2 \alpha + 2 \tan \alpha}{- 2 \tan \alpha}\right) \ln \frac{\bar{t}}{\bar{h}} + 2\bar{\sigma}_f \frac{r_1}{\bar{h}} + \sigma_{yC}$$

with:

$$- \ln \frac{\bar{t}}{\bar{h}} = \ln \frac{\bar{h}}{\bar{t}}$$

and:

$$r_1 = \frac{L}{2} - \frac{\bar{h} - \bar{t}}{2 \tan \alpha} \tag{4.31}$$

We obtain:

$$\sigma_{yA} = 2\bar{\sigma}_f \left[\left(\frac{1 + \tan^2 \alpha + 2 \tan \alpha}{2 \tan \alpha}\right) \ln \frac{\bar{h}}{\bar{t}} + \frac{L}{2\,\bar{h}} - \right.$$

$$\left. - \frac{\bar{h}/\bar{t} - 1}{1 \tan \alpha \cdot \bar{h}/t} \right] + \sigma_{yC} \tag{4.32}$$

The angle α must be such as to give the minimum value for the
stress σ_{yA}, i.e. the deforming material flows to consume the
minimum amount of energy, therefore minimizing (4.32) with
respect to $\tan \alpha$:

$$\frac{\partial\,\sigma_{yA}}{\partial \tan \alpha} = 0 \quad \text{or} \quad \tan \alpha = \pm \sqrt{1 - \frac{\bar{h}/\bar{t} - 1}{\bar{h}/\bar{t} \ln \bar{h}/\bar{t}}} \tag{4.33}$$

The average height of the deformation zone $\bar{h}$ or the ratio $\bar{h}/\bar{t}$
could also be determined by minimizing (4.32) with respect to $\bar{h}$.
This cannot be done analytically.

Numerically it is obtained:

$$\frac{\bar{h}}{\bar{t}} \simeq 0.8 \left(\frac{L}{2\,\bar{t}}\right)^{0.92} \tag{4.20 rep.}$$

for: $\dfrac{L}{2\,\bar{t}} \geq 2$ which is satisfied in most practical forgings.

4.2.3.4 Stresses

The stress calculations are done according to the "slab method" approach (Figure 4.15).

Case (A):

$$\sigma_{max} = \bar{\sigma}_f \left[1 + \mu_1 \frac{D}{h}\right] \tag{4.34}$$

where:

μ_1 = friction coefficient between die and billet,

$\bar{\sigma}_f$ = flow stress according to equation (4.1).

Case (B):

$$\sigma_{z2} = 2\,\mu_1\,\bar{\sigma}_f\,\frac{w}{\bar{t}} + \bar{\sigma}_f \tag{4.35}$$

$$\sigma_{z1} = \frac{K_2}{K_1}\,\ell n\,\left(\frac{\bar{t}}{K_3 + K_1\,r_2}\right) + \sigma_{z2} \tag{4.36}$$

$$\sigma_{max} = \frac{2\,\mu\,\bar{\sigma}_{f_1}}{\bar{h}}\,r_1 + \sigma_{z_1} \tag{4.37}$$

where:

$$K_1 = -2\,\tan\alpha \tag{4.38}$$

$$K_2 = -\bar{\sigma}_{f_2}\,K_1 + 2\,\mu_2\,\bar{\sigma}_{f_2}\,(1 + \tan^2\alpha) \tag{4.39}$$

$$K_3 = \bar{h} - r_2\,K_1 \tag{4.40}$$

$$r_1 = \frac{L}{2} - r_2 \tag{4.41}$$

$$r_2 = \frac{\bar{h} - \bar{t}}{2 \tan \alpha} \tag{4.42}$$

$$\tan \alpha = \sqrt{1 - \frac{(\frac{\bar{h}}{\bar{t}} - 1)}{\frac{\bar{h}}{\bar{t}} \ln (\frac{\bar{h}}{\bar{t}})}} \tag{4.43}$$

$$\mu_2 = 0.577 \text{ (shearing friction)}$$

$$\mu = \mu_1 \quad \text{if} \quad \bar{h} \geq h$$

$$\mu = \mu_2 \quad \text{if} \quad \bar{h} < h$$

w = the flash width calculated as shown in § 4.2.3.5.

4.2.3.5 Flash Width Calculation

From considerations of volume constancy and Figure 4.15, the volume of the billet before deformation is:

$$V = \frac{\pi D_o^2}{4} H \tag{4.44}$$

where:

D_o is the initial billet diameter,

H is the initial billet height.

The volume of the deformed billet with flash can be expressed as:

$$V = V_1 + V_2 \tag{4.45}$$

where:

$$V_1 = \frac{\pi L^2}{4} \cdot h \tag{4.46}$$

$$V_2 = \frac{\pi \, (L + 2\,w)^2}{4}\,\bar{t} - \frac{\pi \, L^2}{4}\,\bar{t} = \pi \, \bar{t} \, [w\,L + w^2] \qquad (4.47)$$

where:

    $\bar{t}$ is the flash thickness,

    w is the flash width,

    L is the die diameter,

    h is the deformed billet height.

Therefore equation (4.45) becomes:

$$V = \frac{\pi \, L^2}{4}\,h + \pi \, \bar{t} \, [w\,L + w^2] \qquad (4.48)$$

Equation (4.48) can be written as:

$$\bar{t}\,w^2 + L\,\bar{t}\,w - \frac{V}{\pi} + \frac{L^2\,h}{4} = 0 \qquad (4.49)$$

This 2nd degree equation gives the solution:

$$w = -\frac{L}{2} + \frac{1}{2}\sqrt{L^2 + \frac{4\,V}{\pi\,\bar{t}} - \frac{L^2\,h}{\bar{t}}} \qquad (4.50)$$

The condition for $w = 0$ gives:

$$0 = -\frac{L}{2} + \frac{1}{2}\sqrt{L^2 + \frac{4\,V}{\pi\,\bar{t}} - \frac{L^2\,h}{\bar{t}}}$$

or:

$$L = \sqrt{L^2 + \frac{4\,V}{\pi\,\bar{t}} - \frac{L^2\,h}{\bar{t}}}$$

and:

$$\frac{4\,V}{\pi\,\bar{t}} - \frac{L^2\,h}{\bar{t}} = 0$$

and finally:

$$V = \frac{\pi L^2}{4} h$$

which verify the solution.

4.2.3.6 Force Calculations

The forces on the die are calculated according to the stress distribution obtained previously and the contact area between die and billet. Therefore the force is [4.7]:

Case (A):

$$F = 2\pi\sigma_f \cdot [\frac{h^2}{4\mu_1^2} \exp (\mu_1 \frac{D}{h}) - 1] - \frac{hD}{4\mu_1} \qquad (4.51)$$

where all the variables have been previously defined.

Case (B):

$$F = F_1 + F_2 \qquad (4.52)$$

where:

$$F_1 = 2\pi\sigma_{z_1} \cdot [\frac{\hbar^2}{4\mu_2^2} \exp (\mu_2 \frac{L}{\hbar}) - 1] - \frac{\hbar L}{4\mu_2} \qquad (4.53)$$

$$F_2 = \sigma_f \cdot \frac{\pi}{4} (2Lw + w^2) (1 + \frac{1}{3} \mu_1 \frac{w}{t}) . \qquad (4.54)$$

F_1 is the average force on the billet and F_2 is the average force on the flash land.

4.2.3.7 Description of the Computer Program

The program consists of a main program and several subroutines for the data input, material data, stress-strain-force calculations, interpolation, data output, plotting, etc.

The input data are interactively given through a terminal. Two versions of the program have been written:

1)   for a refresh graphic display and a printer
2)   for a TEKTRONIX terminal.

In both versions of the program, a summary of the intermediate results is displayed on the terminal screen; in the second version, also the dies, the billet and the stress distribution are displayed at each time interval (user selected) for a continuous control of the process.

The initial step (STEP $\neq$ 0 in Figure 4.16 (a)), and intermediate step (STEP $\neq$ 35 in Figure 4.16 (b)) and the final step (STEP $\neq$ 45 in Figure 4.16 (c)) are given here for illustrative purposes only.

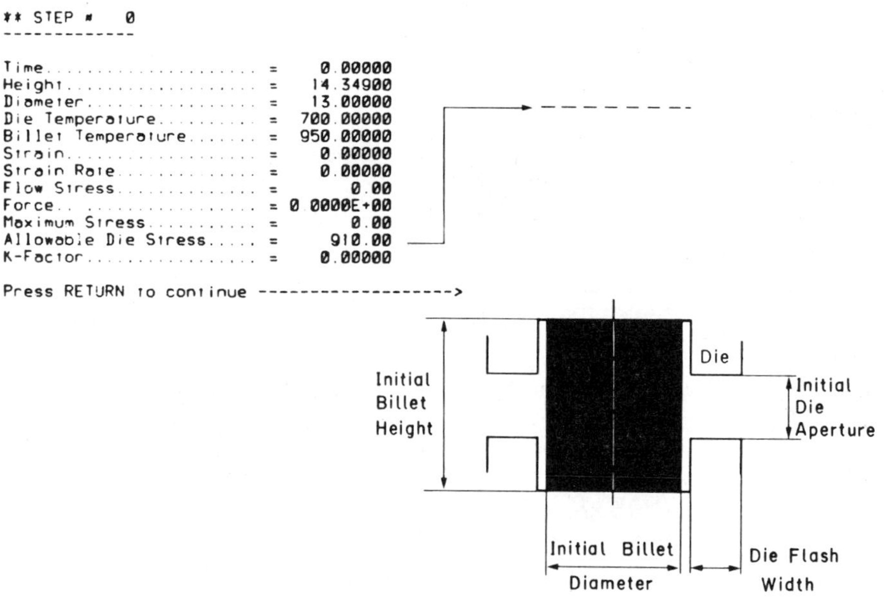

Figure 4.16:   Output display from the process modelling computer
               program.
               a) initial data and configuration.

```
** STEP #  20
--------------

Time...................... =      2.00000
Height.................... =     12.34899
Diameter.................. =     14.01324
Die Temperature........... =    748.34222
Billet Temperature........ =    903.69916
Strain.................... =      0.15011
Strain Rate............... =      0.08098
Flow Stress............... =        89.38
Force..................... =  0.1483E+05
Maximum Stress............ =       109.66
Allowable Die Stress...... =       885.83
K-Factor.................. =      8.07772

Press RETURN to continue -------------------->
```

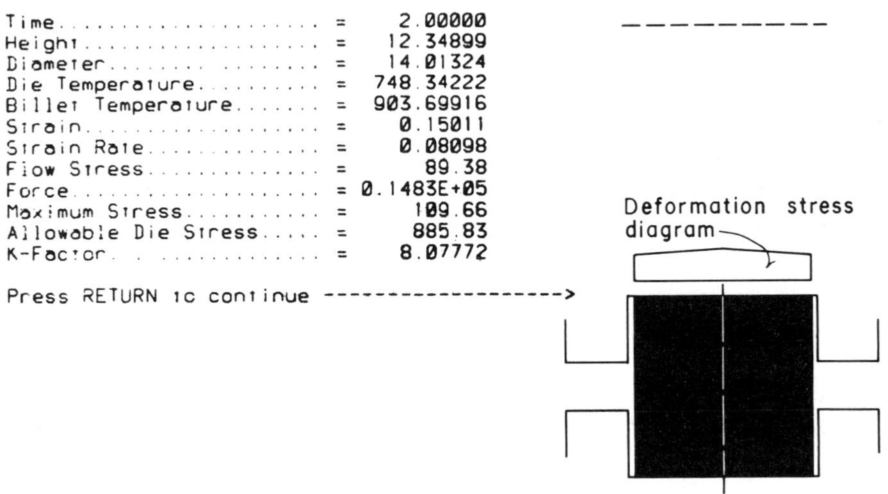

Figure 4.16: Output display from the process modelling computer program.
b)   intermediate results.

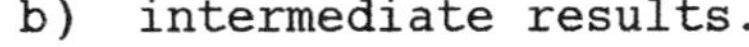

```
** STEP #  40
--------------

Time...................... =      4.00000
Height.................... =     10.34898
Diameter.................. =     15.00001
Flash width............... =      1.10958
Flash thickness........... =      1.34910
--- Flash Parameters ---
Temperature............... =    826.02761
Strain.................... =      0.59487
Strain rate............... =      0.74123
Flow stress............... =       180.79
-- Die Cavity Parameters --
Temperature............... =    854.51288
Strain.................... =      0.30483
Strain rate............... =      0.19118
Flow stress............... =       174.93
Die Temperature........... =    757.07007
Force..................... =  0.8509E+05
Maximum Stress............ =       703.45
Allowable Die Stress...... =       881.46
K-factor.................. =      1.25305

Press RETURN to continue -------------------->
```

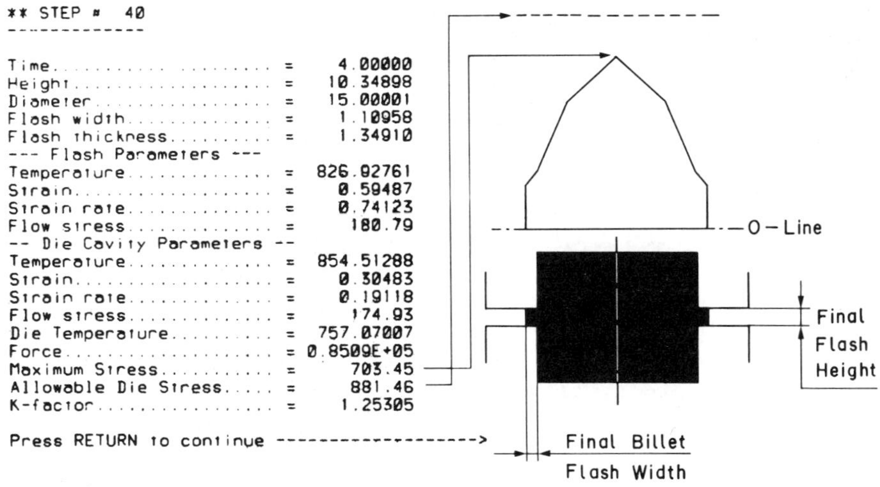

Figure 4.16: Output display from the process modelling computer program.
c)   final results.

4.2.3.8 <u>Experimental Verification of the System Modelling</u>

The theoretical approach is tested by comparing with experimental results of upsetting of cylindrical billets. The material is NIM 80 A. In Figure 4.17, the curves of force versus displacement for isothermal conditions at 200, 800 and 1100°C are given. The theoretical results agree quite well with the experiments.

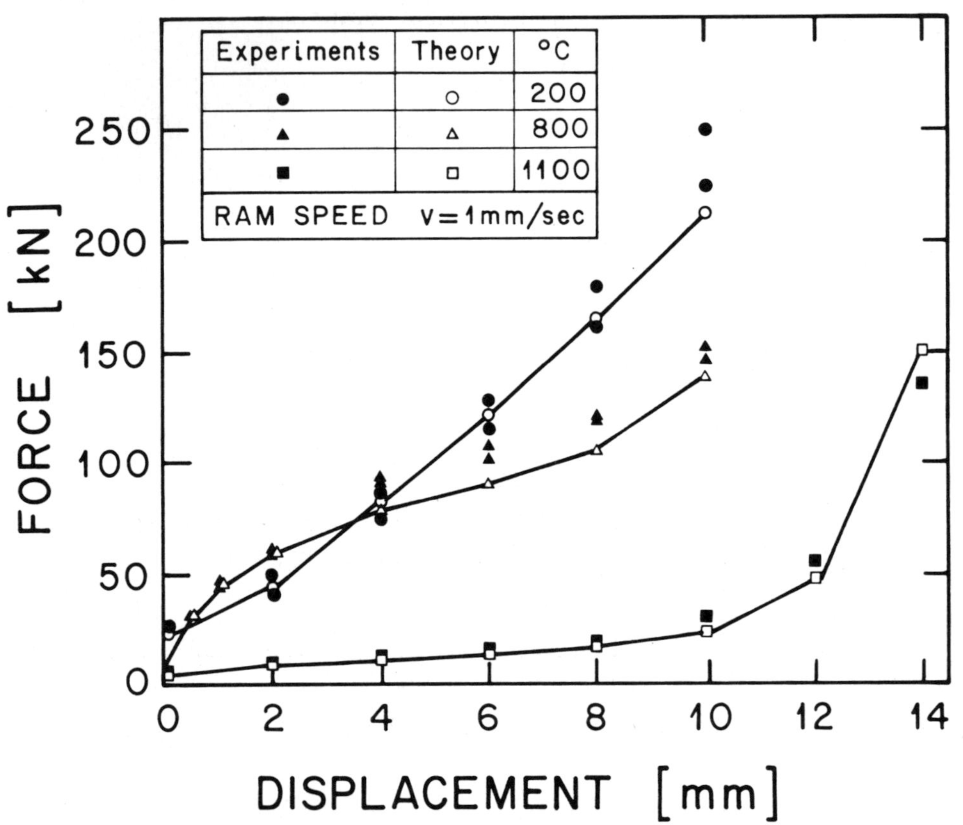

<u>Figure 4.17</u>: Force versus displacement in upsetting NIM 80 A cylindrical billets.

Comparison between experimental results and the theoretical model has been extensive. In Figure 4.18 the experimental results for hot-die upsetting of NIM 80 A billets are reproduced. The initial billet temperature was 1100°C while the initial die tem-

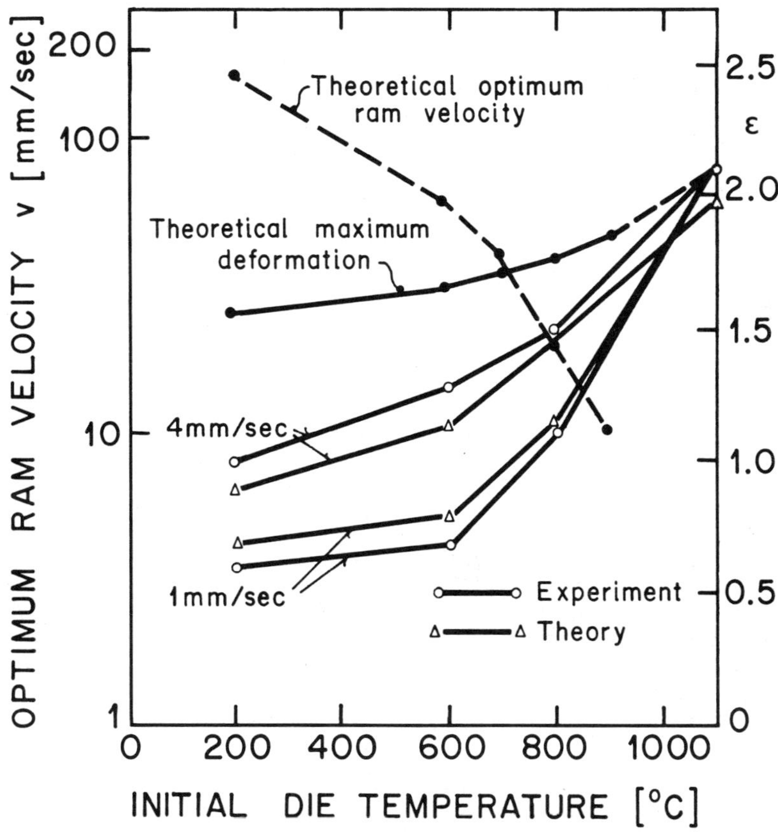

Figure 4.18: Experiment versus theory for upsetting NIM 80 A billets.

perature varied between 200°C and 1100°C (isothermal conditions). A hydraulic press was used with ram speeds of 1 and 4 mm/sec. The test was stopped when the maximum allowable force of 150 kN was reached. The curves of the final maximum deformation (logarithmic) ϕ (or ε) versus initial die temperature are drawn (curves

with open symbols). The total deformation is higher for higher
ram speeds. In this case, the contact time between the die and
the billet is shorter. Such an effect has a larger influence on
decreasing the flow stress than the accompanying increase in
flow stress due to strain rate. The theoretical curves reproduce
the experimental curves quite well.

From Figure 4.18 it can be seen that there exists an "optimum
ram speed" for each die temperature which gives a maximum of deformation for a given force. Also, it can be shown that the increased die temperature increases the possible deformation.

4.2.4 Optimization of Process Parameters

The principal aim of a process model is to simulate and possibly
optimize the process. Several parameters, as described earlier,
were found to influence the optimum forging conditions. Several
optimization criteria can also be selected by the designer for a
specific forging operation; maximum allowable force, maximum
possible deformation, maximum allowable stress, etc.

The following pages will describe the optimization of several
parameters for the upset forging of cylindrical billets in hot-
die and isothermal conditions.

4.2.4.1 Geometrical Parameters

In Figure 4.19, the initial die temperature and initial billet
temperature were fixed, as well as the billet volume and billet
material (NIM 80 A). The maximum allowable force was also fixed
(150 kN). The optimization criterion was the maximum attainable
deformation ϕ (or ε). The ram speed was then the variable to be
optimized. Curves for different billet geometry (height H and
diameter D) are drawn. The conclusions are summarized as follows:

-   the maximum deformation is higher for higher H/D ratios;
    thin billets have a higher contact area with the dies and
    therefore cool more rapidly,

-   there is a ram velocity at each H/D ratio which gives a maxi-

mum final deformation. At this point, the effect of billet
cooling and strain rate hardening on the flow stress are
balanced. Therefore an optimum ram velocity field can be
drawn as in the Figure 4.19.

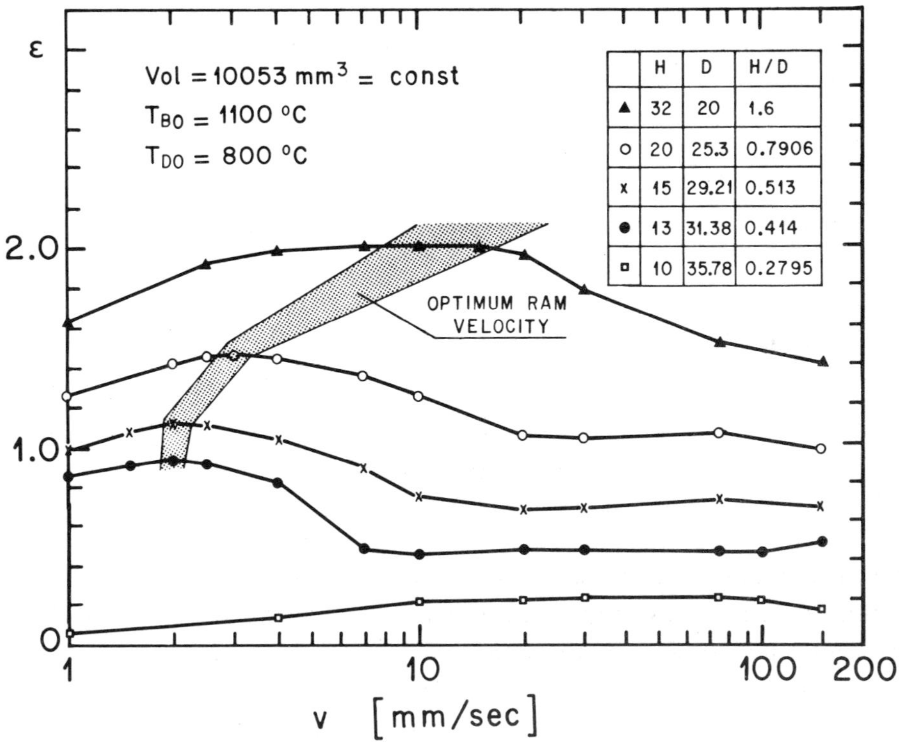

Figure 4.19: Effect of the aspect ratio H/D on the optimum ram
velocity.

In Figure 4.20 the same criterion was applied but the H/D ratio
was kept constant and the volume of the billet was changed. Again
an optimum ram velocity field can be found.

4.2.4.2 Maximum Allowable Die Stress

In a forging operation, it is important to determine when the
load reaches a critical value for the safe utilization of the
die. There are several criteria which can be defined as the

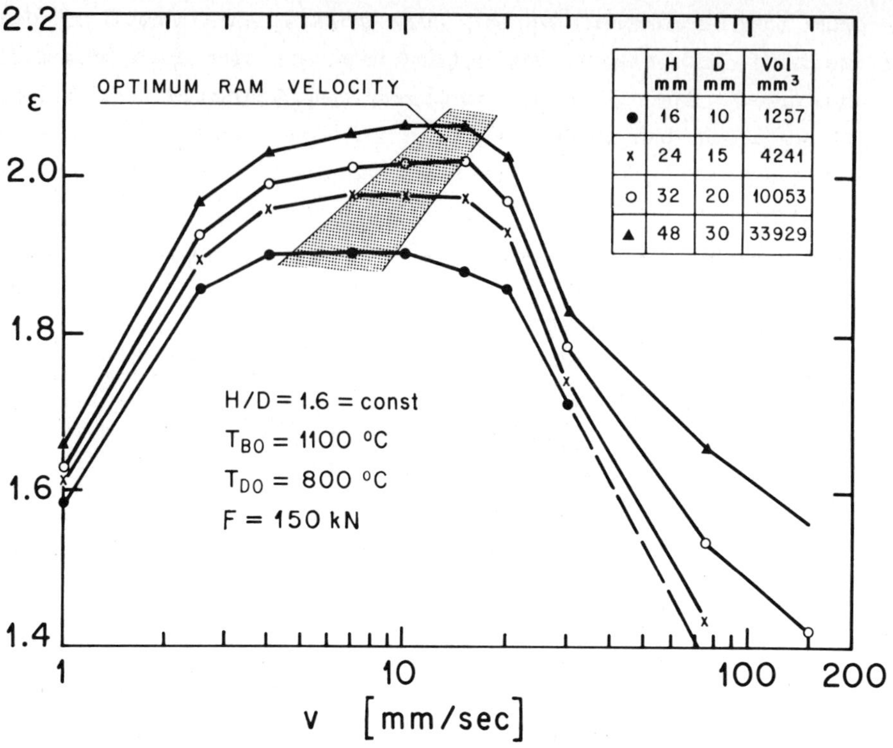

Figure 4.20: Effect of the billet volume on the optimum ram velocity.

safety limit for the forging dies. In the present model, the safety limit was defined as the onset of plastic deformation of the die. To quantify this definition, the following relationship was used:

$$K_I = \sigma_{0.2}/\sigma_{sp} \tag{4.55}$$

where: $\sigma_{0.2}$ is the 0.2 % yield stress of the die material,
        σ_{sp} is the specific pressure defined as:

$$\sigma_{sp} = F/A \tag{4.56}$$

where: F is the forging load,
        A is the die-billet contact area.

The $\sigma_{0.2}$, or maximum allowable, die stress is a function of the die temperature which is calculated in the same way as the billet temperature [4.3, 4.4]. Also the force F, which is a function of the flow stress, depends on the billet temperature, making the problem very difficult to analyze without the aid of a computer. It was experimentally found that the forging die will yield when $K_I < 1.15$ [4.2].

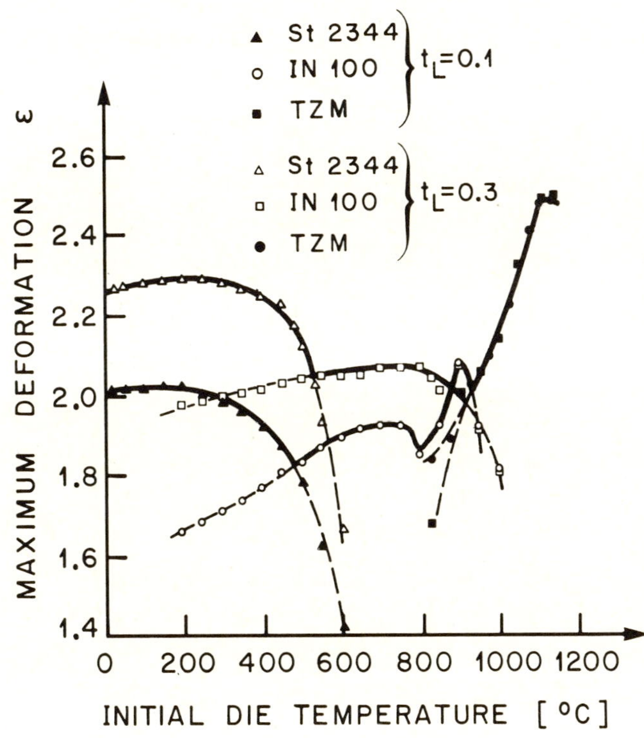

Figure 4.21: Influence of thermal barrier thickness on maximum deformation using different initial die temperatures.

The effect of a thermal barrier was simulated by changing the thickness. The criterion selected was $K_I = 1.15$ as a limiting factor. The results are presented in Figure 4.21 in terms of maximum possible deformation. Three different die materials were considered: a tool steel (St 2344), a nickel-base alloy (IN 100) and a molybdenum alloy (TZM). The effect of the thermal barrier is more important for conventional forging and is obviously insignificant for isothermal forging.

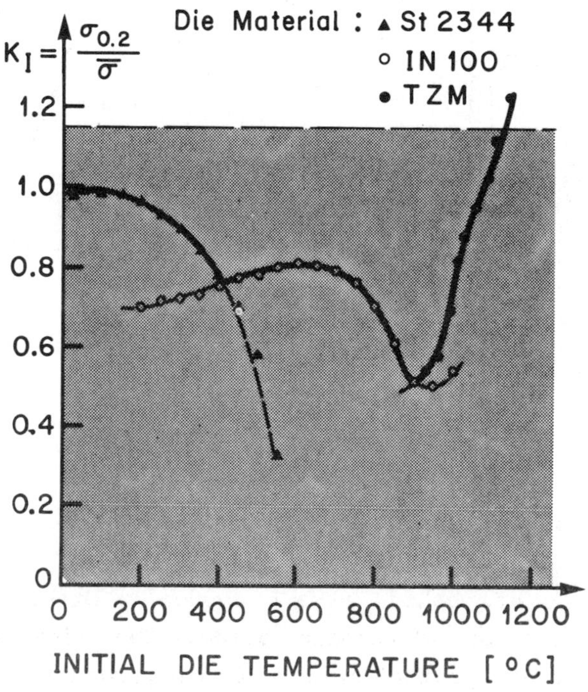

Figure 4.22: K_I-value versus initial die temperature for HOT-DIE forging of NIM 80 A.

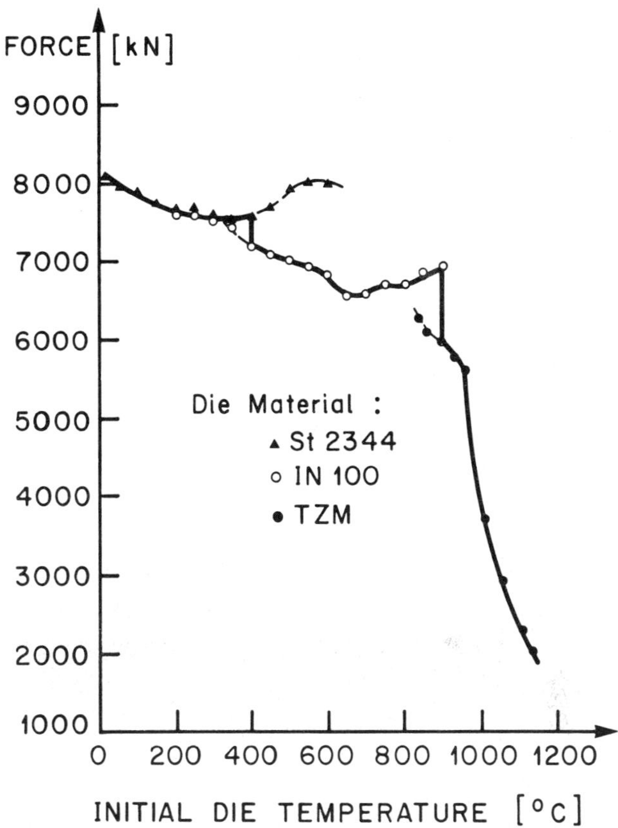

Figure 4.23: Force versus initital die temperature for HOT-DIE forging.

4.2.4.4 Optimum Conditions for Forging NIM 80 A Billets

The dimensions of the billet are chosen in order to simulate the airfoil portion of a turbine blade. The limiting factor for the determination of the optimum forging conditions is the minimum height after forging: 2.5 mm. This height is a typical value in the thin part of the finished airfoil. In Figure 4.22, the stress ratio factor K_I is plotted versus the initial die temperature for the three die materials. Only isothermal conditions using TZM dies give values of $K_I \geq 1.15$. In Figure 4.23, the force at the end of forging versus initial die temperature is given. The advantage of isothermal forging is again shown here: the force needed is four times less than for conventional forging. The optimum ram velocity versus initial die temperature is given in Figure 4.24. The isothermal forging requires low speeds and therefore hydraulic presses can be used.

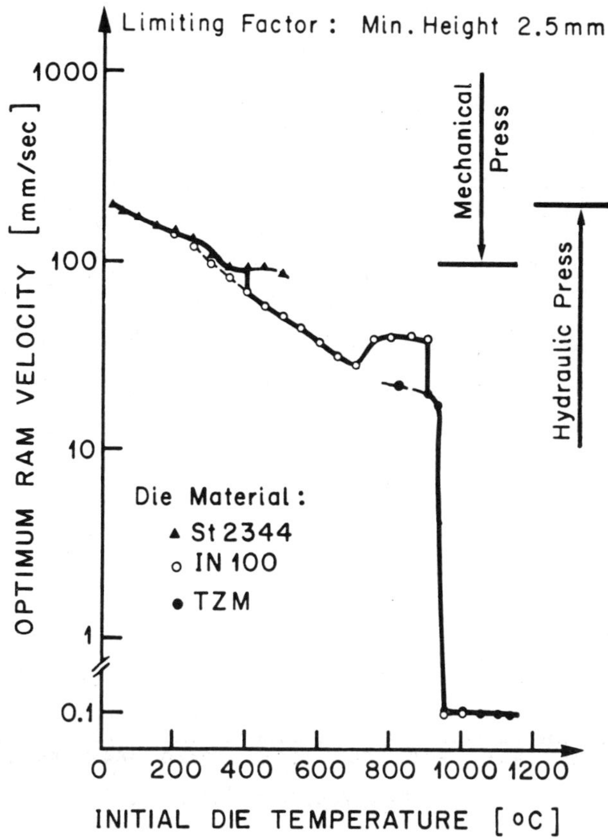

Figure 4.24: Optimum ram velocity versus critical die temperature.

The calculations show that the only method that was possible
to use in this case (in a one-stroke forging) was isothermal
forging. It was also possible to show the optimum die temperature with the model.

4.2.4.5 Isolines of Maximum Deformation

The FORG Process Model program has an added feature consisting of
a post-processor which allows the plotting of the contour lines
representing maximum constant logarithmic deformation. This gives
the designer a quick estimate of the required ram velocity and
initial die temperature required to obtain a certain deformation.

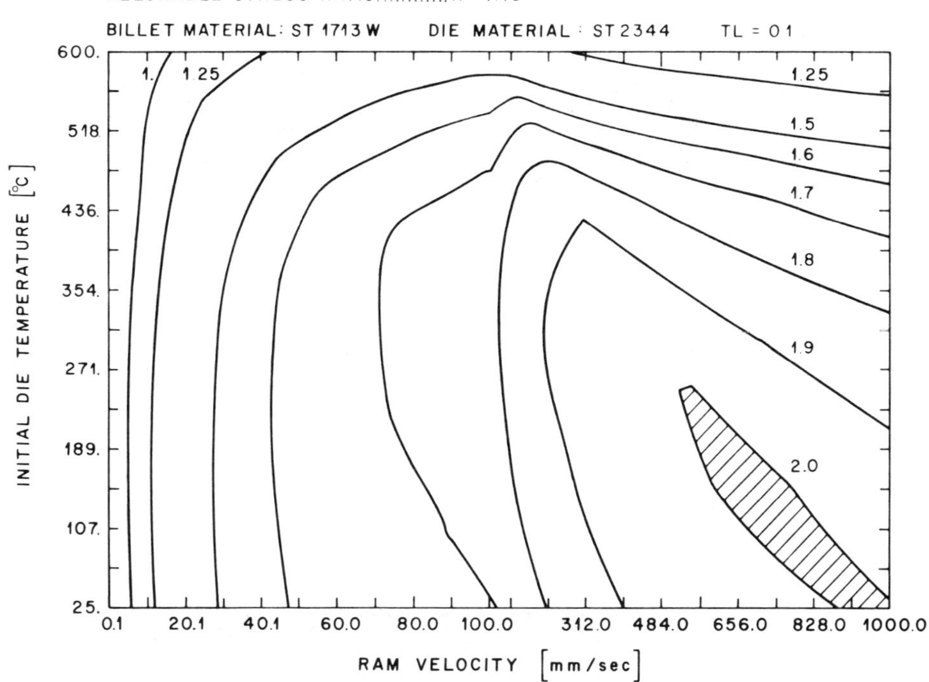

Figure 4.25: Isolines of maximum deformation $\phi = \ln (h_o/h_f)$.

Two figures 4.25 and 4.26 are presented which summarize the results for forging St T 17/13 W billets with steel and IN 100
dies. It is interesting to find, for example, that there is a
region in which it is possible to obtain a high deformation even
with relatively low ram speeds (60 to 120 mm per sec) and initial
die temperature between 650 to 750°C using IN 100 dies.

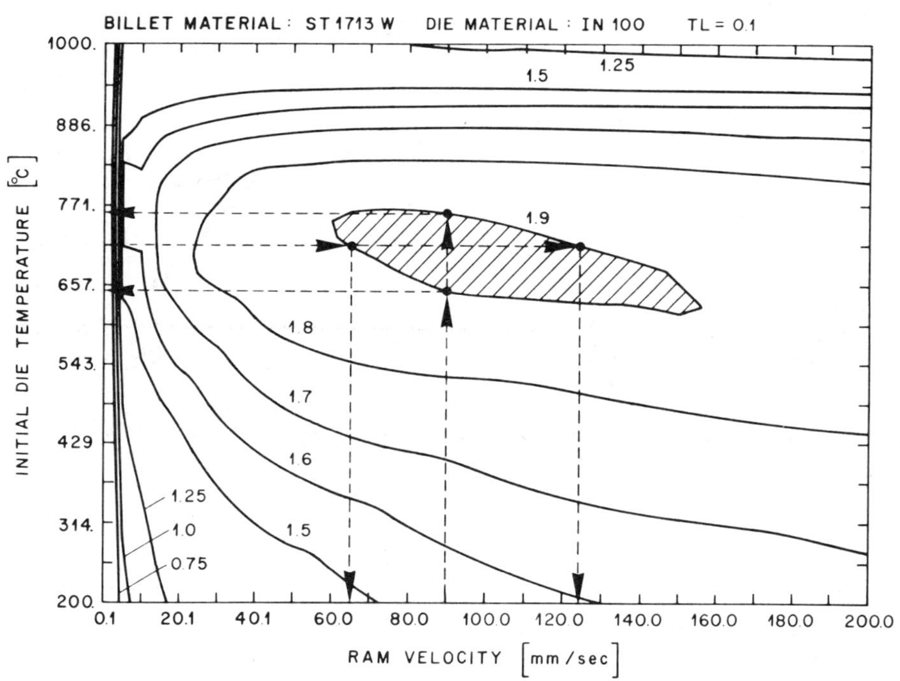

Figure 4.26: Isolines of maximum deformation $\phi = \ln (h_o/h_f)$.

4.2.4.6 Evaluation of Optimum Variable Ram Speed

All the considerations and the results presented so far assume
that the ram speed is constant throughout the stroke of the

press. In fact, it can be shown that if the ram speed is changed
during compression, the specimens can be compressed more than
when the ram speed is constant.

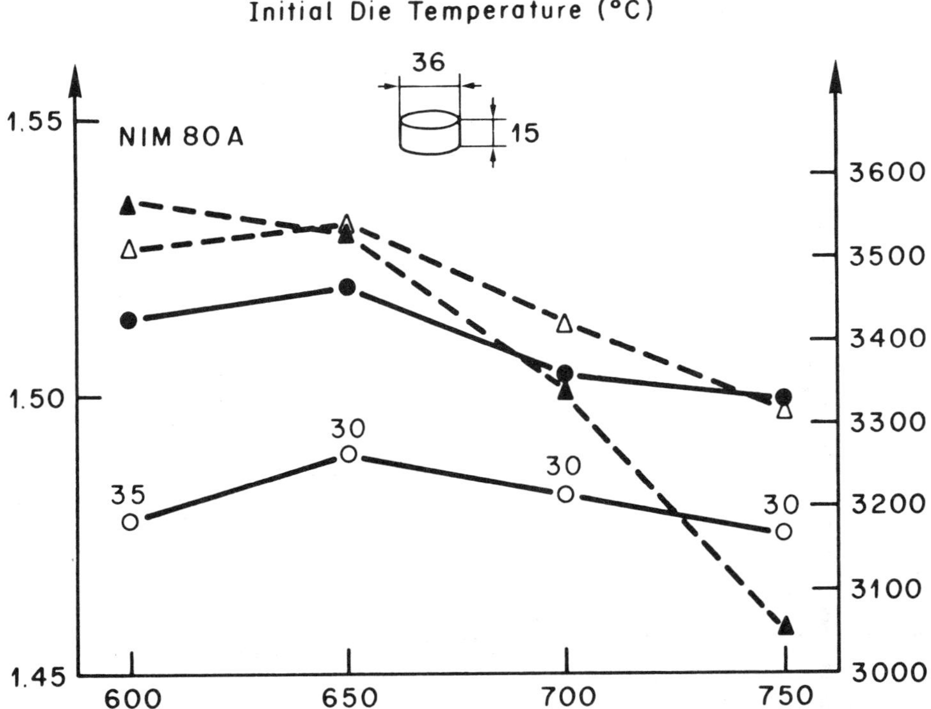

Figure 4.27:   Influence of variable ram speed on the maximum deformation attainable.

In Figure 4.27, we have reported the results for the billet material NIM 80 A. Here the limiting parameter was $K_I = 1.15$. The
maximum deformation obtained is plotted versus the initial die
temperature. The constant ram speed is written near the open
circle: this is the ram speed which gives the maximum deformation
in the range between 0.1 to 100 mm/sec. The line fitting the
closed circles is obtained instead by changing the ram speed
during the stroke. In this case, at each compression step, the
program finds the best ram speed to obtain the maximum deformation without deforming the dies ($K_I = 1.15$). There is an improvement compared with the constant ram speed. Also the required

force is reduced. The improvements are not quantitatively drama-
tic, but the method should be tested further for other billet
geometries (volume, H/d ratios).

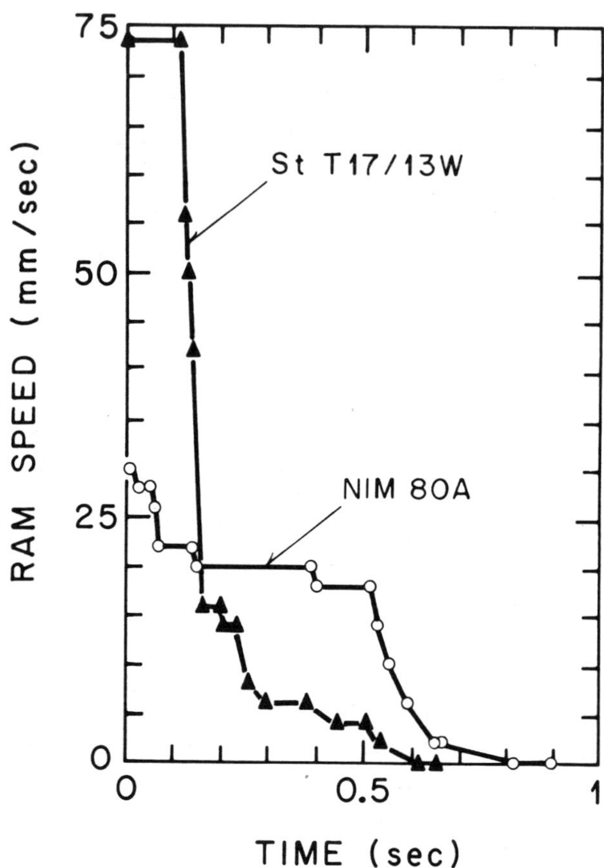

Figure 4.28:   Ram speed schedule for optimum upsetting. Height =
15 mm; Diameter = 36 mm; T_L = 0.1 mm; K_I = 1.15;
Initial die temperature = 700°C; Initial billet
temperature = 1130°C.

The program can produce the ram speed schedule for the process
control algorithm as in Figure 4.28. Finally, the controller of
the press can then be programmed with the ram speed schedule. A
400 kN press was equipped to regulate the ram speed during forging. The regulating is done via a microcomputer and the ram

192

speed interval is 0.1 - 25 mm/s [4.19]. The press can be used to simulate isothermal, hot-die and, in many cases, conventional forging. The regulating system is built around an Apple II which gives the input data to the hardware controler MODACOS (Modular Data and Control System). The Apple II also manages the data acquisition. The press has an automatic billet handling system.

4.2.5 Coupled FEM Analysis of Forging

4.2.5.1 Introduction

The conventional closed die forging technique is a highly non steady-state process:

-   the material flow and the die filling are complex functions of all parameters including friction, shape of dies and pre-form, temperature distribution and velocity. There are parts of the billet with free surface and other parts with surface in contact (sticking or sliding) with the die.

-   The die velocity is a complex function of the press type and also of the forging process. For all press types, but especially for hammer presses and screw presses, there is no inter-connection between machine and process giving the actual die velocity. This is only one non steady-state boundary condition for calculating local parameters such as strain rates.

-   The temperature distribution of the dies and of the billet is characterized in general through a gradient between dies and billet at the beginning of the forging process. An exception is isothermal forging with identical temperatures of dies and workpiece. A simple boundary condition for the conventional closed die forging would be a homogeneous die temperature (200°C) and a homogeneous billet temperature (1100°C) at the beginning. Starting the forging process, the temperature distribution in workpiece and dies becomes inhomogenious as a function of time and location as a result of the temperature gradients and the heat generation of the deformation.

Thus the process simulation of forging in general needs a coupled analysis of plasticity and heat transfer for non steady-state problems with changing boundary conditions.

The Finite Element method process simulation, or a combination of FEM and Finite Difference methods are powerful instruments for solving these problems.

The forging behavior of inductive (radial inhomogeneous) heated billets is cited as an example of the application of a coupled FEM-process modelling analysis of a forging problem. The principal question was the influence of the initial radial inhomogeneity of billet temperature on the force, the material flow and the temperature distribution in dies and billet.

Inductive heating is widely used for the heating of the billets in bulk metal processing like forging, rolling, and hot extrusion. The reason is the quick and clean process which can be integrated smoothly into modern manufacturing systems. A critical point is the inhomogeneous radial temperature distribution caused by heat generation in the skin of the billet in induction heating. The prevention of overheating at the surface is essential. An important question is the degree of inhomogeneity allowable as a temperature gradient between surface and center, Figure 4.29, because a temperature gradient of less than 100°C between surface and core of the billet is expensive to reach, due to an extension of the heating equipment and the heating time.

Thus the question of allowable temperature gradient is important for the manufacturer of induction heating equipment as well as for the user in forging processes [4.8].

4.2.5.2 <u>Principle of Induction Heating</u>

In induction heating, the heat is generated in the billet itself through a current flowing in a surface layer. The thickness of the layer is proportional to $\sqrt{1/f}$ where the frequency is denoted by f - this is called the skin effect [4.9 - 4.11]. The frequency of induction heating equipment is designed or controlled as a function of the size of the billets - low frequency and therefore

a thicker skin of current flow being suitable for heat generation
in larger parts. The function of induction heating equipment
could be compared with that of a transformer. The heating coil
corresponds to the primary winding of the transformer, which
transfers the energy to the billet. This is turn controls the
secondary winding and the magnetic core in a single operation.

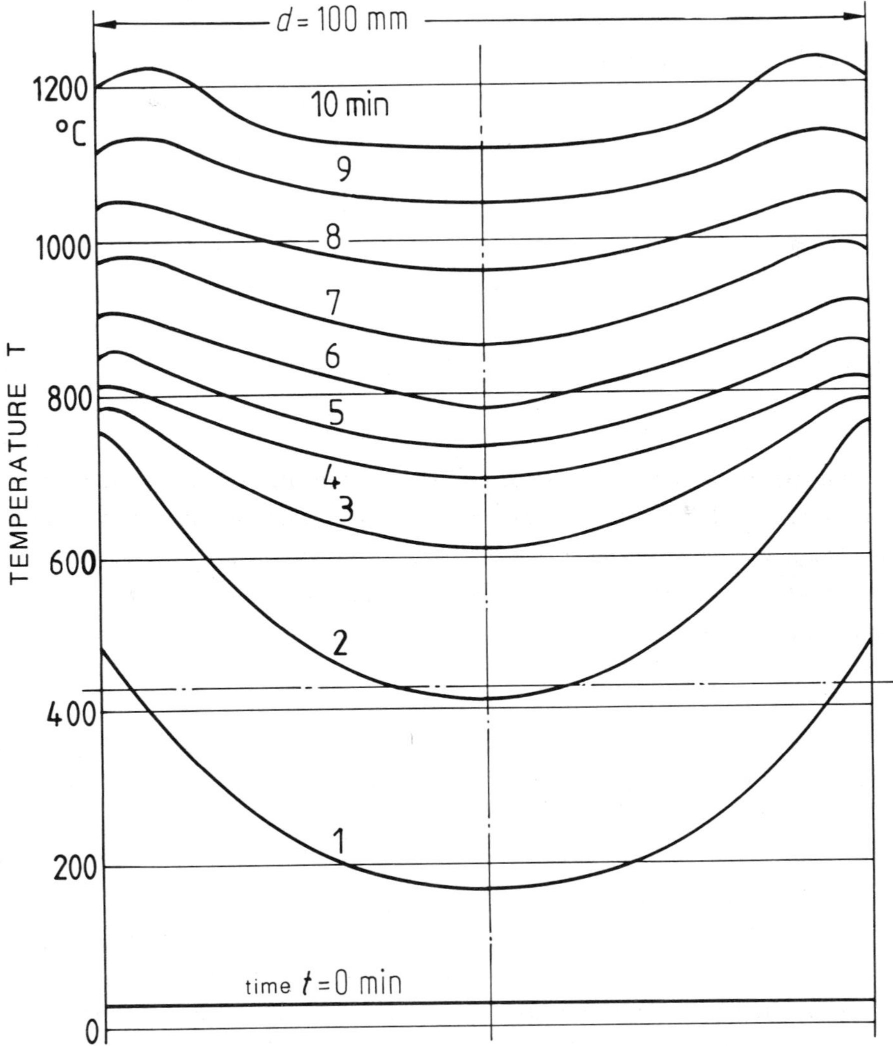

Figure 4.29:   Radial temperature distribution during induction
heating of a steel cylinder as a function of
heating time. Diameter d = 100 mm; Frequency 2 kHz
[4.22].

4.2.5.3 Influence of an Inhomogeneous Temperature Distribution on the Forging Process

In induction heating, the billets are transferred through a series of induction coils which raise the temperature stepwise up to the forging temperature. Because the heat is generated in the skin of the billet, the core temperature is lower than the surface temperature, Figure 4.29. In one or more homogenizing coils, it is possible to get a nearly homogeneous temperature distribution. But a high requirement for homogeneity needs longer and more expensive induction heating equipment; altogether it raises the costs of heating considerably.

Therefore it is of great practical interest to know the allowable inhomogeneity of the temperature in the billet to give a smooth forging process. This needs a consideration of the material flow, the force and the die load. For comparison, the hot upsetting of cylindrical billets is a suitable process due to the simple geometry but possessing all features of the process. A process simulation with coupled Finite Element method program is given.

4.2.5.4 Process Model and Boundary Conditions

A FEM process modelling program described in detail in Chapter 2.6.4 was used to simulate the upsetting process without significant simplifications.

All the important forging parameters were used as input data:

- flow curve $\sigma_f = f(\varepsilon, \dot{\varepsilon}, T_f)$

- density ρ_B, heat capacity C_B, conductivity λ_B of the billet as a function of the temperature

- density ρ_D, heat capacity C_D, conductivity λ_D of the die as a function of the temperature

- starting temperature distribution of the billet (radial)

- starting temperature of the die

- friction coefficient, heat transfer coefficient of the lubricant

- die velocity range for typical forging presses

For the case study, the following conditions were fixed:

- billet material is a carbon steel with 0.45 % C (Ck45), a hardenable steel used widely in the forging industriy.

The billet dimensions were:

- starting diameter d_O = 20 mm
- starting height h_O = 40 mm

When modelling, the reduction in height was considered, up to 50 %.

- The initial temperature distribution in the billet was considered as parabolic with fixing the surface temperature to T_{AO} = 1100°C and a variable initial core temperature T_{KO} = 1100°C, 900°C and 700°C, Figure 4.29 and 4.30.

- The press/die velocity was taken as v = 20, 100, 500, 2000 mm/s, typical for the range of slow hydraulic presses (20, 100 mm/s), mechanical presses (500 mm/s) and hammer or spindle presses (2000 mm/s). Using a constant velocity is a simplification mainly used for better comparison of results of the simulation. The FEM-process model would be able to deal with a time-dependent velocity without difficulty.

With these boundary conditions, a numerical simulation gives the following results:

- temperature distribution in workpiece and dies

- strain distribution locally and the shape change of the billet as an integral measure

- force and power for the forging process

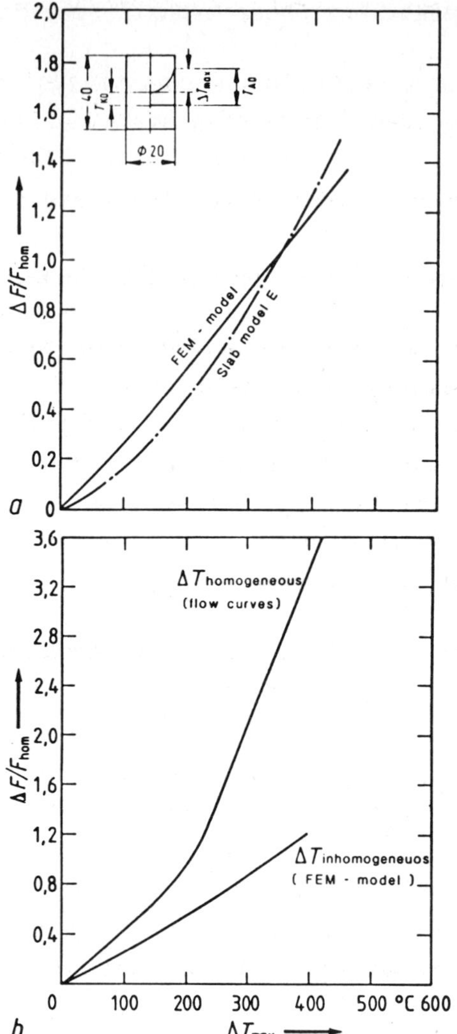

Figure 4.30:   a)   Related force increase $\Delta F/F_{nom}$ for an inhomogeneous temperature distribution with $\Delta T = T_{AO} - T_{KO}$ and F_{nom}: force for homogeneous temperature 1100°C - comparison of FEM and slab model.

       b)   Comparison of an inhomogeneous temperature distribution $\Delta T_{inhomogeneous}$ with a homogeneous temperature decrease $\Delta T_{homogeneous}$.

Starting temperature T_{AO} = 1100°C; die speed v = 20 mm/s; billet diameter d_o = 20 mm; height h_o = 40 mm; material steel Ck45.

-    local die loads, stresses and temperature.

For checking the results in an experimental set up with homogeneous and inhomogeneous induction heated billets, upsetting tests on a 4 kN hydraulic press were performed. The force as a function of the displacement and the temperature was registered.

4.2.5.5 Calculations, Experiments and Results

The compression force F for homogeneous and inhomogeneous temperature distribution was calculated with a slab method model and with the FEM-model as well. For the slab method model "E" in Figure 4.30a, adiabatic conditions were assumed. For the comparison of the effect of an inhomogeneous temperature distribution in Figure 4.30b, the effect of a certain homogeneous temperature decrease is also calculated.

For the approach "E", a parabolic starting temperature distribution $T_O = f(r)$ is assumed:

$$T_O(r) = \frac{T_{AO} - T_{KO}}{R_O^2} \cdot r^2 + T_{KO} \qquad (4.57)$$

with:

T_A: temperature at the surface
T_K: temperature in the core
R: outer diameter of the billet
H: initial height
h: intermediate height
Index o: initial condition

Assuming adiabatic conditions for the slab elements and ignoring heat generation from deformation for a first approach of the slab modelling "E", the height, h, is given as follows:

$$T_O(r,h) = \frac{T_{AO} - T_{KO}}{R_O^2} \cdot \frac{h}{H} \cdot r^2 + T_{KO} \qquad (4.58)$$

According to the procedure of the slab method, the force increment is calculated for a tube element of the whole billet as the product of the area element and the normal stress on the element in the force direction. A first approximation for the normal stress is the flow stress $\sigma_f = f(\frac{h}{H}; \frac{v}{h}; T)$ giving:

$$\Delta F_i \approx 2\pi \cdot r_i^2 \cdot f(\frac{h}{H}; \frac{v}{h}; T) \tag{4.59}$$

and for the whole force:

$$F_i \approx \sum_{i=1}^{n} \Delta F_i \tag{4.60}$$

An improvement of the slab method "E" considering friction and heat generation would easily be possible but is not necessary for this comparison.

From Figures 4.30a and b the following can be deduced:

- Good agreement of force calculations between slab method "E" and FEM-model. (For force calculations "E" is good enough.)

- A temperature inhomogeneity of 100°C increases the relative force $\Delta F/F_{nom}$ about 20 % related to the homogeneous force F_{nom}. A homogeneous temperature decrease of 100°C instead gives a much higher increase of the force of 50 %.

- A (homogeneous) temperature increase of 40°C can compensate a temperature inhomogeneity of 100°C.

  Example: $T_{AO} = 1100°C$; $T_{KO} = 1000°C$ ($\Delta T = 100°C$) gives
  $\Delta F = 20\%$ related to F_{nom} (1100°C).
  $T_{AO} = 1100 + 40 = 1140°C$ and $T_{KO} = 1000 + 40 = 1040°C$ gives the same force as F_{nom} (1100°C)

- In a comparison with the influence of speed on the flow stress, the force shows: a five times higher die speed, giving a force increase of 20 % as a temperature inhomogeneity of 100°C. This means that a temperature inhomogeneity of 100°C has about the same effect as a change from a slow to a fast press type, as from hydraulic- to mechanic- or hammer/screw presses.

4.2.5.6 Local Strain and Temperature Distribution

An inhomogeneous temperature distribution in the billet influences not only the force, but also the local strain and temperature distribution during forging, thus affecting the intermediate material flow.

With the FEM-process model, inhomogeneous temperature distributions were calculated between $\Delta T = 0 \ldots 400°C$ and the influence of the die speed $v = 4 \ldots 2000$ mm/s, according to the band between hydraulic presses and hammer or spindle presses.

Figure 4.31 shows an example of the results. They can be summarized as follows: inhomogeneous temperature distributions with lowest temperature in the core leads to more homogeneous strain distribution and less bulging of the billet during upsetting, so there is some compensation of the friction and cooling effect at the die workpiece interface with the flow stress gradient from the center to the rim. This effect dominates only at high speeds (hammer-, spindle-, mechanical press) but disappears at the low speeds of $4 \ldots 20$ mm/s (hydraulic presses). This is a consequence of the temperature equalization during the longer deformation time.

The computation of the temperature distribution during forging leads to the following results, Figure 4.31:

-   For the low speeds of a hydraulic press ($v = 4 \ldots 30$ mm/s) there is a temperature equalization in the billet during compression. Much more significant is the cooling of the billet at the die surface, which could be up to 200°C.

-   The chilling effect of the die reduces with increasing die speed. From $v = 500$ mm/s upwards, there are nearly adiabatic conditions and the initial radial temperature distribution will be qualitatively unchanged.

- The die load has a stress and a thermal component. The highest normal stresses are around the center of the specimen. In this area, a substantially lower temperature is present for inhomogeneous heated billets with the lower initial temperature in the core. This is obvious for low speeds under hydraulic presses up to 100 mm/s.

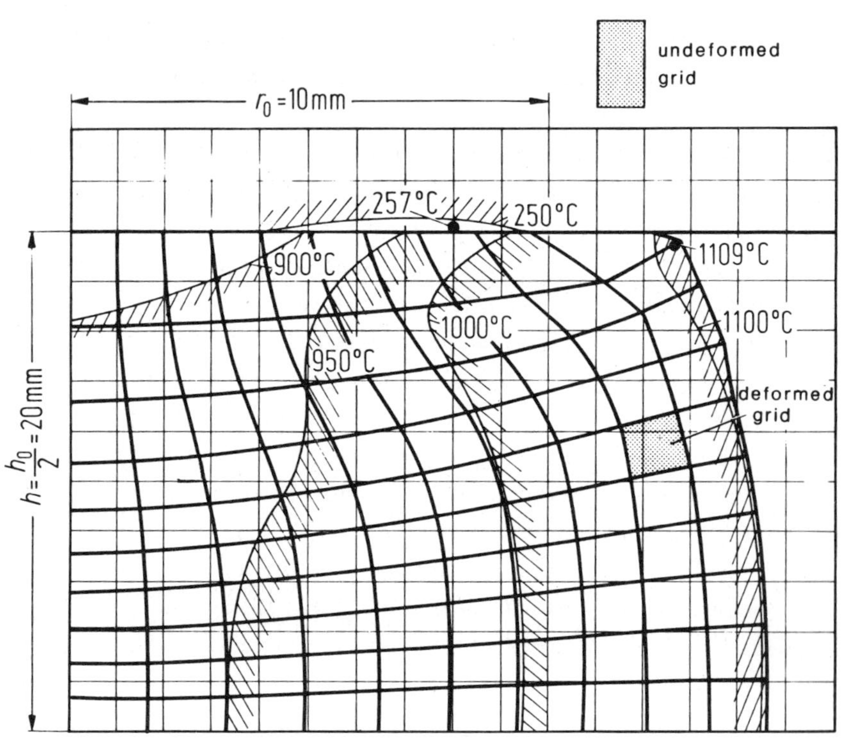

Figure 4.31: Strain and temperature distribution in compression of an inhomogeneous heated billet. Initial temperature distribution T_{AO} = 1100°C surface temperature; T_{KO} = 900 core temperature; die speed v = 500 mm/s; initial die temperature 200°C; billet steel Ck45; d_o = 20 mm; h_o = 40 mm; ε_n = h_o - h/h_o = 50 %.

4.2.5.7 Future Applications of Coupled FEM Process Models

There are specific problems in forging which justify the high cost of such a detailed simulation:

-   forging of parts from titanium- or nickel-base alloys for
    gas turbine and aero-space applications. A detailed knowledge
    of the local strain and temperature history is necessary to
    predict and optimize material properties

-   optimizing properties of parts or dies for the forging of
    parts in large quantities, typically for the automotive
    industry

-   solving principal questions in processing, such as heating,
    process parameters in forging, heat treatment.

It seems that the coupled FEM process modelling in three-dimensional analysis is more a hardware problem than a software problem for users [4.12]. There is hope for the further development of faster and more powerful and cheaper, computers of the next generation.

4.3 Plane Strain Modelling of Closed Die Forging

The slab method can be used very successfully to calculate the
stresses and the loads during forging. The method described in
general terms in Chapter 2.4.2 is applied here to closed die
forging, in particular for forging of turbine and compressor
blades where the assumptions of plane strain holds particularly
true.

4.3.1 Elementary Analysis Approach

In Figure 4.32 the section between the upper and lower dies is
considered subdivided into small deformation zones (between
dashed lines). There is a line (called the neutral line) which
characterizes the direction of flow of the deforming material.
As can be seen, for each deformation zone, the material flow is
essentially perpendicular to the ram motion, while the upper and
lower plates have different inclinations.

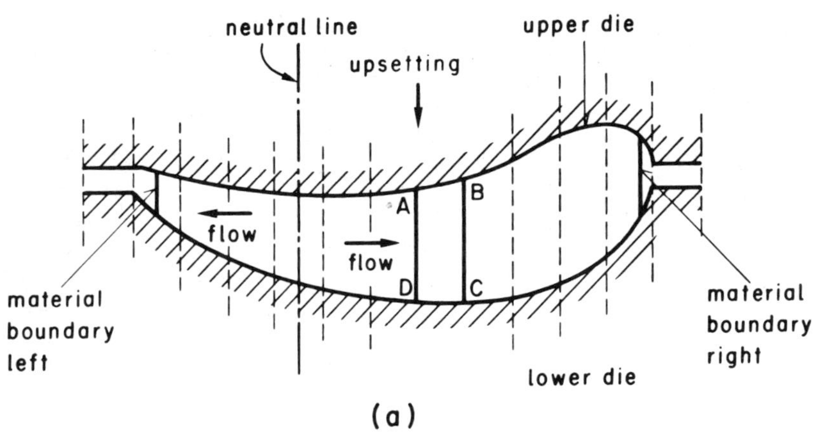

<u>Figure 4.32</u>:   Slab method principle for plane strain closed die
                 forging.

The static equilibrium of each zone is then expressed following
the solution derived in Chapter 2.4.2.2 and 2.4.2.3.

Combining the deformation elements to form the required deformation shape, it is possible to calculate the stress distribution for any complicated plane strain forging processes.

The calculation starts from the left and right material boundaries (see Fig. 4.32): here the stress is known and equal to the flow stress of the material σ_f. The stress distribution is then obtained by calculating element by element, until the two lines cross (see Fig. 4.33): at this point the neutral line has been determined.

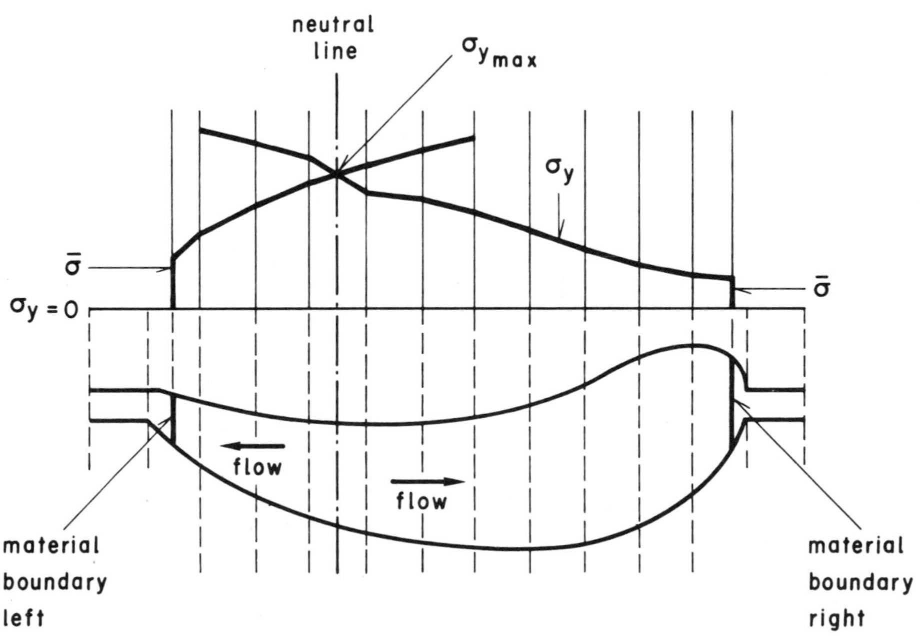

Figure 4.33: Neutral line determination with the slab method.

It should be noticed that the stresses $\bar{\sigma}$ at the boundaries are different in Figure 4.32: this can happen because the flow stress is a function of temperature, strain and strain rate and can therefore be different in each deformation element.

The <u>effective flow stress</u> $\bar{\sigma}$ is given by:

$$\sigma_f = F(\varepsilon, \dot{\varepsilon}, T) \tag{4.61}$$

where F is a function of the material to be deformed but has generally the form of equation 4.1.

The <u>strain</u> ε is expressed as:

$$\varepsilon = -\ln \frac{H}{h_i} \tag{4.62}$$

where:

    H   is the initial preform height
    h_i is the average interval height

The <u>strain rate</u> $\dot{\varepsilon}$ is expressed as:

$$\dot{\varepsilon} = \frac{v}{h_i} \tag{4.63}$$

where: v is the ram speed.

The temperatures in a metal forming operation depend mainly on:

o   the initial material and die temperature

o   the adiabatic heating or heat generated by internal plastic deformation and friction at the die/material interface

o   the heat transfer between the deforming material and the dies and between the material and the environment.

In a process such as forging, the heat loss into the environment can be ignored and only the temperature increase due to deformation and the temperature decrease due to heat transfer between billet and die have to be taken into consideration (as was shown in § 4.2.1.1).

The temperature increase due to deformation, ΔT_d, in a time interval Δt is given by:

$$\Delta T_d \;=\; \frac{\bar{\sigma}\,\dot{\bar{\varepsilon}}\,\Delta t}{c_p\,\rho} \tag{4.64}$$

where:

$\bar{\sigma}$ is the effective flow stress of the deforming material

$\dot{\bar{\varepsilon}}$ is the effective strain rate

c_p is the specific heat of the deforming material

ρ is the specific density of the deforming material

The equation (4.64) can also be written as:

$$\Delta T_d \;=\; \frac{\bar{\sigma}\,\Delta\bar{\varepsilon}}{c_p\,\rho} \tag{4.65}$$

where:

$\Delta\bar{\varepsilon} \;=\; \dot{\bar{\varepsilon}} \cdot \Delta t$ is the effective strain generated during the time interval Δt.

The heat transfer can be calculated considering the deforming material as a thin plate with an average uniform temperature, which is cooled between two infinite dies from both sides. The heat balance is therefore expressed by:

$$-\,c_p\,\rho\,V_a\,dT \;=\; \alpha\,F\,(T - T_{Do})\,dt \tag{4.66}$$

where:

V_a is the average deforming volume under consideration
α is the heat transfer coefficient between the materials
F is the surface area of contact between the material and the die
T is the instantaneous temperature

T_{Do} is the initial die surface temperature

t   is the cooling time

Integration of equation 4.66 gives:

$$T = T_{Do} + (T_{Bo} - T_{Do}) \exp \left[\frac{-2 \; \alpha \cdot t}{h \; c_p \; \rho} \right] \qquad (4.67)$$

where:

h   is the material thickness between the dies

T_{Bo} is the initial material temperature

The final average temperature T_{BF} is given by:

$$T_{BF} = T + \Delta T_d \cdot t_h \qquad (4.68)$$

where: $t_h = \exp \left[\frac{-2 \; \alpha \; t}{h \; c_p \; \rho} \right]$ is the rate of heat transfer.

Therefore the equation 4.68 can be written as

$$T_{BF} = T_{Do} + (T_{Bo} - T_{Do} + \Delta T_d) \exp \left[\frac{-2 \; \alpha \; t}{h \; c_p \; \rho} \right] \qquad (4.69)$$

The temperature T_{BF} is calculated for each deforming element and therefore it is possible to have a temperature distribution across the section.

4.3.2   CAD/CAM and Process Modelling of Closed Die Forging

To obtain a successful forging process, besides the knowledge of metallurgical properties of the part to be forged, it is necessary to evaluate technologically the press, the dies and the workpiece.

This means that in order to optimize the process and reduce the costs, it is necessary to minimize the forces and the stresses during forging, as well as to obtain the best filling of the dies with minimum material scrap.

To minimize the forces and the stresses, it is necessary to find
the optimum position of the finished part in the dies, or, in
other words, it is imperative to find the optimum location of
the dividing line between the upper and lower die. The dividing
line is also the position of the flash in the die.

A good position of the flash will assure that no undercuts are
present in the dies (and it is possible to extract the workpiece
from the dies when the forging process is finished) and that the
loads on the dies are minimized (reducing, therefore, press deformations and assuring better final tolerance).

The best filling of the dies, or, in other words, the knowledge
of material flow during the forging process, is of primary importance. To optimize the material flow, it is necessary to determine the best preform shape and position in the die, as well
as the dimensions (thickness and width) of the flash.

To determine the above parameters, it is necessary to simulate
the deformation process and calculate the stress and temperature
distribution. Furthermore, it should be possible to change and
modify the geometry and position of the preform, as well as the
size of the flash interactively.

The elementary analysis described in Chapter 2.4 and the application to forging shown in Chapter 4.2 is implemented in a CAD/CAM
system in order to help the designer to solve the above technological problems. The program is modular and uses high resolution
graphics for the manipulation of complex 3-D geometry.

The system is coupled with a FEM package for very large plastic
deformation. The software is written according to the description
of Chapter 2.6 and some applications to the forging process are
presented in Chapter 4.3. The FEM program is used when considerable detail of the deformation process is needed.

A third component of the software system consists in the die
stress analysis and the CAM (Computer Aided Manufacturing) of
the dies. The stress analysis of the forging dies is presented
in Chapter 4.4.3 with some typical applications.

4.3.2.1   The DIEDESIGN Software

The first step in design of the forging process starts by examining the finished part drawing (see Figure 4.34).

A preliminary design of the forging dies is then made. Taking
into consideration the equipment available, the material to be
forged and other process parameters, the optimum dividing line
between upper and lower dies is found. The main optimizing factor
is the minimum side loading.

The preform is then designed and the final dimensions of the
flash are established, based again on equipment available and
process parameters. An adequate metal distribution determines
the best design solution.

The dies are then manufactured and the production of blades can
start.

As was previously defined, CAD of forging dies is a system of
computer programs to automate the design and manufacture of
dies for forging (in particular turbine and compressor blades).

In the computer programs, algorithms which simulate the forging
process were implemented. Thus the design of the dies is not
done with the help of _CAD_ alone, as intended in the classical
sense of the term, but rather with the combination of _Process
Modelling_ (for the simulation of the deformation process) and
CAD (for the graphical description of the dies, preform, etc.).

The major functions performed by DIEDESIGN are outlined in Figure
4.35, and may be summarized as:

1. Read in and preprocess the blade geometry, including the
    root. Calculate the necessary parameters, e.g. cross-sec-
    tional area. Give the initial flash geometry, the material
    to be forged and the forging conditions.

2. Determine the flow stress under forging conditions, based on
    the flow stress-strain-strain rate surface built in as func-

tions in the program.

3. Determine the forging plane, based on minimum side loading of the dies during forging. The turbine blade is rotated around its stacking axis in small increments until the position with minimum resultant horizontal force is determined. This, in turn, minimizes die shift during forging, resulting in improved tolerances on the product.

4. A simulation of the forging process is carried out in order to determine the best form and position of the preform in the die cavity. This position not only ensures die cavity fill during forging, but also flash losses may be minimized by using the minimum dimensions indicated.

   The flash design can be performed interactively to obtain the best metal deformation and minimum forging stresses.

5. The results are summarized for each cross-section and for the whole blade.

6. A 3-D display of the die surfaces is provided, together with the flash land as designed by the computer programs. (This step is performed with Part I of DIEDESIGN).

Following the use of DIEDESIGN, other steps are possible in order to reach the final manufacturing of the dies (see Figure 4.35):

a. If a more accurate description of the metal deformation is required along with the strain distribution in each section, the DIEDESIGN results can be used as input data to the Finite Element program for large plastic deformations (SINTER, MRPNEW).

b. Due to relatively high loads and complex geometry, uneven local elastic deflections of the die are unavoidable. These deflections must be taken into consideration in order to compensate with material addition and to remain in the specified tolerances. When the user is satisfied with the design, the information regarding the geometry and the stress state is given as inputs to a Finite Element program. Where necessary, the local elastic deflections may be superimposed on the gross deflections of the press. Using this information, the die surface coordinates are corrected for the expected deflection during forging.

c. The output from the corrections due to elastic deflections are used by part programming routines for Numerical Control (NC) machining.

The DIEDESIGN package has been written as a series of modules to facilitate the introduction of new blocks and algorithms during the research and development phase. Each module works and operates around data bases that can be common to each module as shown in Figure 4.36. One data base that is in common with all modules is the data file on the x, y, z coordinates, describing each section of the object.

The communication between the different computer hardware and software does not constitute a problem as long as the interfaces are standardized and the information data base has a simple and constant structure, as in our case with the x, y, z section's coordinates.

The DIEDESIGN program can be used to design a die for isothermal or hot-die forging. Testing on real problems has shown the feasibility and accuracy of the software.

The major improvements with respect to previous programs are:

-   design with variable position of the flash

-   possibility of simulating the sliding of the preform in the die

-   temperature distribution in each section of the deforming billet (not only average temperature)

-   as a consequence of the above point there is also a better representation of the stress distribution

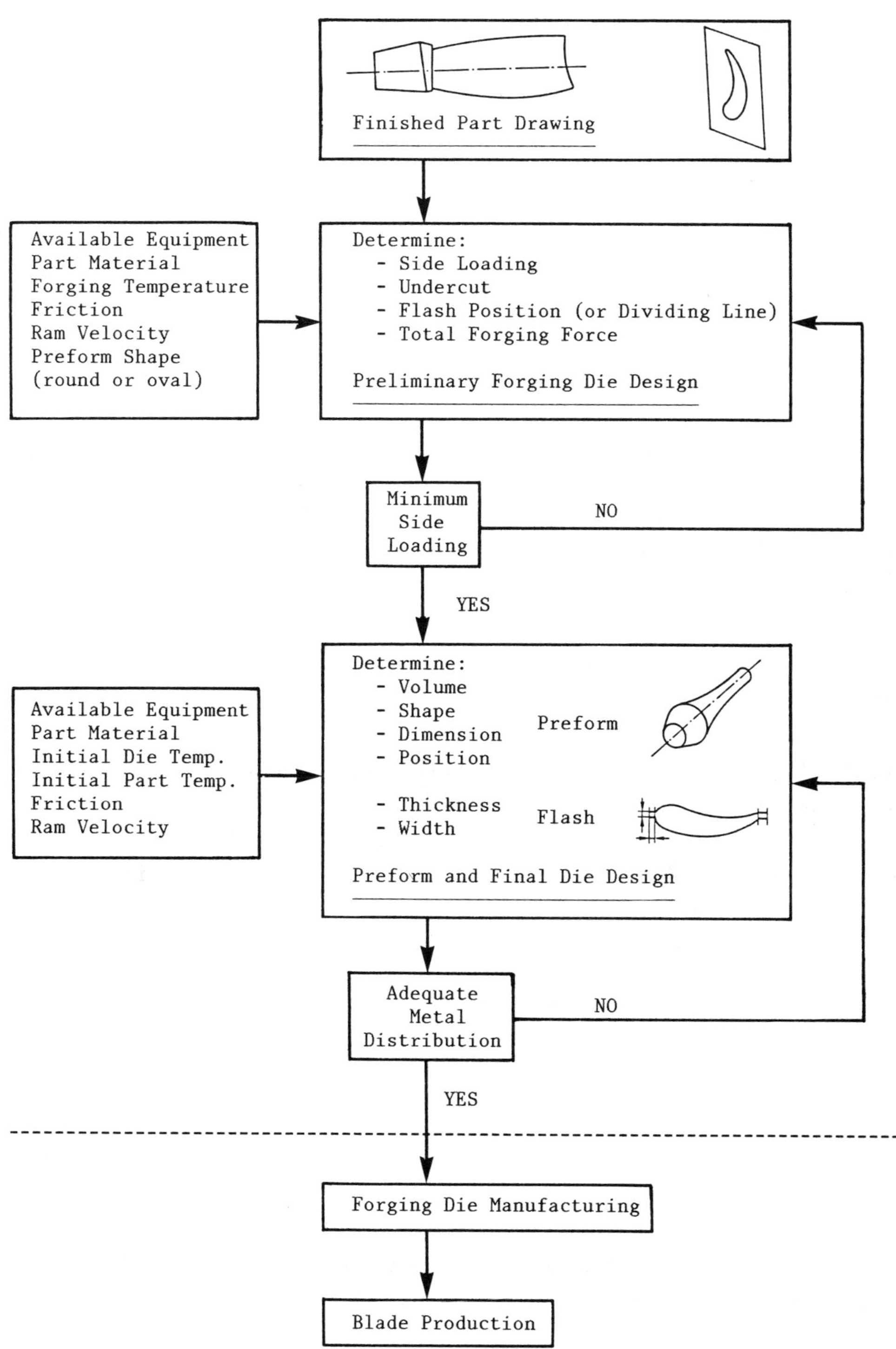

Figure 4.34: Design of forging dies and preform.

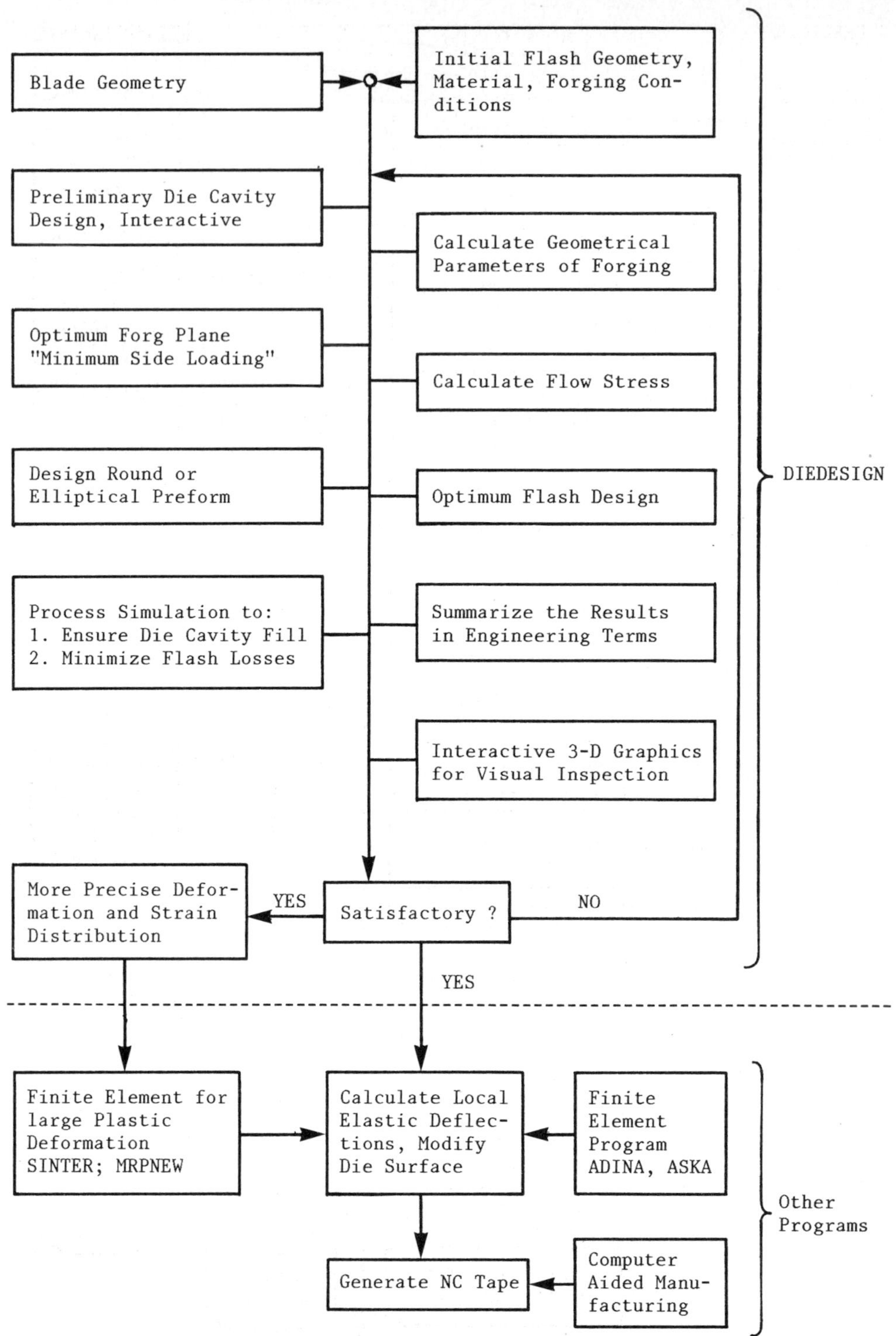

<u>Figure 4.35</u>: CAD and process modelling for forging die design.

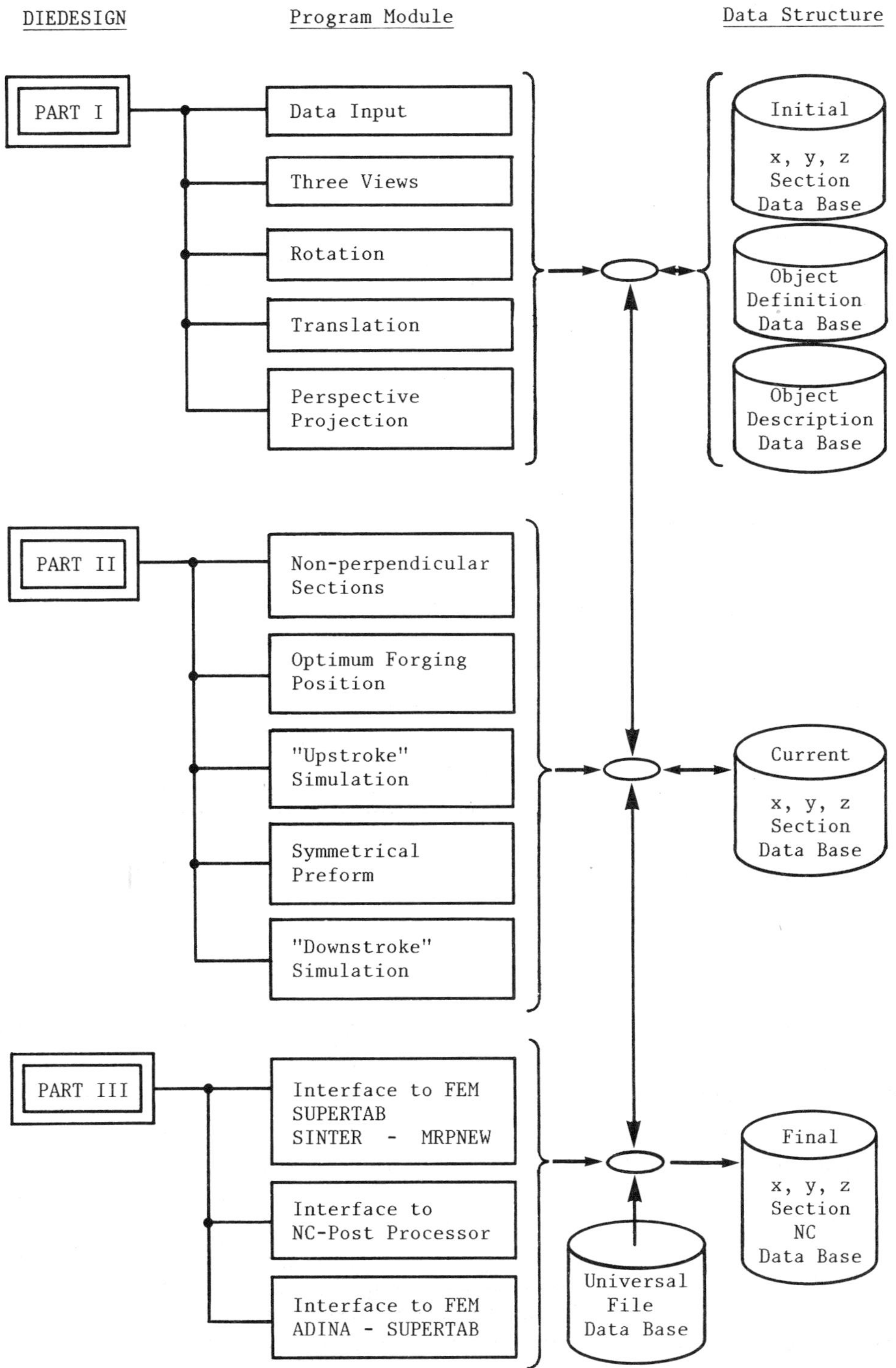

Figure 4.36: DIEDESIGN modules and data bases.

215

- calculation of round or oval (with ratio of large to small axis) preform

- possibility of fixing or changing the position of the preform in the die

- use of a microcomputer instead of a large mainframe for many of the calculations involved.

4.3.2.2 Geometrical Input Data Description

The data describing the geometry of the part are given as a series of x, y, z coordinates. These coordinates are organized in sections perpendicular to one of the main axes of the object. This is the normal way of description of, for example, turbine blades as shown in Figure 4.37. Several sections describe the airfoil shape (Sec. A-A, B-B, ... F-F), while the root is given by its dimensions. The program can only treat objects whose sections are perpendicular to one of the main axes. Therefore a special algorithm has been written and implemented in the computer model, which can handle any section not perpendicular to such an axis. The algorithm subdivides the inclined surface into a given number of sections perpendicular to the axis: the higher the number of sections, the better the approximation to the true surface.

4.3.2.3 Process Modelling of the Forging Process

The heart of DIEDESIGN is the _modelling of the forging process_. The algorithm on which the model works is based on the "slab method", or elementary analysis. This method gives a relatively good estimation of the stresses and loads involved during the process, as was demonstrated in Chapter 4.2.

The description of the algorithms and generalized application to any plane strain situation is given in Chapter 4.3.1.

The present method assumes that the three-dimensional object is subdivided into cross-sections parallel to each other, as was previously described. Each section is then treated separately, assuming that there is no deformation in the direction perpendicular to the cross sections: this assumption works particularly well in the case of the turbine blade.

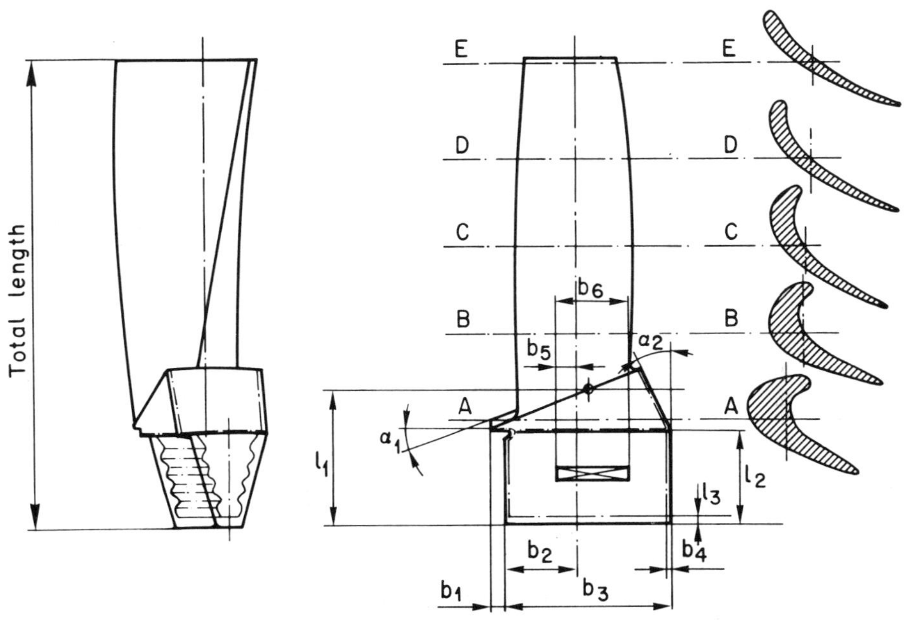

Figure 4.37: Geometrical description of a blade.

In addition to the slab method, other algorithms were implemented in order to improve the performance and accuracy of the calculations. For example, a relatively good description of the temperature distribution has been given in order to take into consideration both the heat transfer phenomena between billet and dies (as in the case of hot die forging), as well as the adiabatic heating generated in the deforming material.

4.3.2.4 Forging Loads Minimization and Flash Positioning

One problem in forging a part (blade) is to know how the part should be positioned in the dies. The orientation of the blade itself defines where the flash will be. The initial position of the blade is shown in Figure 4.38 (a).

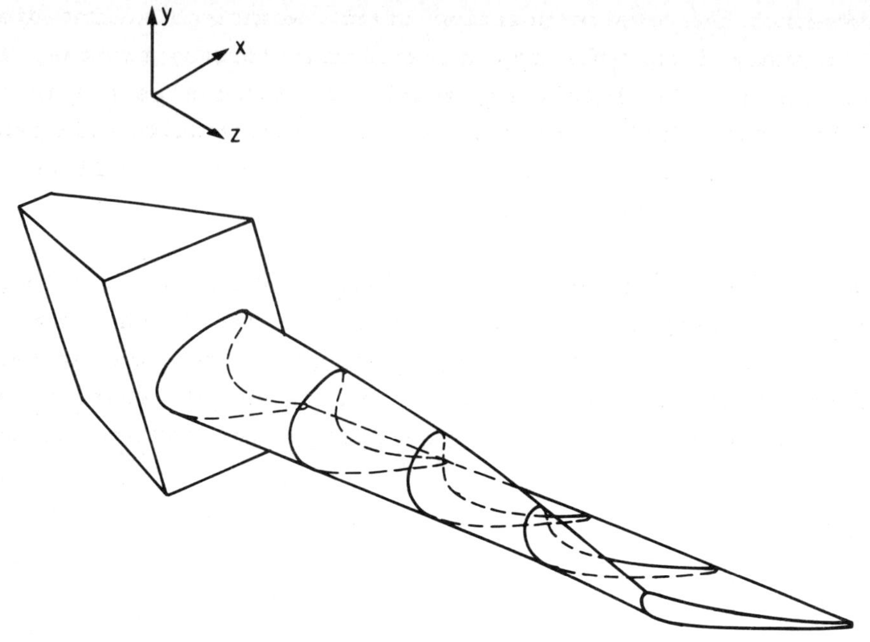

<u>Figure 4.38 a)</u>:   Initial blade position.

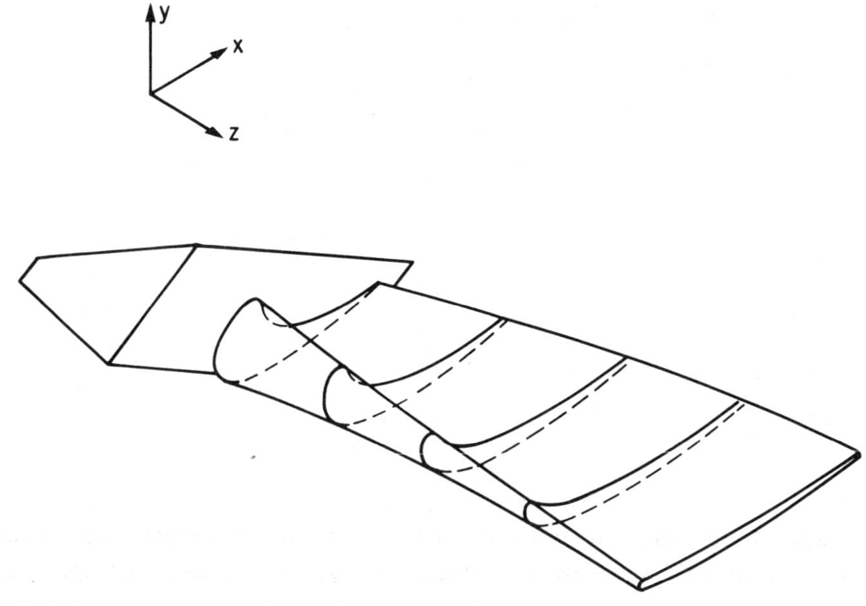

<u>Figure 4.38 b)</u>:   Optimum blade position.

218

The determination of the position of the workpiece in the dies
is of primary importance for a successful forging process. If
the workpiece is as highly asymmetric and twisted as a turbine
blade, the determination of the position becomes quite different
because the aim is to minimize the lateral force as well as to
avoid undercuts in the dies.

In Figure 4.39 (a), a section of a blade is shown with the forging direction perpendicular to the plane of the flash. In this
position, the blade cannot be forged because one portion (shaded)
is in undercut. The section has to be rotated, at least, by an
angle α until the line A is parallel to the forging direction.

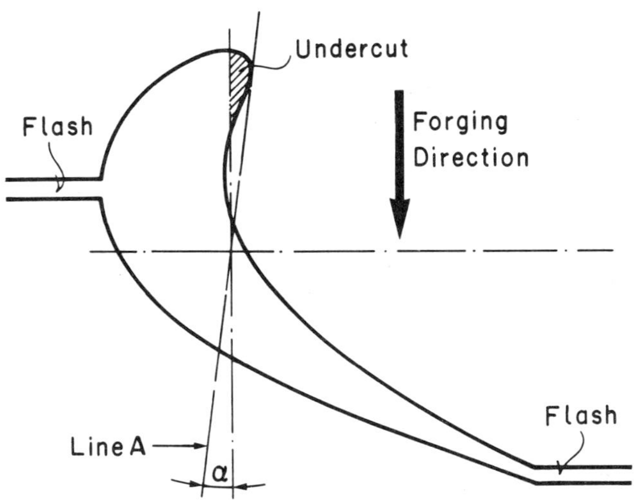

Figure 4.39 a): Undercut detection.

In Figure 4.39 (b), a section of a blade is shown subdivided
into a number of deformation elements. With the help of the slab
method, it is possible to calculate not only the forging force,
but also the horizontal forces acting on the upper and lower
dies.

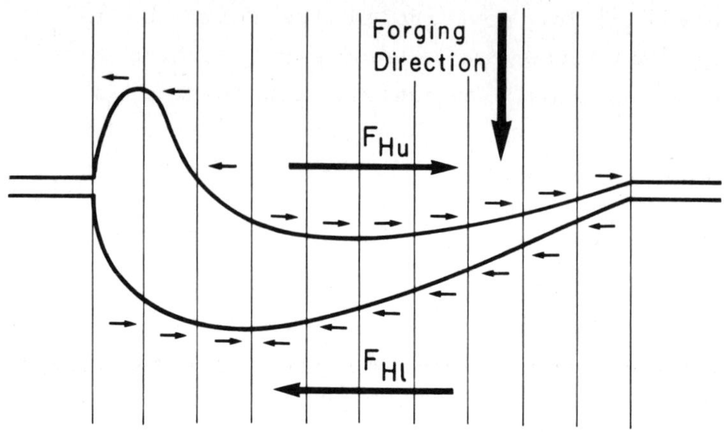

<u>Figure 4.39 b)</u>:   Optimum forging direction detection.

In each section i the horizontal forces are given by:

$$F_{u_i} = L_i \sum_{j=1}^{m} F_{u_{i,j}}$$ (4.70)

$$F_{l_i} = L_i \sum_{j=1}^{m} F_{l_{i,j}}$$ (4.71)

where:

L_i is half the distance between two sections i and i+1

$F_{u_{i,j}}, F_{l_{i,j}}$ are the elementary horizontal forces per unit length.

m is the number of elementary deformation elements

The total horizontal forces on the workpiece (or dies) are therefore:

$$F_{H_u} = \sum_{i=1}^{n} F_{u_i} \qquad\qquad (4.72)$$

$$F_{H_l} = \sum_{i=1}^{n} F_{l_i} \qquad\qquad (4.73)$$

where:

    n is the number of sections representing the workpiece.

The total resultant force is therefore given by:

$$F_{TOT} = [F_{H_u} - F_{H_l}] \qquad\qquad (4.74)$$

The total resultant force is a function of the rotation angle α and the best orientation of the workpiece is obtained when:

$$F_{TOT}(\alpha) = minimum \qquad\qquad (4.75)$$

An example is shown in Table 4.2, where, for the given conditions, the resulting force in x has a minimum for $\alpha = 42°$.

Interactivity between the computer and the user is possible in the first phase of initializing the search process for the minimum.

The bounded angle of search (lower and upper limit) with an angle increment and the number of points are used to determine the search. Data such as temperature, friction coefficient and ram velocity are necessary to calculate stresses and forces for a certain material (see Table 4.2). A round or oval preform can be specified.

The search process can be divided into the following steps:

1 - rotate sections

2 - divide sections into upper and lower parts

3   -   check for undercut in lower part and overcut in upper part; if undercut or overcut is detected: new angle and continue at point 1

4   -   trim upper and lower part with cubic spline function into an equally-spaced data set

5   -   determine stress distribution and forces with the slab method

6   -   the total forces (vertical and horizontal) for the whole object define one forging position

7   -   new angle and continue at point 1

The blade rotated in the optimum position is shown in Figure 4.38 (b). In this position, the sum of the horizontal forces has a minimum value.

TABLE 4.2

Calculations are done with the following parameters :

```
Input   data file name ............: [KLR16.DIE.DATA]BLADE.DAT
Starting angle ....................:    30.00000
Limiting angle ....................:    50.00000
Angle increment ...................:     1.00000
Ratio width/height ................:     1.00000
Number of intervals ...............:    20
Material ..........................: NIM80
Temperature .......................: 1100.00000
Friction coeffecient ..............:     0.20000
Ram velocity ......................:     1.00000
```

All forces are calculated in kN

ANGLE	RESULTING FORCE IN X	FORCE IN Y	FORCE IN X UPPER	FORCE IN X LOWER
30.000	OVERCUT IN SECTION : 3 AT POINTS :135 136			
31.000	OVERCUT IN SECTION : 3 AT POINTS :135 136			
32.000	OVERCUT IN SECTION : 3 AT POINTS :136 137			
33.000	1.575432E+003	3.858227E+003	7.835513E+002	-7.918804E+002
34.000	1.455738E+003	3.882891E+003	7.202931E+002	-7.354453E+002
35.000	1.332156E+003	3.893423E+003	6.620187E+002	-6.701377E+002
36.000	1.218088E+003	3.906790E+003	6.052589E+002	-6.128290E+002
37.000	1.094150E+003	3.915671E+003	5.429824E+002	-5.511673E+002
38.000	9.591906E+002	3.925905E+003	4.819688E+002	-4.772218E+002
39.000	8.448735E+002	3.934698E+003	4.173634E+002	-4.275101E+002
40.000	7.209365E+002	3.936801E+003	3.593171E+002	-3.616194E+002
41.000	5.978454E+002	3.953472E+003	3.048900E+002	-2.929554E+002
42.000	4.800434E+002	3.954186E+003	2.411642E+002	-2.388792E+002
43.000	3.549402E+002	3.961231E+003	1.754057E+002	-1.795346E+002
44.000	2.152625E+002	3.962629E+003	1.015129E+002	-1.137496E+002
45.000	9.603321E+001	3.960658E+003	4.488404E+001	-5.114917E+001
46.000	4.280045E+001	3.954175E+003	-2.582024E+001	1.698021E+001
47.000	1.590133E+002	3.953960E+003	-8.899065E+001	7.002267E+001

4.3.2.5 Preform and Flash Definition

The shapes of the closed dies correspond to the final shape of
the blade after the forging process. The next difficulty is to
get an idea of the starting material size and distribution
which allows a forged blade to be obtained.

Procedure

Considering that the metal volume remains unchanged and the metal
flows away from the neutral surface towards both sides, the geometry of the forging is determined after each step.

On the supposition that the material does not flow lengthwise
("plane strain"), the area in any vertical cross-section remains
unchanged.

The following steps are performed:

1) Each section has to be provided with flash on both sides.
    The user can dimension the flash width and thickness. It is
    possible to choose the increment of the ram stroke and the
    ratio width/height which defines the type of preform (round
    or oval). Indications as to temperature, friction coefficient
    and ram velocity represent the process data for a certain
    material that is forged.

2) The upper die - each section - is raised in small steps
    (upstroke simulation). After each step, the new material
    boundaries on both sides of the neutral surface are calcu-
    lated. The "slab method" gives results for the stress
    distribution, the new location of the neutral surface and
    the forces. The die is raised until the area is approxima-
    tely a square or a rectangle (for round or oval preform).

3) - From all sections is received:
                        - starting die opening
                        - location of the neutral surface
                        - area distribution left and right

- In order to get a machinable shape, the locations of the
centers of the preform sections have to be situated on a
straight line. This is obtained by linear regression (see
Figure 4.40).

- At the beginning of the forging process, the die has to be
opened to the maximum.

- These changes apply for each section:
  - adjust die opening to maximum
  - new location of neutral surface
  - new material distribution at left
    and right

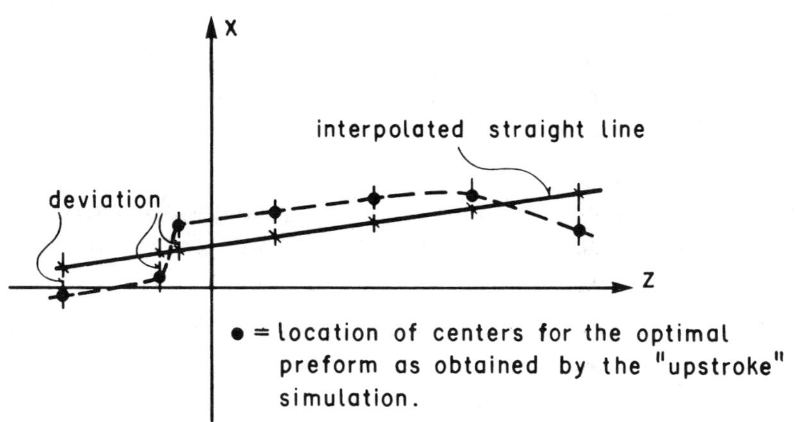

Figure 4.40:   Linear regression for straight preform.

4) The forging process can now be simulated by closing the upper
die in steps (downstroke simulation) as in 2) until the final
form of a section is reached ("bottom dead center"). At each
step, the material boundaries are shown, and the stress and
temperature distribution are displayed. Four part areas for

224

the left hand and right hand side give additional control,
even in cases where the material exceeds the die boundaries.
The forging force, as well as the lateral forces on the upper
die and on the lower die, is calculated and displayed.

If, after the last step, a section is not totally filled (see
Figure 4.41 (a)) the user has the possibility of "adding"
material to the section and the forging process is simulated
again (see Figure 4.41 (b)).

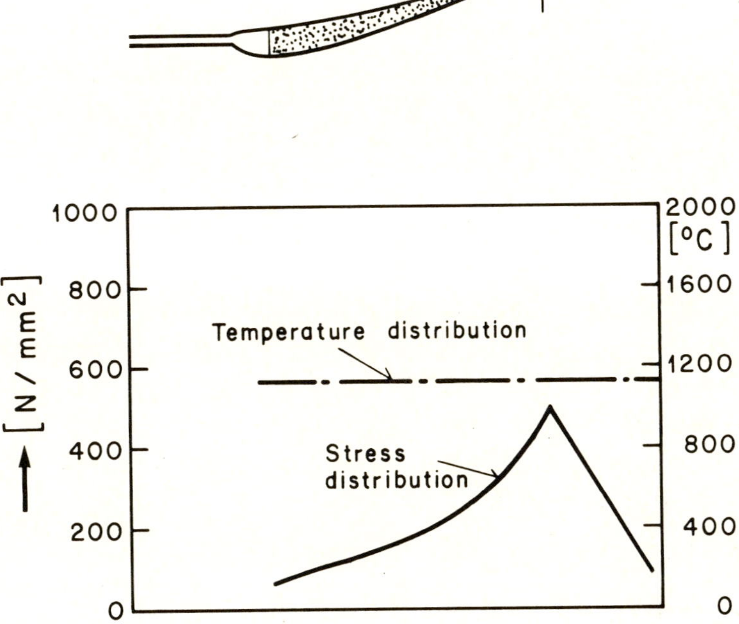

<u>Figure 4.41 a)</u>: Downstroke simulation: "bottom dead center"
position (Section G).

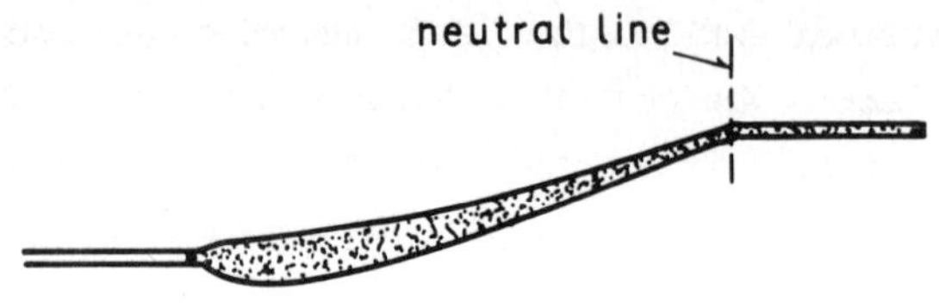

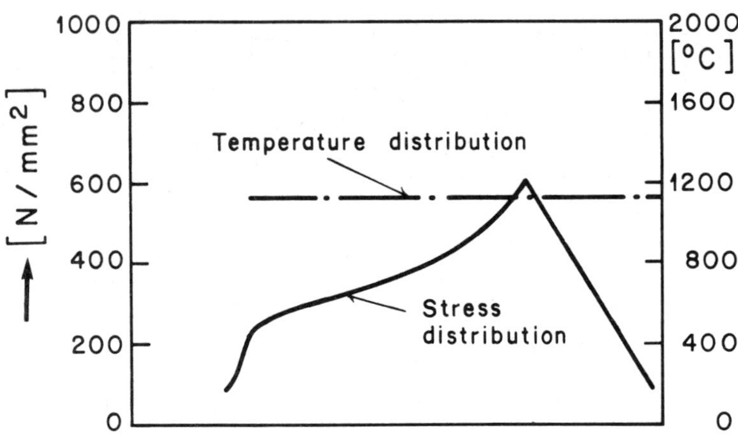

Figure 4.41 b):   Interactive preform dimensions changed.

5)   After all sections have been processed and material flow is
     optimized, the program displays the lower die and the calcu-
     lated form and position of the preform in the die (see Figure
     4.42). Furthermore, the center of loading is also calculated
     and displayed (see Chapter 4.3.2.6). This is the position of
     the forging axis corresponding to the axis of the press,
     where horizontal forces and torque are minimized, reducing,
     therefore, lateral deflection of the frame of the press and
     increasing the forging precision.

6)   After a successful forging, the calculated preform is
     sketched where all the sections are of circular shape.

     It is also possible to draw the upper and lower dies with the
     flash part. The final coordinates are interfaced into the

first part of DIEDESIGN and can be plotted as in Figures
4.43 (a) and 4.43 (b).

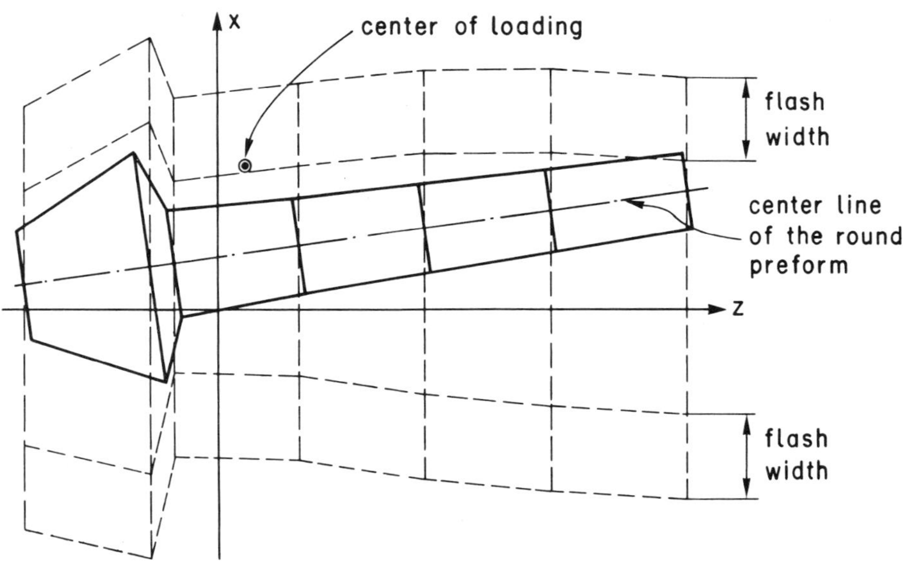

<u>Figure 4.42</u>: Preform position and center of loading.

4.3.2.6 <u>Center of Loading</u>

In a forging process, it is very important to be able to locate
the position of the dies on the press bed that gives least
distortion of the press structure. If such a location is found,
the tolerance of the finished product will also improve.

To find such a <u>center of loading</u> is not easy if the workpiece is
as highly asymmetric as a turbine blade. With the discretization method used in DIEDESIGN, such a location is easier to
obtain.

The center of loading is obtained by finding the equilibrium of
the moments with respect to two axes.

If we consider the workpiece in Figure 4.44 with the forging axis

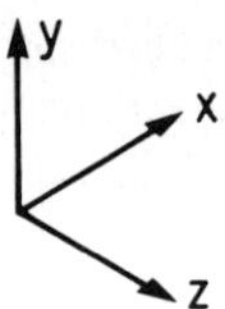
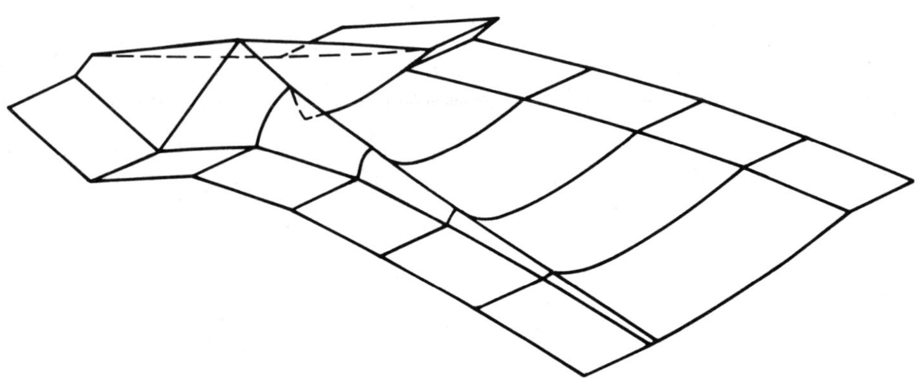

Figure 4.43 a): Upper forging dies.

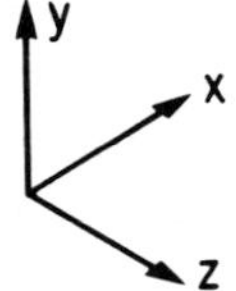
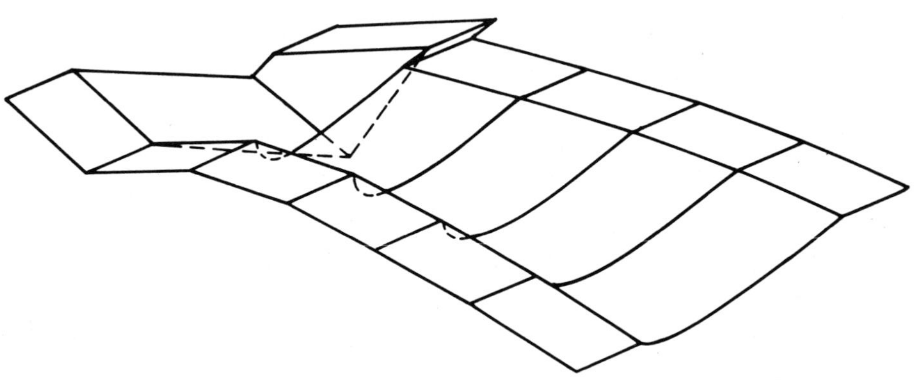

Figure 4.43 b): Lower forging dies.

perpendicular to the sheet, it will be necessary to calculate the two distances R_x and R_y: the center of loading will be at the intersection of the two lines.

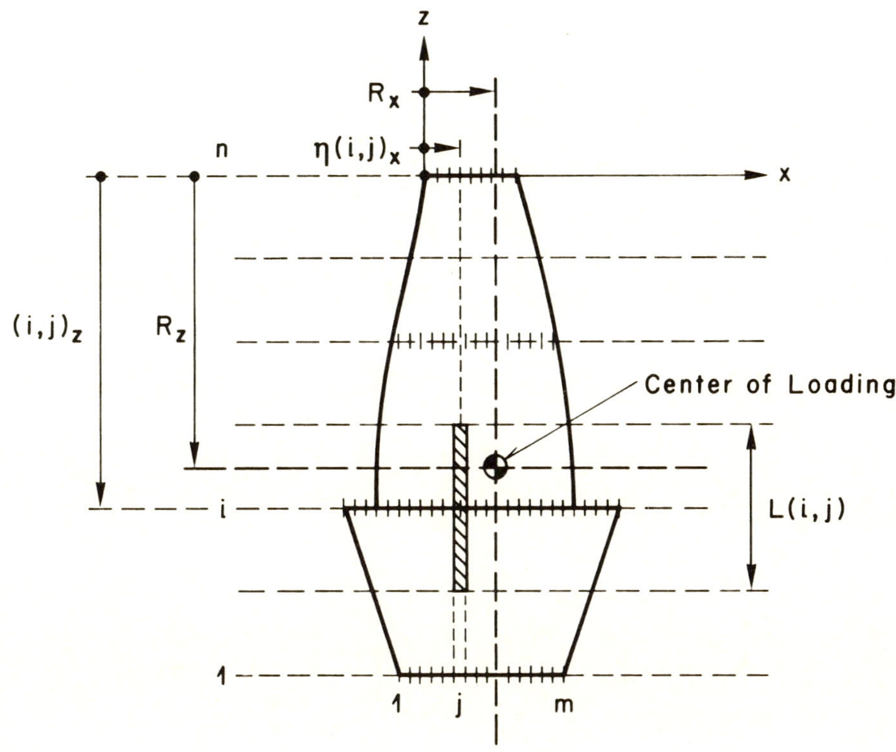

Figure 4.44:   Center of loading determination.

Taking an element (dashed in Figure 4.44) of length L (i,j) and the force per unit length F (i,j) calculated with the slab method, the moment with respect to a z axis passing by the center of loading is written as:

$$F\,(i,j) \cdot L\,(i,j) \cdot (R_x - r\,(i,j)_x) = M_{ij} \qquad (4.76)$$

where:

R_x is the polar axis distance from the reference axis Z

$r(i,j)_x$ is the distance of the element from the reference axis Z.

The summation over all the sections i=1, n and all deforming elements j=1, m imposing a total moment equal to zero will give:

$$\sum_{i=1}^{n} \sum_{j=1}^{m} [F(i,j) \cdot L(i,j) \cdot (R_x - r(i,j)_x)] = 0 \qquad (4.77)$$

Solving for R_x and considering that the length $L(i,j)$ is equal for every deforming element in a section i, is obtained:

$$R_x = \frac{\sum_{i=1}^{n} [L(i) \sum_{j=1}^{m} F(i,j) \cdot r(i,j)_x]}{\sum_{i=1}^{n} [L(i) \sum_{j=1}^{m} F(i,j)]} \qquad (4.78)$$

If the same consideration is given for the equilibrium with respect to the X axis, this produces:

$$R_z = \frac{\sum_{i=1}^{n} [L(i) \cdot r(i)_z \sum_{j=1}^{m} F(i,j)]}{\sum_{i=1}^{n} [L(i) \sum_{j=1}^{m} F(i,j)]} \qquad (4.79)$$

where, in this case, the distance $r(i)_z$ is also constant on a section i.

4.3.3 Application of CAD/CAM to Forge a Blade in a Nickel-Base Alloy

The DIEDESIGN program (see Chapter 4.3.2) was used to evaluate the filling of the die and the forces and stresses required to forge a blade in a nickel-base alloy.

The optimum position of the profile to minimize lateral forces on the upper and lower dies was previously obtained using the same DIEDESIGN program.

The final position with the flash design is shown in Figure 4.45.

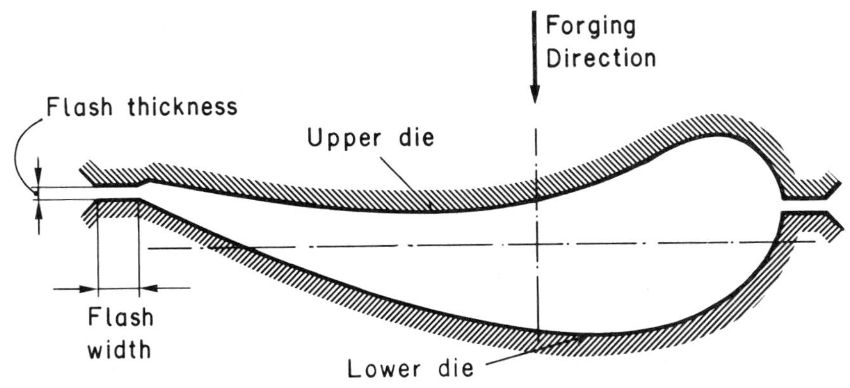

Figure 4.45: Blade forging: die geometry.

The material properties of the nickel-base alloy are given below as:

$$\dot{\varepsilon} \; = \; K \cdot D_{Ni} \; \left(\frac{\bar{\sigma}}{E}\right)^{n} \tag{4.80}$$

where:

K, n   are experimentally determined

E = E (T)   is the Young's modulus

T   is the forging temperature

$$D_{Ni} \; = \; 1.5 \times 10^{-4} \, \exp\left(\frac{-3.43 \times 10^{4}}{T_{K}}\right)$$

T_{K} =   temperature in °K.

The program performs an upstroke simulation to determine the form and position of the preform. In this case an elliptical preform with diameter ratio 1.786 is obtained.

The slab method used simplifies the geometry of the preform by straight lines at the free surfaces and by the boundary of the dies. The form of the profile is such that the flash has to be designed correctly to obtain a good filling of the die.

The object of the simulation was to predict the necessary force and the stress distribution in order not to overload the press and not plastically deform the dies.

A downstroke simulation provides the force versus die opening curves as well as other information such as the stress distribution, die filling, neutral line position, horizontal forces, etc. (the forces are per unit length (or mm)).

The influence of the friction is quite noticeable. For the calculations it was decided to use $\mu = 0.2$ as a reasonable value. A higher forging temperature reduces the final forging force.

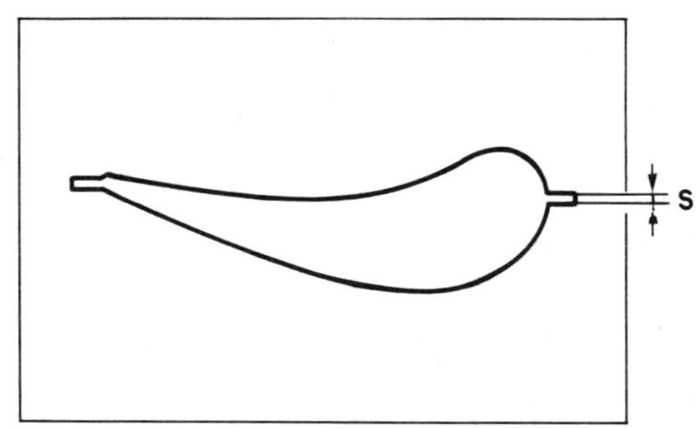

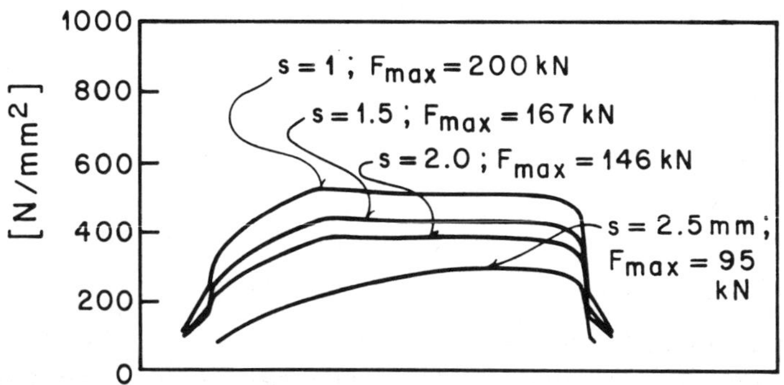

<u>Figure 4.46</u>:   Blade forging: stress distribution   and   maximum forging force in function of die opening s.

The influence of the ram velocity (or deformation velocity) was also studied.

The material has filled the die and extruded through the flash for a die opening between 1.5 and 1 mm (see Figure 4.46). The required force is between 95 and 146 kN/mm length. To obtain the correct profile, it will further be necessary to squeeze the material with an unnecessarily high force (see Figure 4.46).

All the above calculations were made for isothermal conditions.

4.3.4 FEM Plane Strain Analysis of Blade Forging

The forging of a turbine or compressor blade is a rather complex operation due to the properties of the material under deformation, which is generally made of tough steel or titanium or a nickel-base alloy. Furthermore, the operation is performed at high temperatures, with considerable structural and thermal stresses applied to the dies. The problems connected with the heat transfer between the billet and the dies and the environment were discussed in Chapter 4.2.1 which deals with the modelling of hot-die upsetting.

The forging of airfoil shapes is considered one of the more difficult problems of precision forging, due to the rather complex geometry involved and large dimensional differences along the three principal axes of the blade. The problem can generally be simplified, assuming that the deformation along the major axis of the blade is negligible with respect to the other two axes. In this chapter, attention is focussed therefore on the plane strain deformation of one section perpendicular to the major axis of the blade.

The forces and stresses during forging are calculated using the elementary analysis described in Chapter 4.3.1. The same geometry of the section is used in this chapter to analyse the deformations, the strains and strain rates by the application of the Finite Element method. The stress distribution is then compared with that calculated with the application of the slab method.

The forging process is assumed to be isothermal. The material properties follow the law:

$$\sigma_f = 800\ \dot{\varepsilon}^{0.714}$$

and the friction coefficient is m = 0.5.

The preform geometry is an oval, as shown in Figure 4.47 and is positioned as shown in the same figure.

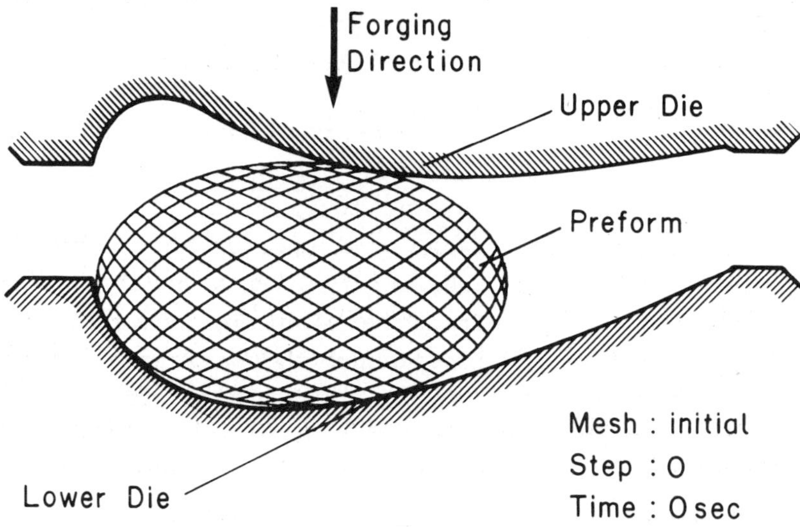

Figure 4.47: Initial die position and preform for closed die forging of a blade.

4.3.4.1 Deformations

The deformed grid is shown at different stages of deformation in Figures 4.47 to 4.51 (a). In Figure 4.49 the material has started to enter the flash at the left-hand side of the dies, but neither the upper portion of the upper die, nor a large part of the die cavity on the left has yet been filled. It is also necessary to remesh the original grid if the material has to enter into flash smoothly. This is done in Figure 4.50 (a), which shows the particular arrangement of the elements in the zone around and in

234

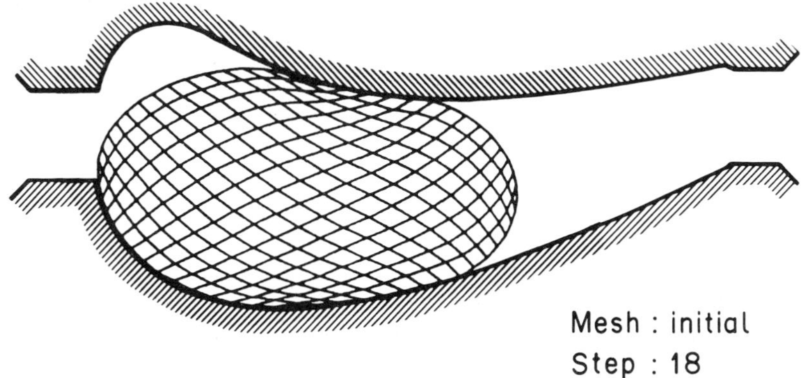

Figure 4.48:   Preform deformation after 0.20 sec.

the left flash. The filling of the die is mainly due to the beneficial influence of the left flash, which allows the material to flow into the rather thin aperture only a little at a time. In Figure 4.51 (a) the material has reached the right flash and the process is completed. It should be noticed how the initial straight vertical lines are deformed in the final stages, due to the influence of the friction at the interface die-billet.

4.3.4.2 Effective Strains and Strain Rates

The strain and strain rate distribution are very important for the prediction of the final microstructure of the material after the forging process has taken place. The effective strain (or von Mises strain) distributions are shown in Figures 4.49 (b) to 4.51 (b). The values of the contour lines are reported in the figures. The time from the beginning of the forging shown in Figures 4.49 (b) and 4.50 (b) is the same (0.48 sec). This is when the remeshing is taking place. The distributions are very similar, indicating that the interpolating routine (see Chapter 2.6.6) is working quite well. A very high strain gradient is developed at the corners of the left flash. Values of $\varepsilon > 2$ are shown in Figure 4.51 (b).

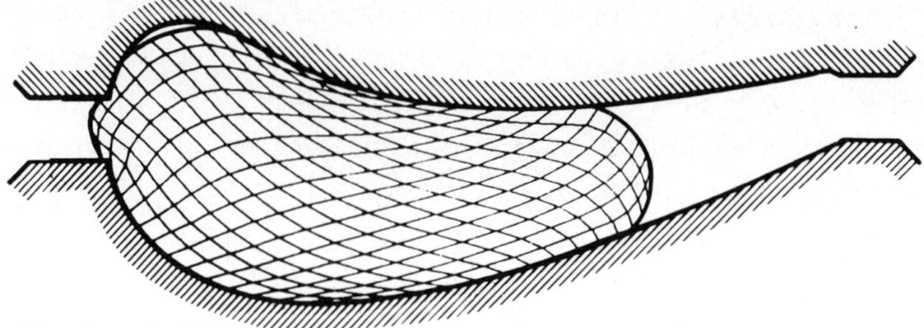

Figure 4.49 a):   Deformation at time 0.48 sec.

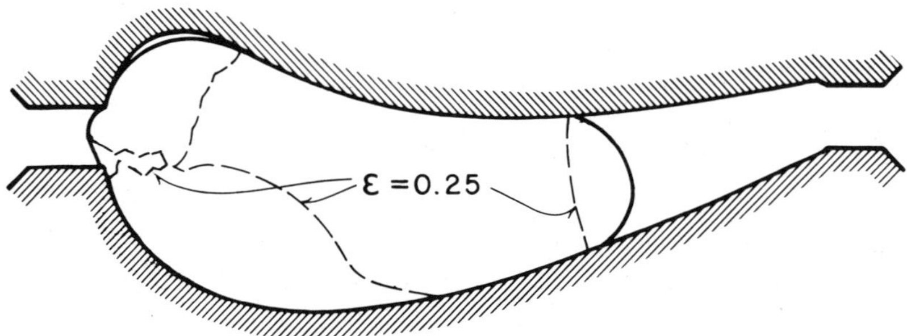

Figure 4.49 b): Effective strain at time 0.48 sec.

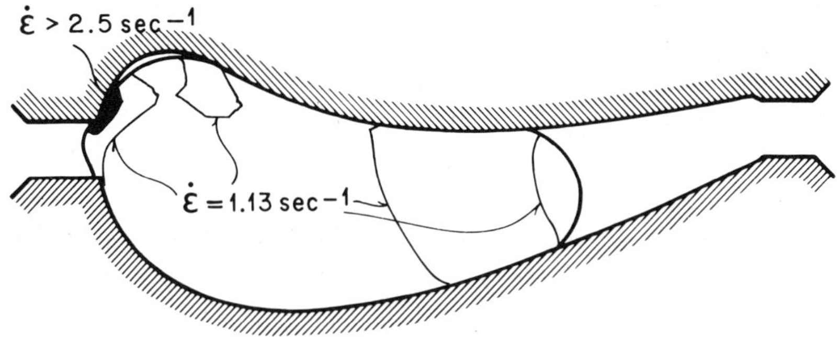

Figure 4.49 c):   Effective strain rate at time 0.48 sec.

The strain rate distribution is shown in Figures 4.49 (b) to 4.51
(b). Very high values of strain rate develop when the material is
entering the flash, as shown in Figure 4.49 (c). The gradient of
strain rate values along the x-axis can give rise to modification
in the microstructure of the material after forging.

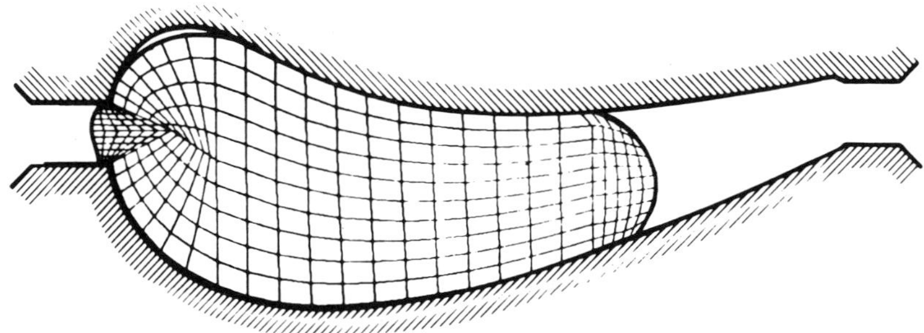

Figure 4.50 a): Remesh at time 0.48 sec.

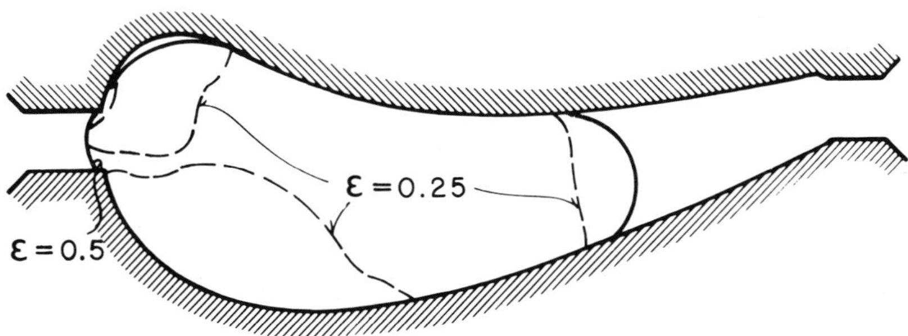

Figure 4.50 b): Interpolated strain distribution after remesh.

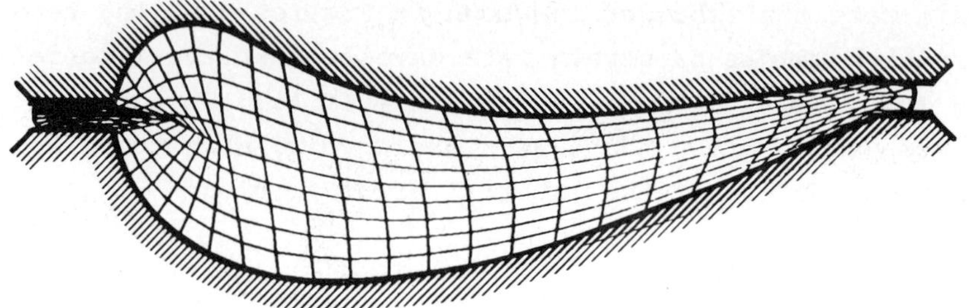

Figure 4.51 a):   Deformations at time 0.70 sec.

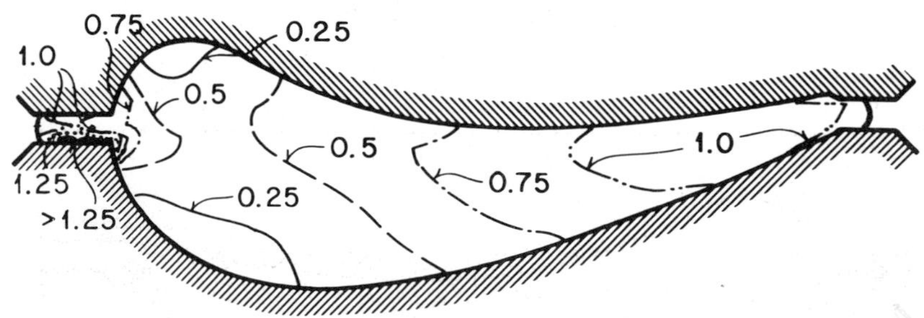

Figure 4.51 b):   Effective strain distribution.

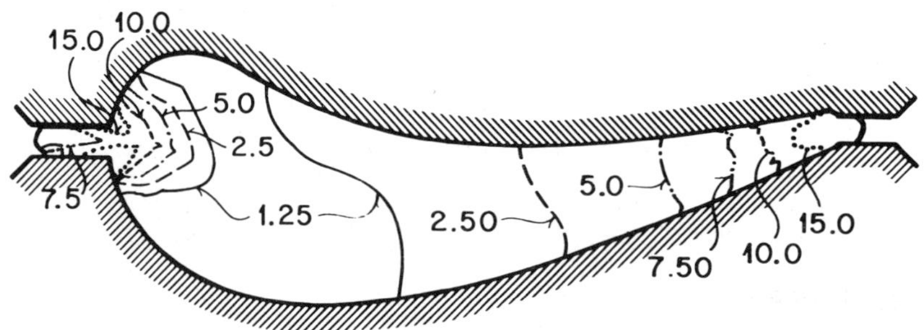

Figure 4.51 c):   Effective strain rate distribution.

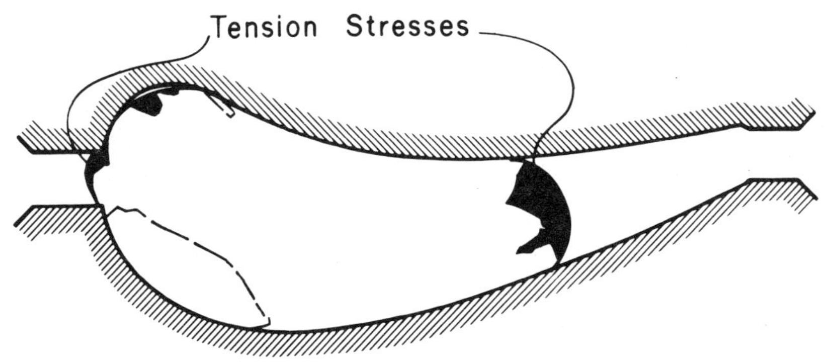

Figure 4.52 a): Stress σ_y distribution at 0.48 sec.

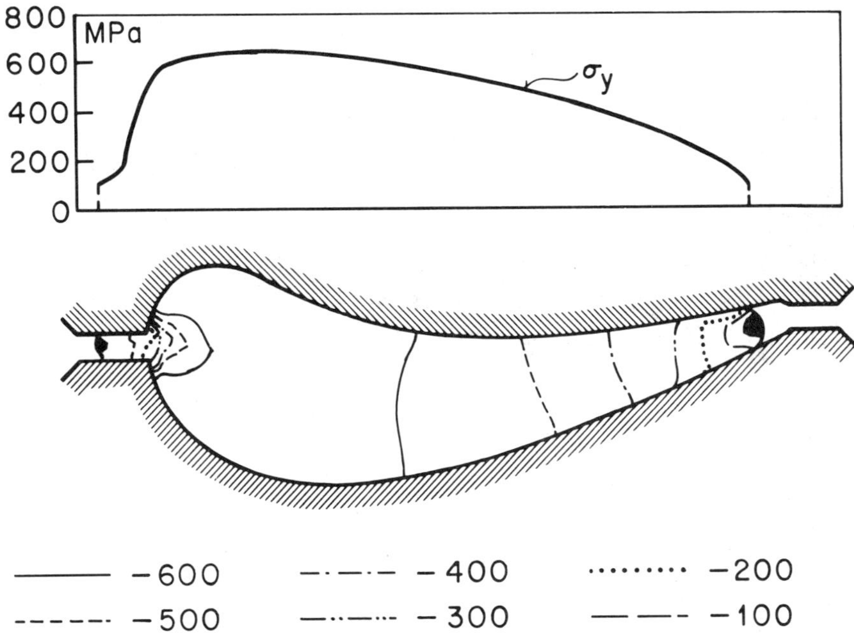

Figure 4.52 b): Stress σ_y distribution at time 0.68 sec.

4.3.4.3 Stresses

In Figures 4.52 (a) and 4.52 (b), the σ_y (parallel to the forging direction) distributions are shown. The negative values are com-

pressive. The dashed area at the free boundaries zone are where
the stresses are tensile and there is a danger of possible
cracking, as was determined by experiments. In Figure 4.52 (b)
the stresses are reported in a diagram that can be compared with
the stress distribution obtained by the slab method (see Chapter
4.3.1).

4.3.4.4 Friction

The influence of friction on closed die forging can be very important. This is shown in Figure 4.53. In Figure 4.53 (a), the
geometry of the dies and the position of a circular preform is
given [4.20].

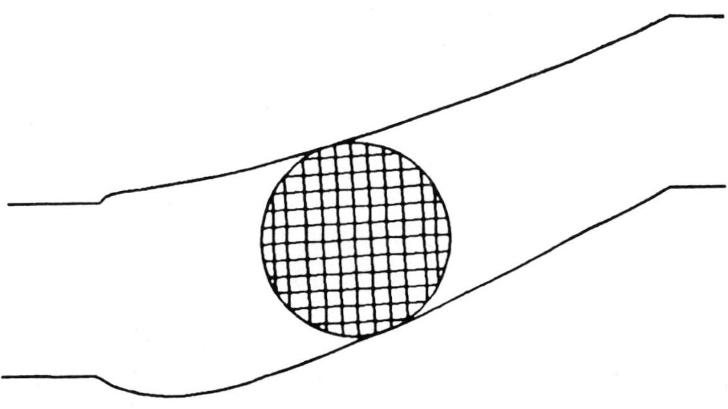

Figure 4.53 a): Problem geometry.

The differences between the deformed grids for m = 0.56 (Fig.
4.53 (c)), m = 1, (Fig. 4.53 (d)), and sticking friction, (Fig.
4.53 (e)), are hardly noticeable without superimposing the grids.
However, a close comparison reveals that the calculations with
m = 1 give the pattern that most closely simulates the experimental conditions. It should be noticed that sticking friction and
m = 1 do not produce exactly the same grid deformation. As can
be seen, the problem becomes quite insensitive to the friction
factor used above a certain value. The strong shearing in the
center, together with almost no deformation in the neutral zones

and on the free surfaces, is very well simulated.

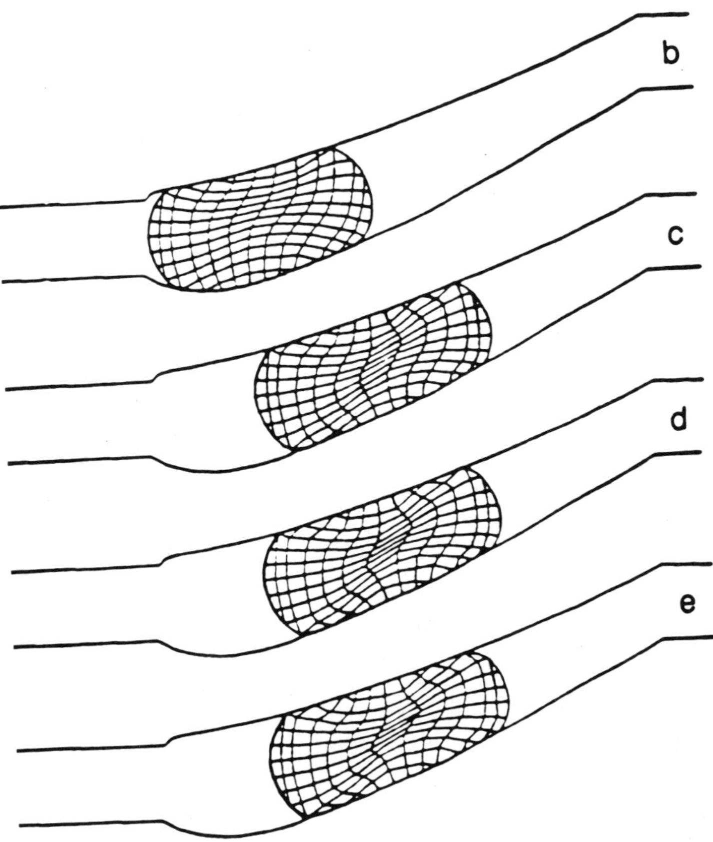

<u>Figure 4.53</u>:   Calculated grid distortions after 31.5 mm compression.
b) m = 0.0; c) m = 0.56; d) m = 1.0; e) sticking.

It should be mentioned that, at least during the first deformation stages, the problem is geometrically very ill-defined. The workpiece lies between almost parallel dies, without having any restrictions in tangential displacement; and at the same time the stiffness of the system is very insensitive to large variations in these tangential displacements. Therefore, except for the cases in which only essential boundary conditions are imposed, the convergence character of the problem is quite poor,

241

and considerable care has to be taken when choosing the strategy
to be adopted.

In Figure 4.53 (b), a large translation of the preform is possible, due to the low friction value and the divergent geometry
of dies at the point of contact with the preform.

The effect of friction in metal forming has been extensively
studied, but only very little is well understood. The mathematical methods used to evaluate the influence of friction show
the same deficiency. The interface between the workpiece and the
die has been simulated in a variety of ways but none has the
capability of taking into account the complex interactions. A
greater effort is required in the future to simulate the interface die-billet in terms of lubrication, friction, die wear,
etc.

4.4 Plane Strain Modelling of Thin Rib Forging

The isothermal forging of nickel and titanium alloys to produce
near-net-shape components has proved to be a successful industrial process that can lead to reduced component costs compared
with conventional closed die forging and machining techniques
[4.13].

It is important to study the metal flow behavior during forging
because of the need to ensure good die-filling, uniformity of
deformation and control of local strain rates in the deforming
workpiece.

The study of the mechanics of the flow during the isothermal
forging of a radial impeller is important, especially in rib
areas. Previous studies [4.14] have shown that die-filling can be
assured by an optimum design of the preform geometry. The results
obtained were for a geometry of the ribs with no taper and therefore easier to fill. If the blade geometry presents a taper angle
of a few degrees (5 - 10°), the filling of the die becomes increasingly difficult and can be likened to a complex form of
extrusion. To reach the filling of the rib portion, it would be
necessary to increase the applied forging pressure (normally
around 120 N/mm^2), but this would lead to increased die stresses,
especially to the tip of the rib where a stress concentration
occurs [4.15]. The problem of die stresses is examined in Chapter
4.4.3.

In some extreme cases, it is necessary to modify the rib geometry
of the die in order to get good filling. The workpiece will
sometimes have to be machined afterwards to obtain the required
geometry. The modifications to be made must therefore be evaluated, with a careful cost analysis.

4.4.1 Elementary Analysis Approach

In order to evaluate better the effects of the die modifications
in terms of better rib filling, a mathematical model was elaborated which can simulate the forging of a part with ribs, such as
airplane structures or impellers.

The part geometry can be given in simplified form while the rib geometry is more accurately prescribed in terms of rib angle, rib width entrance, rib base radius, distance between ribs, etc.

The influence of applied pressure, friction and material flow stress can also be studied.

Using the model, a parametric analysis can be quickly made in order to determine the influence of the various effects on the forming of the rib. Costly and lengthy experimental analysis can therefore be reduced.

The model has been tested and experimentally verified for different sizes and different materials.

4.4.1.1 Forging Stress Distribution

The stress distribution during the forging of a part such as an impeller (see Figure 4.54) is rather complex. In Figure 4.55, the portion of the disk which includes a rib is simplified in order to give an idea of the stresses. Figure 4.55 (c) is the plan view, Figure 4.55 (a) is the cross-sectional view B-B and Figure 4.55 (d) is the cross-sectional view A-A. Looking at the section A-A, the σ_y stress (vertical or axial) increases from the border to the axis of symmetry. In a section B-B the stress will reach a maximum on the neutral lines and a minimum at the entrance of the rib. This stress distribution is valid if the material has not yet reached the flash zone and the value of σ_{ye} is equal to the flow stress of the material. If the material has entered the flash, σ_{ye} has to be estimated in some other way.

For the purpose of estimating the stress distribution along the A-A section, the geometry of the disk (Figure 4.56 (a)) was simplified as in Figure 4.56 (b). The sections H_1^* and H_2^* in Figure 4.56 (a) become H_1 and H_2 in Figure 4.56 (b). The elementary analysis for axisymmetrical forging was applied to the two elements A and B.

Figure 4.54:   Forged turbocharger impeller.

The following assumptions are made:

a)   the elastic deformations are ignored,

b)   the material is incompressible and isotropic,

c)   the inertia forces are ignored,

d)   the deformations are homogeneous or, in other terms, the plane surfaces in the material remain planar,

e)   the flow stress ($\bar{\sigma}$) is constant in the deformation zone considered,

f)   the friction shear stress (τ) is expressed by a constant or friction factor:

$$\tau = \mu \cdot \bar{\sigma} \qquad \text{where} \quad \mu = \frac{m}{\sqrt{3}} \quad \text{(see Chapter 2)}$$

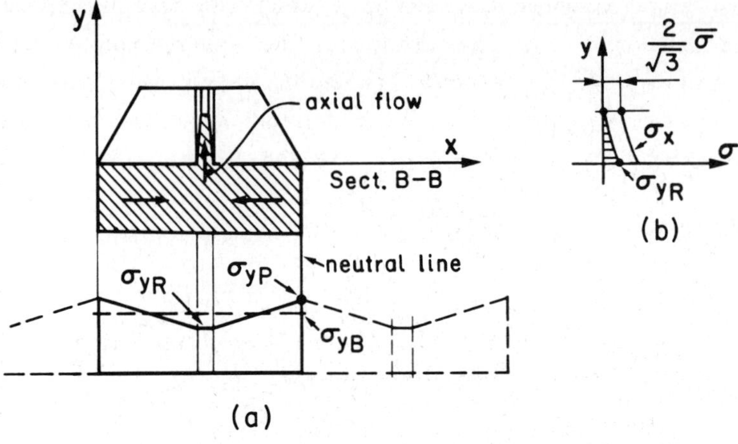

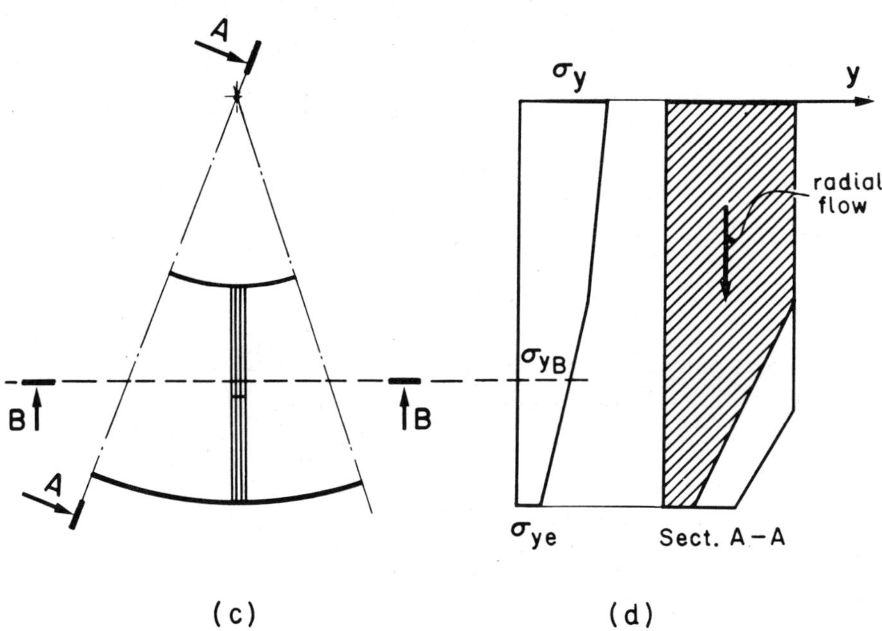

Figure 4.55: Forging stress distribution in a circular plate with a thin rib.

The assumptions a), b) and c) will not have any effect on the results if we model slow, isothermal forming. The assumption d) is certainly not fulfilled in our case where boundary friction is an important factor. The results are therefore approximate but close to the experimental results and very useful in understanding the mechanics of the process. The assumption e) is only limited to the specific deformation zone being studied, but it does not necessarily have the same value in other deformation zones in the forging (like the disk and the rib).

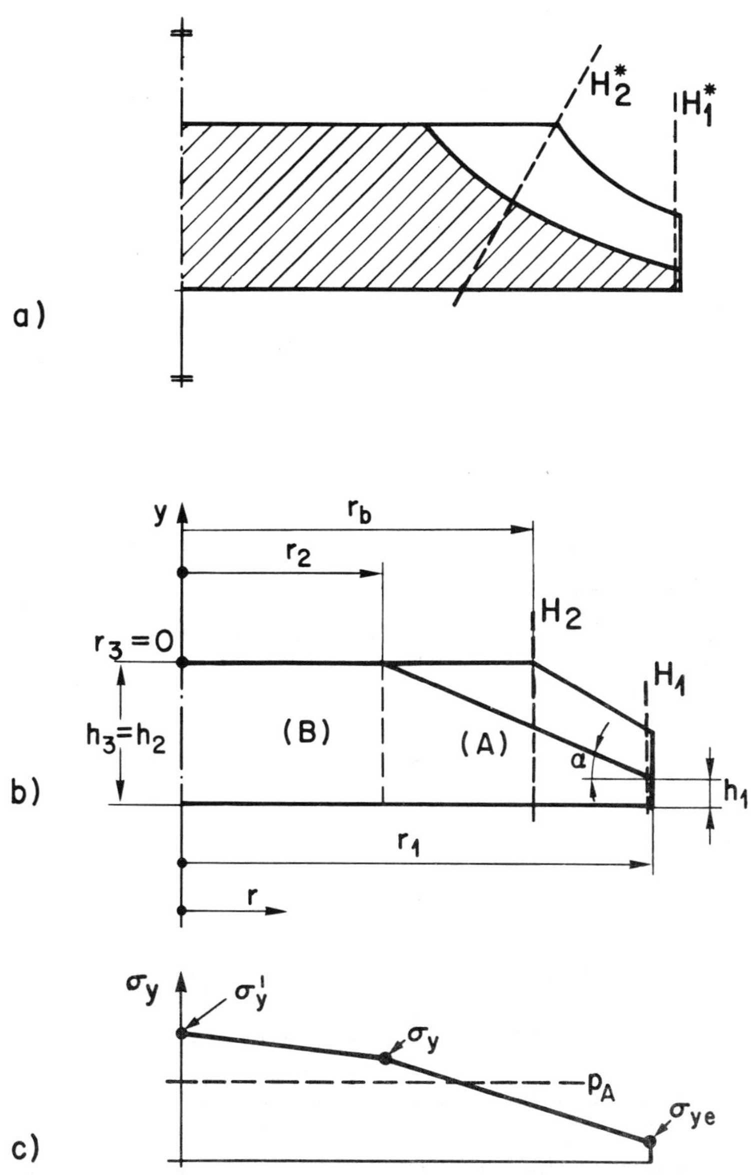

Figure 4.56: Simplified geometry of the disk.

The stresses in element A are calculated according to the slab method following the description given in Chapter 2.4:

$$\sigma_y = \frac{K_2}{K_1} \ln \left(\frac{h_e}{K_3 + K_1 r} \right) + \sigma_{ye} \tag{4.81}$$

where:

$$K_1 = \tan \alpha + \tan \beta \tag{4.82}$$

$$K_2 = -\bar{\sigma} \cdot K_1 + \mu\bar{\sigma} \left(\frac{1}{\cos^2 \alpha} + \frac{1}{\cos^2 \beta} \right) \tag{4.83}$$

$$K_3 = h_b - r_b \cdot K_1 \tag{4.84}$$

In the element B, the stresses are calculated according to:

$$\sigma'_y = \frac{2\tau}{h} (r_e - r) + \sigma_y \tag{4.85}$$

The stress distribution will be as shown in Figure 4.56 (c) and is expressed as in function of the initial stress σ_{ye}.

To evaluate the initial stress σ_{ye}, it is assumed that the equilibrium of the forces will be expressed by:

$$\bar{\sigma}_y \cdot A_B + \bar{\sigma}'_y \cdot A_A = p_A \cdot A \tag{4.86}$$

where:

$A_B = r_2^2 \pi$ is the axial area of element B

$A_A = (r_1^2 - r_2^2)\pi$ is the axial area of element A

$A = r_1^2 \pi$ is the axial area of the disk

$\bar{\sigma}_y = (\sigma_{ye} + \sigma_y)/2$ is the average stress over the area A_B

$\bar{\sigma}'_y = (\sigma_y + \sigma'_y)/2$ is the average stress over the area A_A

p_A is the average applied pressure over the total area A

But the stresses are also given by (4.81) to (4.84) for $h_e = h_1$, $r = r_2$, $h_b = h_2$, $r_b = r_2$:

$$\sigma_y = \frac{K_2}{K_1} \ln \left(\frac{h_1}{K_3 + K_1 r_2} \right) + \sigma_{ye} \qquad (4.87)$$

and by (4.85) for $r_e = r_2$, $r = r_3$

$$\sigma_y' = \frac{2\mu\bar{\sigma}}{h_3} (r_2 - r_3) + \sigma_y \qquad (4.88)$$

where:

$$K_1 = \tan \alpha + \tan \beta \qquad (4.89)$$

$$K_2 = -\bar{\sigma} K_1 + \mu\bar{\sigma} \left(\frac{1}{\cos^2 \alpha} + \frac{1}{\cos^2 \beta} \right) \qquad (4.90)$$

$$K_3 = h_2 - r_2 K_1 \qquad (4.91)$$

If:

$$Z_1 = \frac{K_2}{K_1} \ln \left(\frac{h_1}{K_3 + K_1 r_2} \right) \qquad (4.92)$$

$$Z_2 = \frac{2\mu\bar{\sigma}}{h_3} (r_2 - r_3) \qquad (4.93)$$

Substituting equations 4.87 to 4.93 in equation 4.86:

$$\frac{\sigma_{ye} + \sigma_y}{2} A_B + \frac{\sigma_y + \sigma_y'}{2} A_A = p_A \cdot A$$

$$\frac{A_B}{2} (\sigma_{ye} + Z_1 + \sigma_{ye}) + \frac{A_A}{2} (Z_1 + \sigma_{ye} + Z_2 + Z_1 + \sigma_{ye}) = p_A \cdot A$$

Solving for σ_{ye}, is obtained:

$$\sigma_{ye} = \left[p_A \cdot A - Z_1 \left(\frac{A_B}{2} + A_A \right) + Z_2 \frac{A_A}{2} \right] \frac{1}{A_B + A_A} \qquad (4.94)$$

The average stresses in the sections h_1 and h_2 can then be calculated using equations 4.81 and 4.85.

4.4.1.2 Plate-Rib Forging Model

The model includes the description of the plate and the rib [4.16] in a plane strain situation. The geometry is given in Figure 4.57. If the coefficient of friction at the contact plane of the rib and the border of the plate is the same and constant, and if the shear stresses have a maximum value at all points of the contact plane of the rib and border ($\tau = \frac{1}{\sqrt{3}}\,\bar{\sigma}$), then the specific pressure at the section A is determined by:

$$q_A \;=\; \frac{2}{\sqrt{3}}\,\bar{\sigma}\,(1 + \frac{1}{\alpha}\,\ln\frac{A}{a}) \tag{4.95}$$

where the geometrical variables are given in Figure 4.57.

The specific pressure at the section of the plate of length c is determined by:

$$q_C \;=\; \frac{2}{\sqrt{3}}\,\bar{\sigma}\,(1 + \frac{1}{\alpha}\,\ln\frac{A}{a} + \frac{c}{4h}) \tag{4.96}$$

The average specific pressure on the length of the element B = A + C is found by:

$$q \;=\; \frac{q_A A + q_C C}{B} \;=\; \frac{2}{\sqrt{3}}\,\bar{\sigma}\,(1 + \frac{1}{\alpha}\,\ln\frac{A}{a} + \frac{c^2}{4Bh}) \tag{4.97}$$

This formula determines the maximum possible value of the specific forging pressure for a ribbed panel.

Solving equation 4.97 for a:

$$a \;=\; A \cdot \exp^{-1}\left[\alpha\,(\frac{\sqrt{3}q}{2\bar{\sigma}} - \frac{c^2}{4Bh} - 1)\right] \tag{4.98}$$

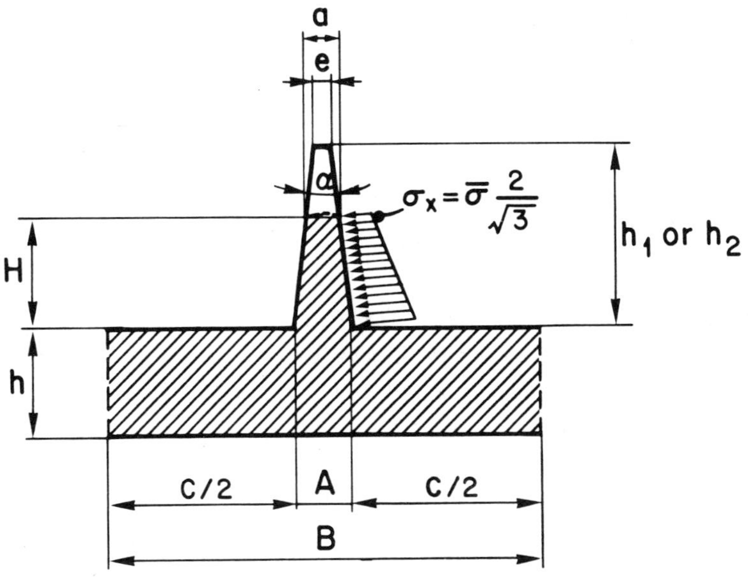

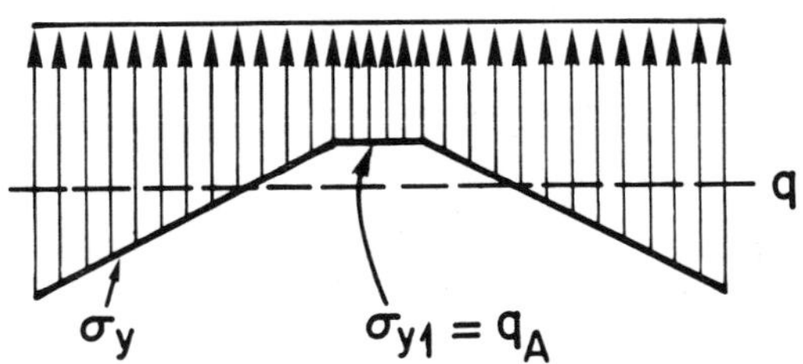

<u>Figure 4.57</u>:   Plate-rib forging model.

But, from geometric considerations it is also:

$$a \;=\; A - 2H \tan \frac{\alpha}{2} \tag{4.99}$$

(N.B.:   This is true only if we assume that the material flowing into the flash is negligible)

Substituting 4.99 in 4.98 and solving for H :

$$H \;=\; \frac{A}{2\,\tan\frac{\alpha}{2}}\; \{1 - \exp^{-1}\,[\alpha\,(\frac{\sqrt{3}q}{2\bar{\sigma}} - \frac{c^2}{4Bh} - 1)]\} \qquad\qquad (4.100)$$

Equation 4.100 gives the height of the rib H as a function of:

1. rib entrance width   A
2. angle of the rib α
3. distance between two ribs   B
4. height of thickness of plate   h
5. applied specific pressure   q
6. flow stress of the material $\bar{\sigma}$

All the results obtained with this model are for a coefficient of friction corresponding to sticking.

4.4.1.3 Rib Forging Model

A model which includes only the rib but takes into consideration also the friction can be applied to determine the height of the forged rib [4.7]. The description of the elementary analysis of this model is given in Chapter 2.4.2.4.

The geometrical variables and the stress distribution are given in Figure 4.58.

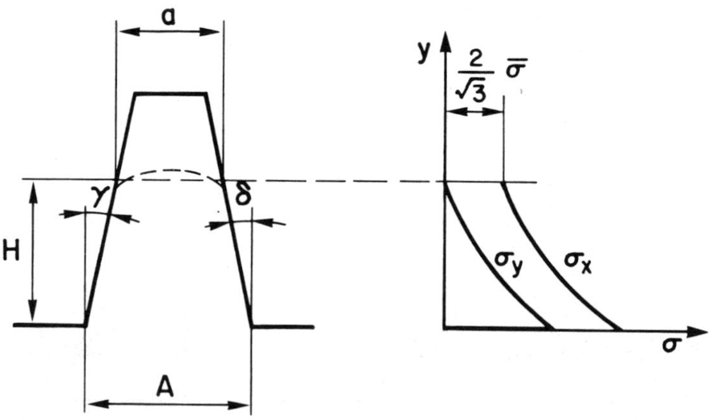

Figure 4.58: Rib forging model.

The σ_y stress is given by equation (see Chapter 2.4.2.4):

$$\sigma_y = \frac{K_2}{K_1} \ln \left(\frac{A + K_1 H}{A} \right) \qquad (4.101)$$

From equation 4.101 it is possible to obtain the height of the forged rib:

$$H = \frac{A}{K_1} \left[\exp \left(\frac{K_1}{K_2} \sigma_y \right) - 1 \right] \qquad (4.102)$$

Equation 4.102 can be used to calculate the influence of friction (as well as of the other parameters) on the height of the rib. The value of the stress σ_y is obtained by equation 4.95 where $q_A = \sigma_y$.

4.4.1.4 Rib Forging Model with the Effect of the Radius

If the effect of the radius at the base of the rib has to be in-cluded in the model, the rib has to be discretized into elements as in Figure 4.59.

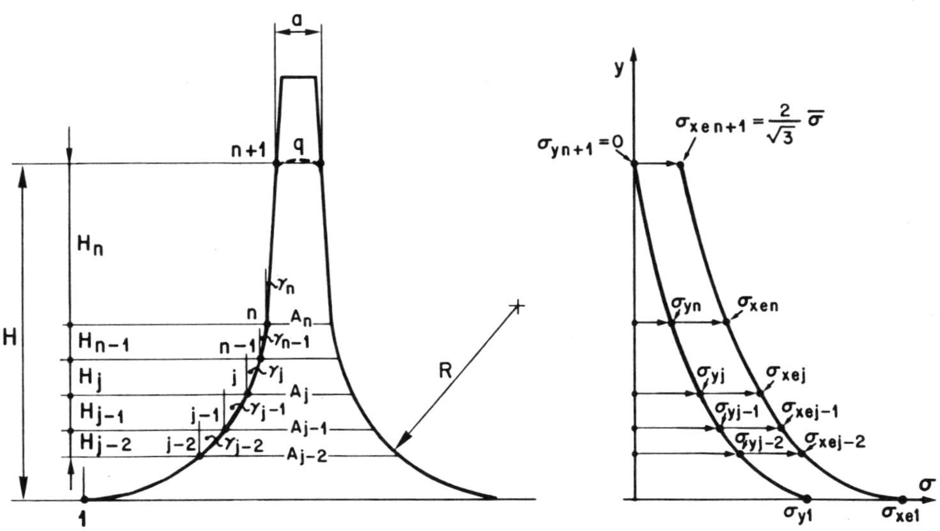

Figure 4.59: Rib forging model with the effect of the radius.

The height of the rib is then:

$$H = \sum_{i=1}^{n} H_i \qquad (4.103)$$

where n is the number of elements.

The width of the rib at the height H is expressed by:

$$A_{n+1} = A_n + K_{1n} H_n \qquad (4.104)$$

where:

$$K_{1n} = - (\tan \gamma_n + \tan \delta_n) \qquad (4.105)$$

It is also:

$$K_{2n} = - \frac{2}{\sqrt{3}} \bar{\sigma} K_{1n} + \mu\bar{\sigma} (2 + \tan^2 \gamma_n + \tan^2 \delta_n) \qquad (4.106)$$

For $y = H$ is:

$$\sigma_{xe\,(n+1)} = \frac{2}{\sqrt{3}} \bar{\sigma} \qquad (4.107)$$

$$\sigma_{y\,(n+1)} = 0 \qquad (4.108)$$

For $y = \sum_{i=1}^{n-1} H_i = H - H_n$ is:

$$\sigma_{xen} = \frac{K_{2n}}{K_{1n}} \ln \left(\frac{A_{n+1}}{A_n + K_{1n}\, y} \right) + \sigma_{xe\,(n+1)} \qquad (4.109)$$

$$\sigma_{yn} = \sigma_{xen} - \frac{2}{\sqrt{3}} \bar{\sigma} \qquad (4.110)$$

254

Substituting equation 4.107 and 4.104 into 4.109 gives:

$$\sigma_{xen} = \frac{K_{2n}}{K_{1n}} \ln \left(\frac{A_n + K_{1n} H_n}{A_n + K_{1n} \sum_{i=1}^{n-1} H_i} \right) + \frac{2}{\sqrt{3}} \bar{\sigma} \tag{4.111}$$

Finally, substituting equation 4.111 into 4.110 and solving for the height of the rib H, produces:

$$H = \sum_{i=1}^{n-1} H_i + \frac{1}{K_{1n}} \left[\left(A_n + K_{1n} \sum_{i=1}^{n-1} H_i \right) \exp \left(\frac{K_{1n}}{K_{2n}} \sigma_{yn} \right) - A_n \right] \tag{4.112}$$

In equation 4.112, the only unknown variable is the stress σ_{yn}.

If an interval $j-1$, j in Figure 4.59 is examined, the following equation can be written:

$$\text{for} \quad y = \sum_{i=1}^{j-2} H_i$$

$$\sigma_{xe(j-1)} = \frac{K_{2(j-1)}}{K_{1(j-1)}} \ln \left(\frac{A_j}{A_{j-1} + K_{1(j-1)} \sum_{i=1}^{j-2} H_i} \right) + \sigma_{xej} \tag{4.113}$$

$$\sigma_{ye(j-1)} = \sigma_{xe(j-1)} - \frac{2}{\sqrt{3}} \bar{\sigma} \tag{4.114}$$

Therefore the stress σ_{xej} can be derived from equation 4.113 as:

$$\sigma_{xej} = \sigma_{xe(j-1)} - \frac{K_{2(j-1)}}{K_{1(j-1)}} \ln \left(\frac{A_j}{A_{j-1} + K_{1(j-1)} \sum_{i=1}^{j-2} H_i} \right) \tag{4.115}$$

where:

$$K_{1(j-1)} = - (\tan \gamma_{j-1} + \tan \delta_{j-1}) \tag{4.116}$$

$$K_{2(j-1)} = -\frac{2}{\sqrt{3}} \bar{\sigma} K_{1(j-1)} + \mu\bar{\sigma} (2+\tan^2\gamma_{j-1}+\tan^2\delta_{j-1}) \qquad (4.117)$$

Assuming that the two rib angles are the same, then:

$$\gamma_{j-1} = \delta_{j-1} = \text{arc tan} \left(\frac{A_{j-1} - A_j}{2\,H_{(j-1)}}\right) \qquad (4.118)$$

Equations 4.113 to 4.118 are valid for any symmetric geometry of the rib. The stress distribution, and therefore the final height of the rib, can be determined starting from the base of the rib where the applied pressure is known or for $j = 1$:

$$\left.\begin{array}{l} \sigma_{y1} = q_A \\[1em] \sigma_{xe1} = \sigma_{y1} + \dfrac{2}{\sqrt{3}} \bar{\sigma} \end{array}\right\} \qquad (4.119)$$

Expressions 4.119 are the initial conditions for an iteration based on the equations 4.115 to 4.118. At the end of the loop, the σ_{yn} is obtained and the height of the rib H can be calculated with equation 4.112.

4.4.1.5 Experiments and Results

To test the theoretical model, model-parts of the impeller-type were forged from different materials.

The height of two critical sections in each rib was measured and the maximum and minimum values obtained were compared with the theoretical model.

The two critical rib sections chosen to determine the influence of the different parameters on the forging of the ribs are indicated as H_1 and H_2 and their locations on the disk are shown in Figure 4.56 (a).

The forging parameters that were investigated are summarized in Table 4.3. All the results are presented as the height of the forged rib versus the applied pressure or another parameter. In each figure, the experimental results determined on a forged

impeller with an applied pressure p_A = 120 MPa are shown as a
bar. The required height of the rib is also shown as a horizontal
line.

Table 4.2

Forging parameters for the computer calculation and experiments

Section H_1

Figure ≠	Rib radius [mm]	Flow stress [MPa]	Friction Coeffi- cient	Rib Entrance [mm]	Rib Angle [degree]	Applied Pressure [MPa]
4.60 a	0	20	0.1...0.5	3.6	10	0...200
4.61	0	5,10,20,30	0.4	3.6	10	120
4.62 a	0	20	0.1...0.5	3.6	5,10	120
4.62 b	0	20	0.1...0.5	3.6,4.6,5.6	10	120
4.62 c	0,3,5	20	0.1...0.5	——	10	120
Experiment	3	20	∿ 0.2÷0.4	3.6	10	120

Section H_2

Figure ≠	Rib radius [mm]	Flow stress [MPa]	Friction Coeffi- cient	Rib Entrance [mm]	Rib Angle [degree]	Applied Pressure [MPa]
4.60 b	0	20	0.1...0.5	6	11.4	0...200
4.61	0	5,10,20,30	0.4	6	11.4	120
Experiment	3	20	∿ 0.2÷0.4	6	11.4	120

Influence of friction

Figure 4.60 shows the influence of friction on the forming of
the rib. The influence is definitely very large and the spread
of the experimental results can be explained by the different
lubrication conditions encountered in each rib cavity, as was
apparent by examining the die surfaces a posteriori. From this
figure it is also possible to determine the friction factor
that can be used to calculate the influence of the other parameters effectively.

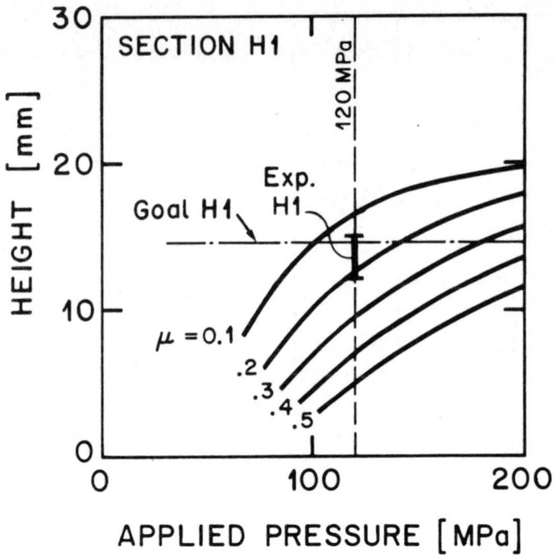

Figure 4.60:   Influence of friction on the forged rib height.

Influence of flow stress

Examining Figure 4.61, it is possible to show that the flow
stress has a large influence on the height of the forged rib. If
a maximum applied pressure of 120 MPa is assumed, it will be necessary not to exceed a flow stress of 10 MPa to meet the required designed values for the rib.

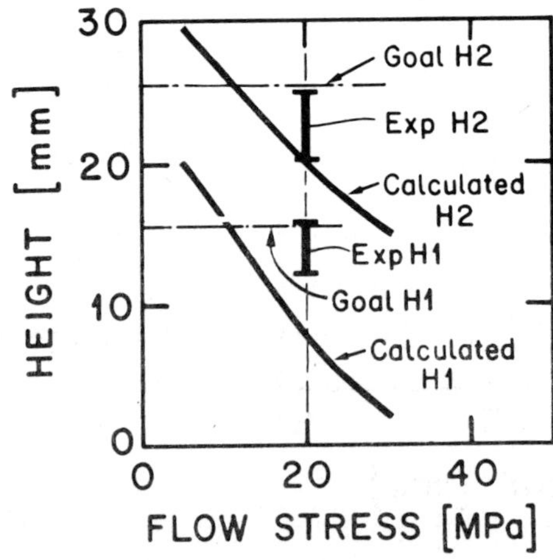

Figure 4.61:   Influence of flow stress on the forged rib height.

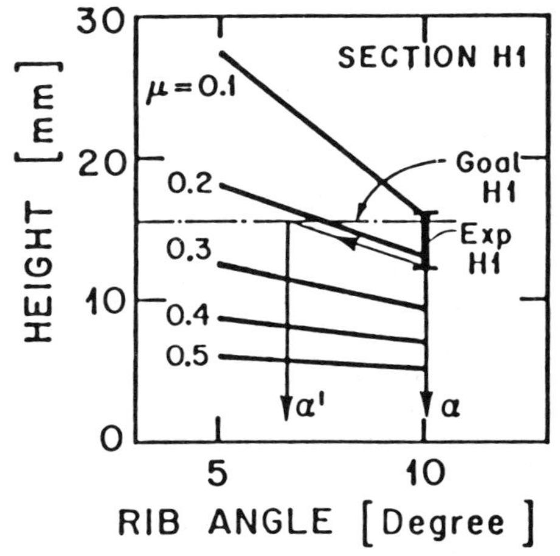

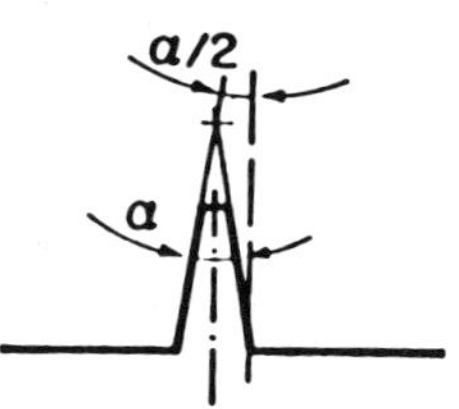

Figure 4.62 a):

Influence of rib angle.

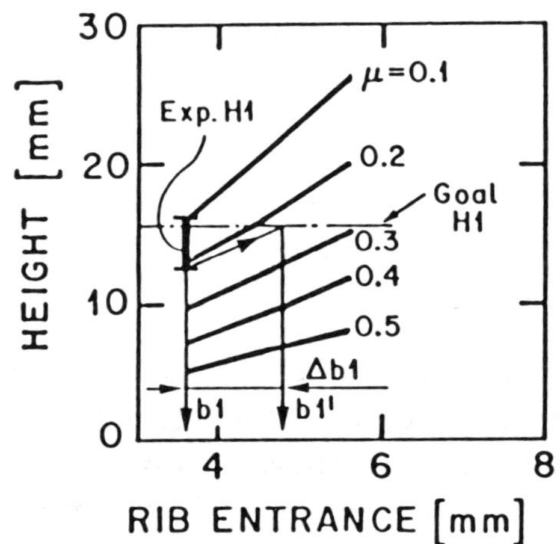

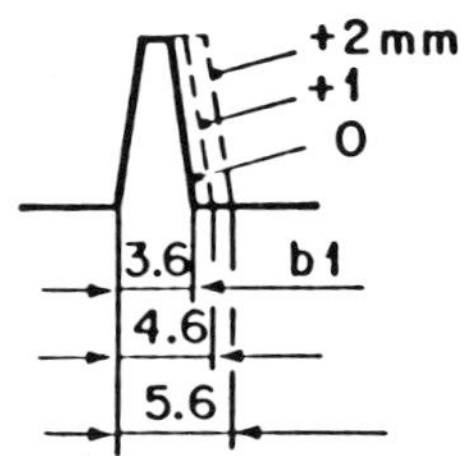

Figure 4.62 b):

Influence of rib entrance.

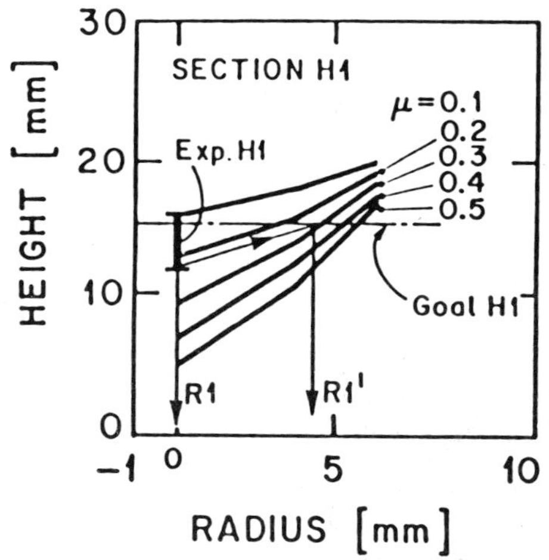

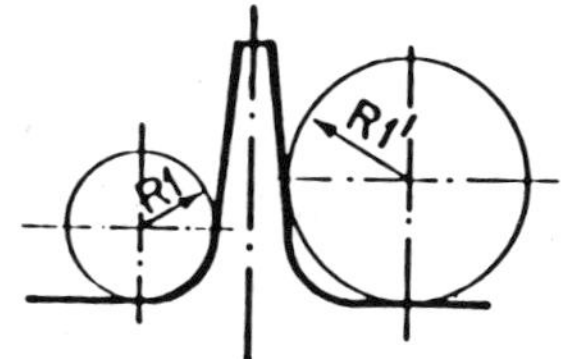

Figure 4.62 c):

Influence of rib radius.

Influence of the rib angle
==========================

In Figure 4.62 (a), are summarized the results that show the influence of the rib angle for the sections H_1. A smaller angle
has a larger benefit only for a low coefficient of friction.

Even with poor friction conditions, a small increase of the
pressure will increase the height of the forged rib above the
required values if the rib angle is decreased by half. To obtain
the required height, as indicated in Figure 4.62 (a), it is
necessary to reduce the angle α to a value α', considering the
worst experimental case.

Influence of the rib entrance
=============================

An increase of the width of the rib entrance (see Figures 4.62
(b)) has a definite influence on the increase of the forged rib.
Without exceeding a pressure of 120 MPa, it would be possible
to reach the design goals by increasing the rib entrance of
1.5 ∿ 2 mm (from b_1 to b_1' in section H_1).

Influence of the radius at the rib entrance
===

A larger radius at the base of the rib has a large influence for
the section H_1 (Figure 4.62 (c)).

4.4.1.6 Conclusions

A mathematical model to simulate the forging of structural parts
with very thin ribs has been described. It is possible to give
the geometry of the rib in details such as angle, thickness and
radius at rib entrance.

The comparison with experiments shows very good agreement. The
results obtained by applying the model were verified experimentally and the predictions regarding the influence of the rib
thickness proved to be correct.

The simulation model was found very useful in evaluating different die design modifications to obtain good filling of the rib.

260

The model does not give any information about the deformation and strain distribution during forging. A FEM analysis is required in order to determine a map of strains during forging (see 4.4.2).

The results obtained can be summarized:

a) friction has a great effect on the filling of the dies and the model can quantify this result and be used to determine the friction factor in combination with some experimental results,

b) the flow stress of the material also has a large influence, showing that for strain rate sensitive material, the iso-thermal forging is advantageous,

c) the rib angle has a large influence only at a low friction factor,

d) the rib thickness has a large influence for any friction condition and the subsequent machining operation to bring the workpiece to the required dimensions is easier and less expensive than a change in rib angle,

e) the radius (or the shape) at the base of the rib has an influence that cannot be ignored and should be taken into consideration when refining the model.

4.4.2 FEM Plane Strain Analysis of Thin Rib Forging

In Chapter 4.4.1 a mathematical model based on elementary analysis was used to simulate the forging of a part with ribs.

The influence of geometrical parameters such as rib angle, rib base radius and distance between ribs could be analyzed with the model, as well as the influence of physical parameters such as applied pressure, friction and material flow stress.

The main interest was to study the conditions that determine the filling of the rib. The model could not provide any information on material deformation or strain distribution.

In this chapter some results are presented on the analysis of the rib forging using the Finite Element method. The influence of the friction is shown on the metal deformation and its effect on the strain distribution. Comparison is also made with the experimental results and the previous analysis [4.17].

The Finite Element program used in this investigation has been previously described in Chapter 2.6.

4.4.2.1 The FEM Model for the Rib Forging

The FEM model of the rib forging process is shown in Figure 4.63. It consists of two dies, one fixed and the other moving with a constant speed (irrelevant in this case because the material model is based on constant flow stress and therefore no strain or strain rate effect is taken into account), simulated by a series of nodes connected by linear elements and a preform described by a mesh of 20 x 20 elements. For reasons of symmetry, only half of the geometry is considered, with the nodes indicated by the triangles bounded to move only in the vertical direction. The other nodes on the periphery are allowed to move following the die geometry according to the friction parameter selected. The grid is more concentrated in the region where larger deformations and strains are to be expected.

The material has a constant flow stress of 20 MPa. The frictionless case (m = 0) was examined, as well as a case with a friction factor m = 0.4.

4.4.2.2 Metal Deformation

The deformation of the preform is shown in Figure 4.64 (m = 0) and Figure 4.65 (m = 0.4). The grid distorts very rapidly, but in the frictionless case, it is possible to simulate the process without mathematical instability until the rib is filled.

Friction m = 0. In Figure 4.64, three deformed configurations are shown corresponding to three different displacements of lower die. The elements on the axis of symmetry are squashed by the movement of the material around the radius of the rib. Due to the

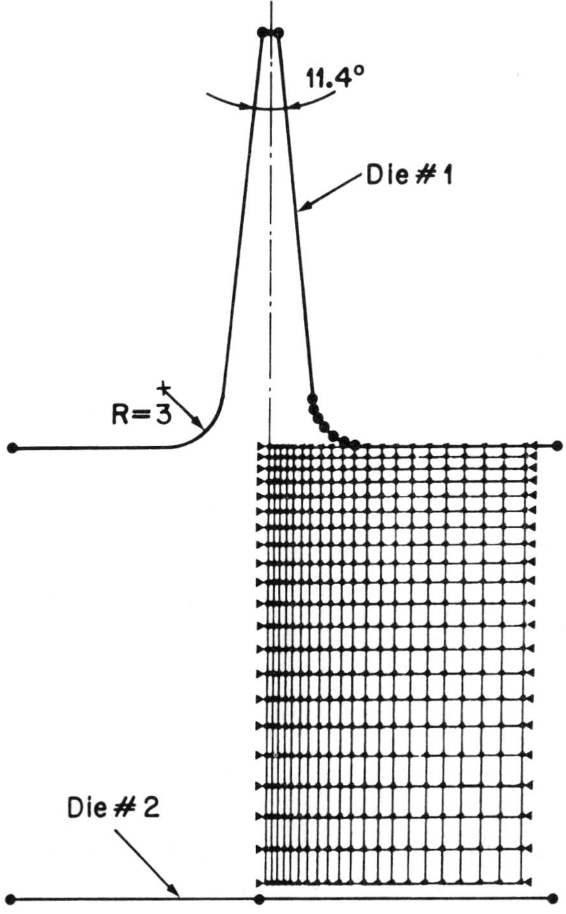

Figure 4.63:   FEM model of the rib forging process.

absence of friction, the material in contact with the die tends
to slide towards the entrance, helping to fill the rib. A large
portion of the metal remains undeformed.

Friction m = 0.4. The material becomes more distorted than in the
frictionless case. Somewhere between (a) and (b) the distortion
is such that the model becomes unstable and the solution does
not converge. Figures 4.65 (a) and 4.64 (a) can be compared: the
material in contact with the die ≠ 2 can no longer slide as before.

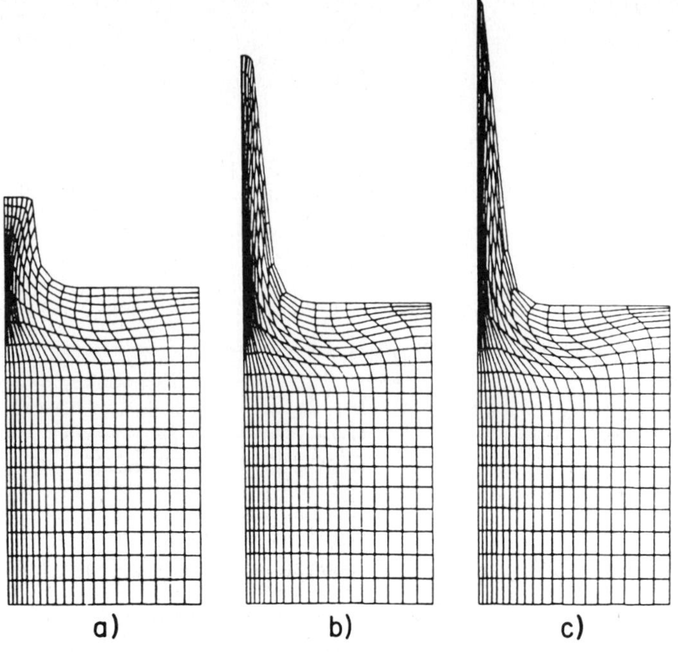

Figure 4.64: Preform deformation (Friction coefficient m = 0).

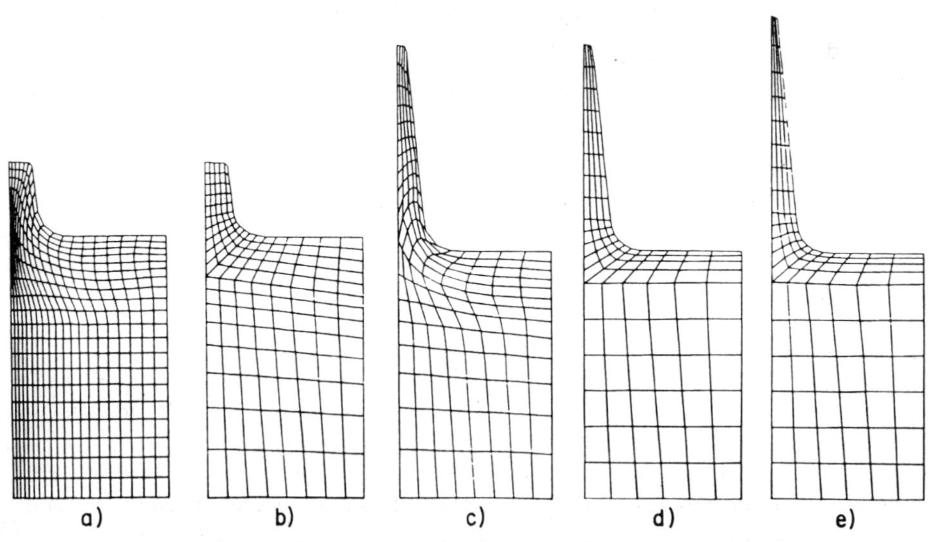

Figure 4.65: Preform deformation (Friction coefficient m = 0);
(b) and (d) remeshed grid.

Material below the surface in contact with the die is moving towards the rib entrance: the shearing effect is higher and, as will be shown later, a higher force is required.

The radius at the base of the rib acts as a punching tool on the deforming material and the elements tend to be stretched - one side going into the rib, and the other away from it. The grid needs to be remeshed, as is shown in Figure 4.65 (b). The simulation process can be continued, but the deformed meshes can no longer be compared with the previous cases.

4.4.2.3   Strain Distribution

The most important feature of the FEM in metal forming is the possibility of calculating locally the strain, strain rate and stresses, allowing for the determination of material properties after forging. The plotting of strain distribution allows the comparison between different friction cases, even if the remeshing has been done at different stages, or, as in the frictionless case, not at all. In Figures 4.66 and 4.67 the strain distributions for friction $m = 0$ and $m = 0.4$ respectively are presented.

The regions with higher strains are larger in Figure 4.67 ($m = 0.4$) and it shows that material folding can take place. This phenomenon can, in turn, create crack problems.

4.4.2.4   Forging Pressure

The applied forging pressure is obtained from the FEM, dividing the force by the projected area. In Figure 4.68, the applied pressure is plotted versus the height of the forged rib.

The FEM results are compared with the results of a model described in Chapter 4.4.1. Some experimental results were available for a structural part (an impeller) with a rib of the same geometrical cross-section. The material flow in such a part is more complex than plane strain, but the results can be used for purposes of comparison.

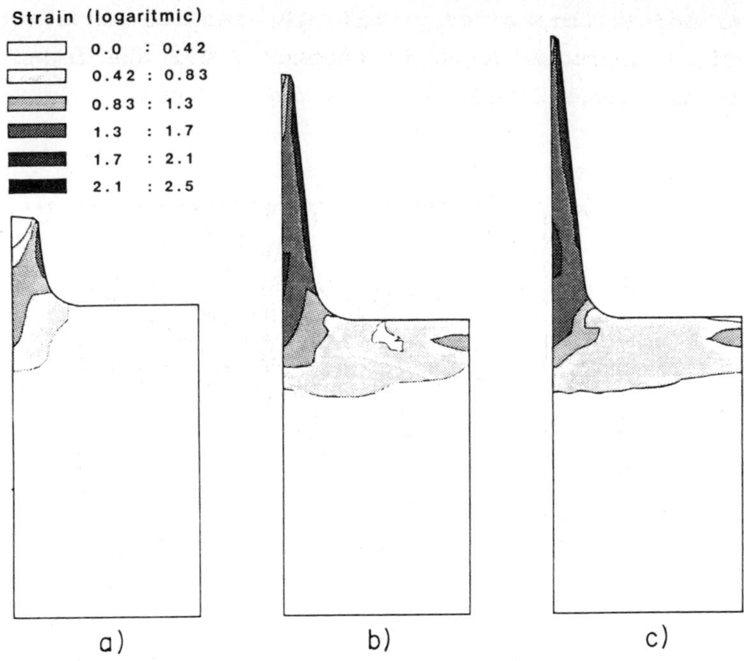

Figure 4.66: Logarithmic strain distribution for friction coefficient m = 0.

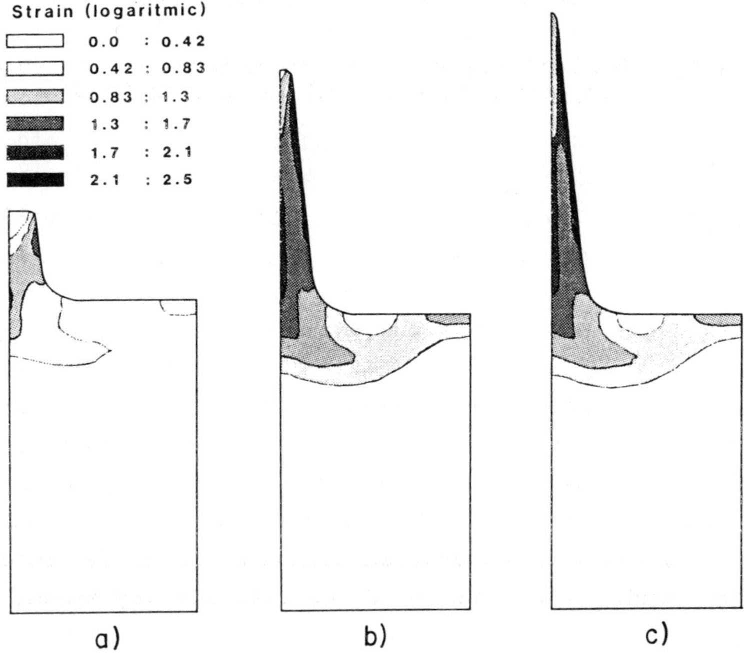

Figure 4.67: Logarithmic strain distribution for friction coefficient m = 0.4.

The two models compare very well: the results obtained with the theoretical approach given in Chapter 4.4.1 are lower than the FEM results, as expected.

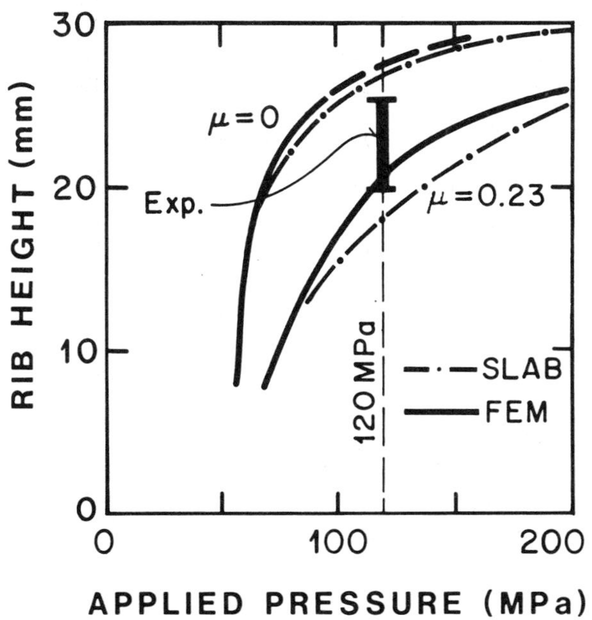

Figure 4.68: Applied pressure versus height of the forged rib: comparison between FEM and slab method.

The experimental results were obtained for a final applied pressure $p = 120$ N/mm^2 and are scattered, as shown in Figure 4.68, in a rather large band. Friction can be the cause of such a scatter in the experimental results.

In Figure 4.69 is shown a cross-section of a rib with similar geometry used in the calculation. The material was macroetched to show the flow at the end of the forging operation. The figure can be compared with the deformed mesh of Figure 4.64. A direct, one-to-one, evaluation cannot be done, due to different friction conditions, but good qualitative agreement can be seen.

The FEM can be applied to simulate a thin rib forging process,

Figure 4.69: Macroetched cross-section of an experimentally
              forged rib.

but remeshing capabilities must be included in the algorithm in
order to study the very large deformations, particularly in the
case when friction plays an important role.

4.4.3 Stress and Strain Analysis of Forging Dies

The Finite Element method is used to analyze the stresses and
strains in a die for isothermal forging of an impeller. The real
model is idealized in a 3-D structural model composed of pentahedral and hexahedral elements. Different loading cases are

considered in order to simulate and study the effect of different
die design arrangements. The results are presented in terms of
stress conditions and deflections.

This study analyzes the stress-strain behavior of a die for isothermally forging a turbocharger impeller, Fig. 4.54. Different
die designs were considered for the forging operation.

The use of a systematic approach (Chapter 4.2) has indicated
the range of conditions for the optimum forging in isothermal or
hot die conditions. The influence of the initial die temperature
could be evaluated with this method. The approach was very useful
for evaluating several design parameters quickly, but had the
limitation of being restricted to a rather simple axisymmetric
geometry. When the range of optimum conditions has been determined with a systematic approach, a detailed approach to the
deformation mechanics is necessary in order to find the most
realistic distribution of the strains, strain rates and flow
stresses (Chapter 4.2). It is also necessary to evaluate the
stress distribution in dies accurately, particularly at points
where there is a danger of stress concentration. The present
chapter is intended to show such an approach in order to evaluate
different die designs.

4.4.3.1 The Finite Element Model

The computer software package for structural analysis by the
Finite Element method is directly coupled to the CAD/CAM and
Process Modelling of Forging (Chapter 4.3.2). The present analysis is three-dimensional and therefore solid elements are
used (a solid pentahedral element with 18 nodes and a solid
hexahedral element with 27 nodes).

Figure 4.70 is the mesh of elements for the forging die.

Two systems of coordinates are included in this study:

- a global coordinate system
- a local coordinate system

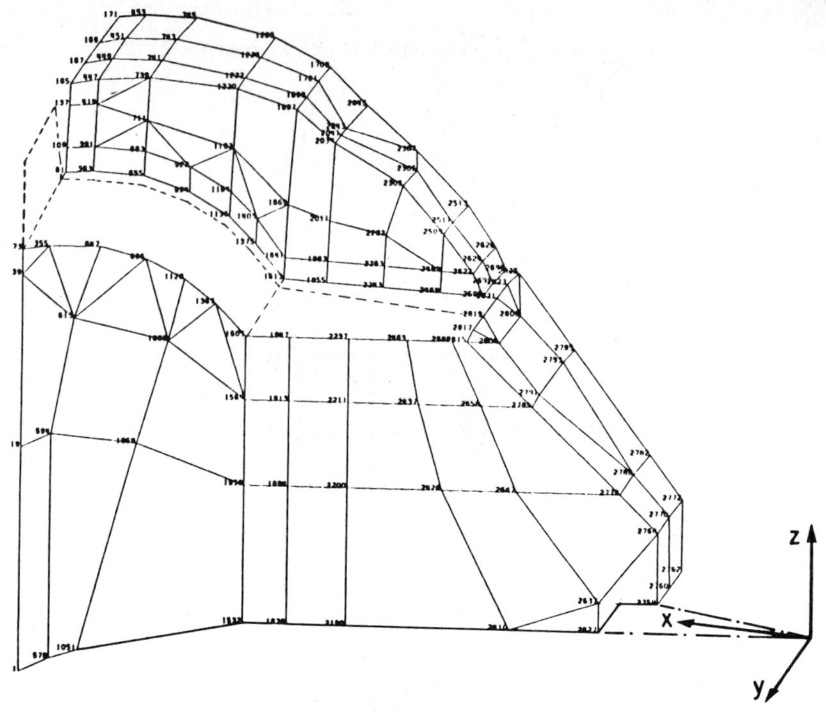

<u>Figure 4.70</u>:   The Finite Element model of the forging die.

The global coordinate system is a fixed system of reference for
the definition of all node coordinates. This system is given in
Figure 4.70 with X, Y, Z axes. The local coordinate system is
defined for displacements and loads at certain nodes. This is
useful when the natural boundaries are not parallel with the
global axes. In Figure 4.71, the local coordinate system is
given as X', Y', Z'.

The kinematic boundary conditions are described in Figures 4.71
and 4.72. Figure 4.71 is a view in the Z negative direction of
the Finite Element model. Supports along the two boundaries are
necessary to simulate material continuity. These supports have
zero displacement in the Y' direction in the local coordinate
system. Figure 4.72 is a view in the X negative direction of the
Finite Element model. Two types of support are shown. The supports on the vertical boundaries are the same as the ones in Fi-

270

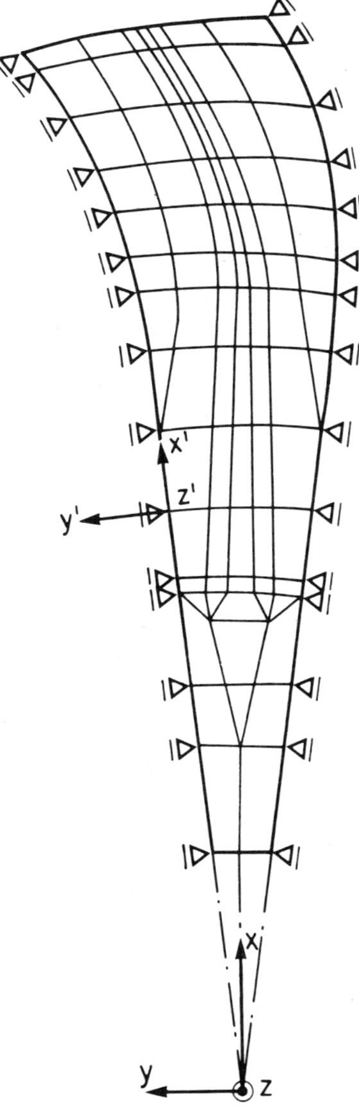

<u>Figure 4.71</u>: Boundary conditions (z-view).

gure 4.71. The supports on the base are the simulation of the
contact surface between the die and the table of the press. The
displacements in three directions are zero: it is therefore
assumed that there is full friction.

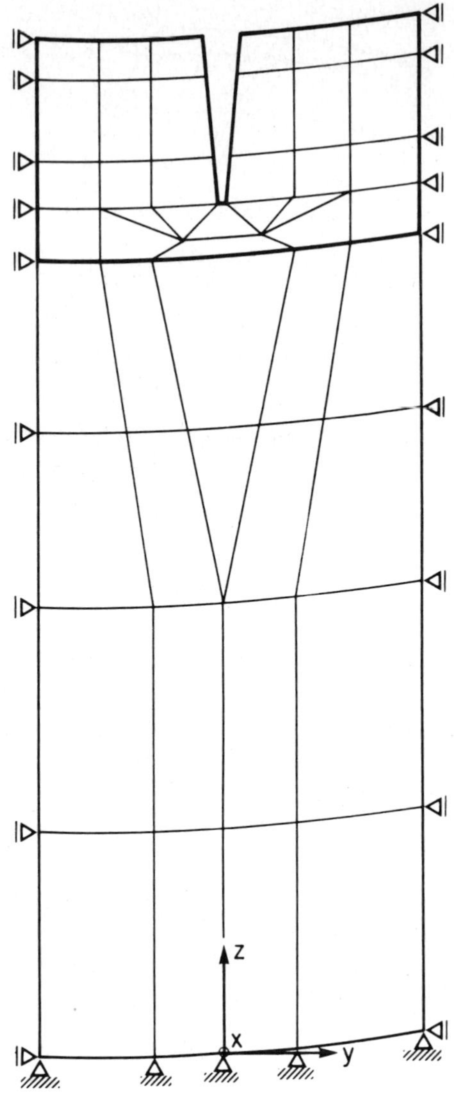

<u>Figure 4.72</u>:   Boundary conditions (x-view).

It was assumed that the contact forces between die and workpiece during the forming operation can be represented by a hydrostatic pressure $p_i = 120$ N/mm^2.

The calculations are done in isotropic conditions with material parameters:

-    Elasticity modulus $E = 214 \cdot 10^3 \ \text{N/mm}^2$
-    Poisson's ratio $v = 0.32$

4.4.3.2 Deformations at the Nodal Points

In Figure 4.73, the deformations due to the internal pressure
are plotted. The dashed lines are the contour plots of the undeformed structure; the continuous line is the plot of the deformed
shape. A maximum total deformation (u_x) of 0.387 mm at node 165
(see Figure 4.70) is obtained.

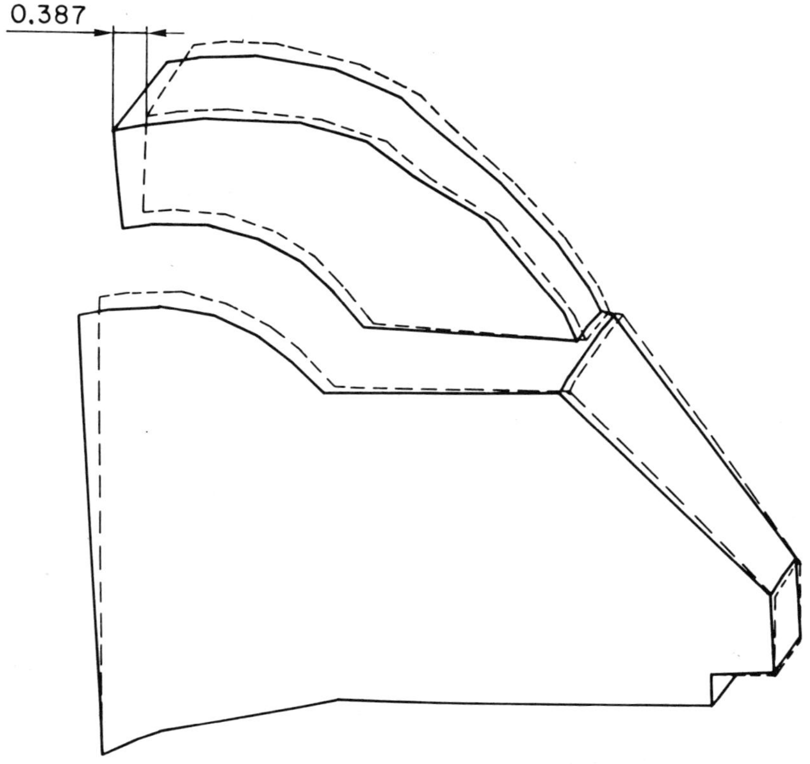

Figure 4.73: Deformations.

4.4.3.3 <u>Stress Distribution</u>

Two kinds of stress plots were produced:

a. Element stresses at the nodal points

b. Nodal stresses that are the average stresses of the element
    stresses at the nodal point.

Negative stresses are compressive and positive stresses are tensile. The stresses are all expressed in the global coordinate
system.

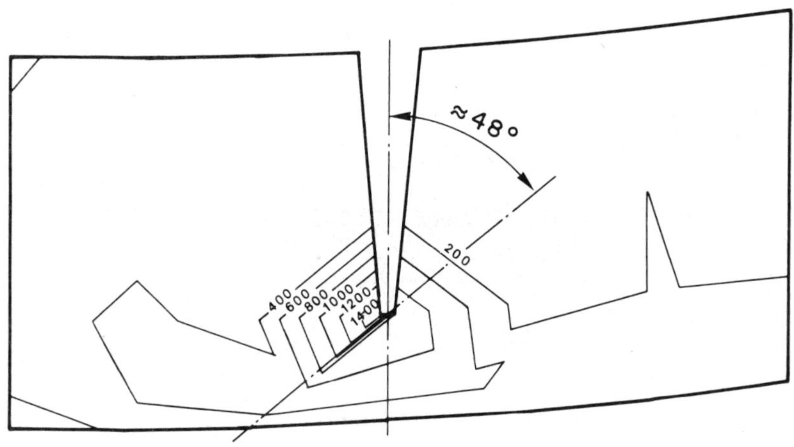

<u>Figure 4.74</u>: Von Mises stresses.

Contour plots of the von Mises stress are given in Figure 4.74,
on the outside radius of the die. If this figure is examined
only qualitatively, it is possible to see that the point most
highly stressed is at the corner of the slot where a stress
concentration factor is present. Furthermore, the direction perpendicular to the maximum stress gradient is at an angle α^* with
respect to the Z-axis. Comparing this result with Figure 4.75,
where the picture shows the same portion of the die as Figure

4.74, a crack in the die starts in a direction which has an angle
similar to the one predicted by FEM.

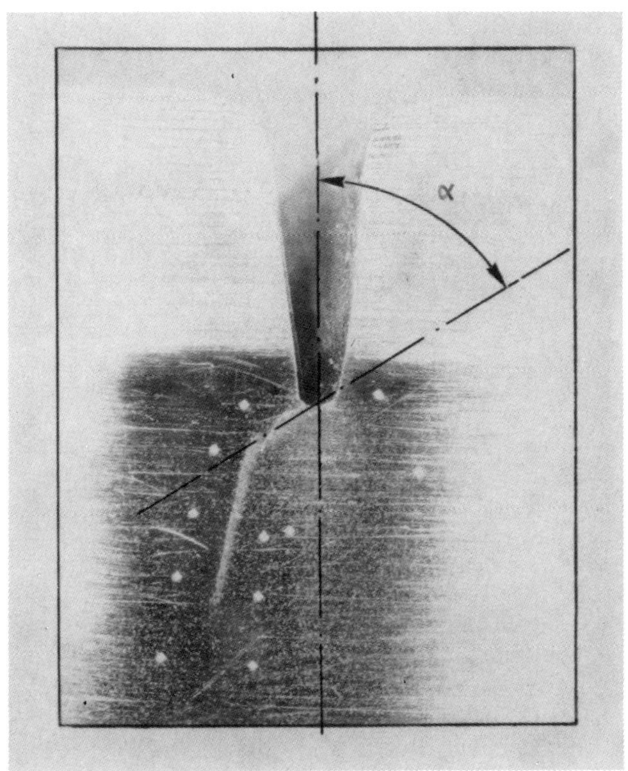

Figure 4.75:   Crack initiation and propagation.

This particular study represents only a partial view of the
applications of the FEM to die design of turbocharger impellers,
from which the following points are worth mentioning:

1. The present approach gives a better understanding of the non-
    symmetric stresses and deformations otherwise impossible to
    obtain due to the geometrical and loading complexity of the
    die.

2. The effects of different loading conditions are easily and
    inexpensively assessed. The stiffness matrix remains un-
    changed.

3.   Geometrical modifications are more difficult and more expen-
     sive to analyze because the mesh must be changed and the
     stiffness matrix has to be calculated again.

4.   The results can be interpreted easily with the examination
     of the plots of the stresses and deformations.

Furthermore, the following should be noted:

-    the fillet radius in the slot can be modelled satisfactorily
     with the use of second order elements and the potential un-
     real stress concentrations at the sharp corners thus reduced;

-    a two-dimensional analysis would reduce the calculation time
     but the non-symmetrical loading due to the curvature of the
     slot would not be reproduced;

-    the analysis was purely elastic; the tip of the slot can
     undergo some plastic deformation with consequent redistribu-
     tion of the stresses and a lowering of the maximum stress.
     An elastic-plastic study may be done, but it was outside the
     scope of the analysis;

-    an experimental fracture mechanic approach was used to
     establish design criteria for loading and unloading cycles
     and to determine the stresses which will initiate and pro-
     pagate cracks [4.6].

4.5 Finite Element Analysis of a Complex Axisymmetric Shape

The forging of axisymmetric shapes is one of the most common processes in metal forming. The geometry of the workpieces is three-dimensional, but the analysis of the process can be performed in two dimensions if the mathematical model is written accordingly.

It is therefore possible to study very complex shapes, using a two-dimensional approach. In the following pages an example is shown to illustrate the advantages and the difficulties of a FEM analysis of a real, complex forging shape. Only the deformations and the strain distribution will be examined, and the metal flow problem, which for the particular example chosen is the most difficult problem, will be concentrated on.

The geometry of the dies and of the preform are shown in Figure 4.76. The difficulty of this shape lies in the two circular ribs: one thinner and longer than the other one. The problem to solve consists of the filling of the two ribs with the lowest strains possible and without exceeding the allowable stresses in the dies.

For reasons of symmetry, only half of the die-workpiece is analyzed.

4.5.1 Deformations

One of the most important issues in a closed die forging operation is the deformation of the billet when pressed between the dies and the determination whether complete failure is achieved or not. The FEM is an important help for the designer in the selection of the correct geometry for the dies and the preform. The initial contour of the preform is deformed following the die geometry and depending on the material properties, ram velocity, friction, die and billet temperature.

Figures 4.77 (a) to 4.80 (a) show the deformed mesh at different die openings. In Figure 4.77 (a), the preform has touched the right wall of the lower die and from now on, the workpiece will be forged-extruded in the two ribs. The deformation shown in Fi-

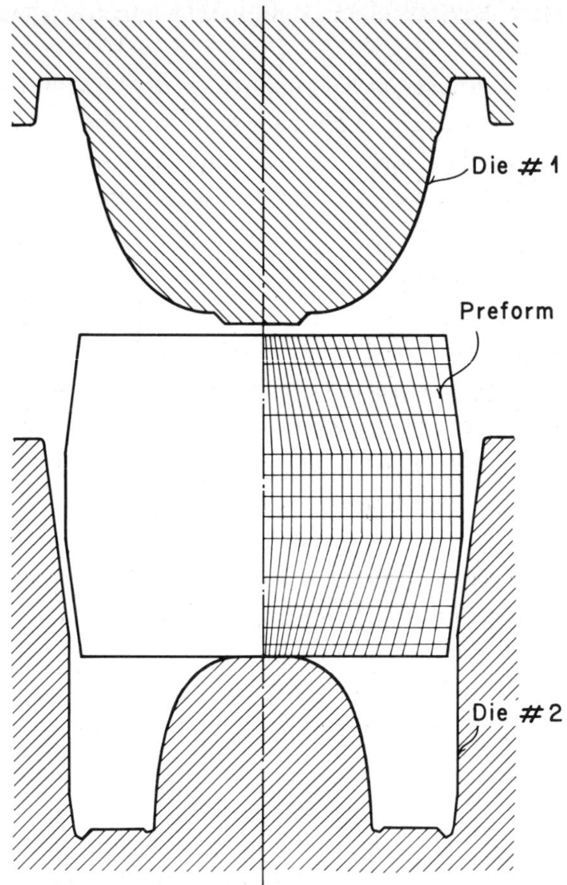

Figure 4.76: Preform and die configuration for the axisymmetric forging FEM analysis.

gure 4.78 (a) is obtained after a first remeshing of the initial mesh. The workpiece has reached the bottom of the lower die, but the rib is not yet filled. At this stage, a new remesh is necessary in order to follow the geometry of the lower rib with a good degree of accuracy. Figure 4.79 (a) shows the stage of filling of the rib while the upper portion of the workpiece has reached the flash zone. At this point, the beneficial influence of the flash can be seen: the material has more difficulty in entering the flash than in filling the die and, in Figure 4.80 (a), the process is terminated with the filling of the rib.

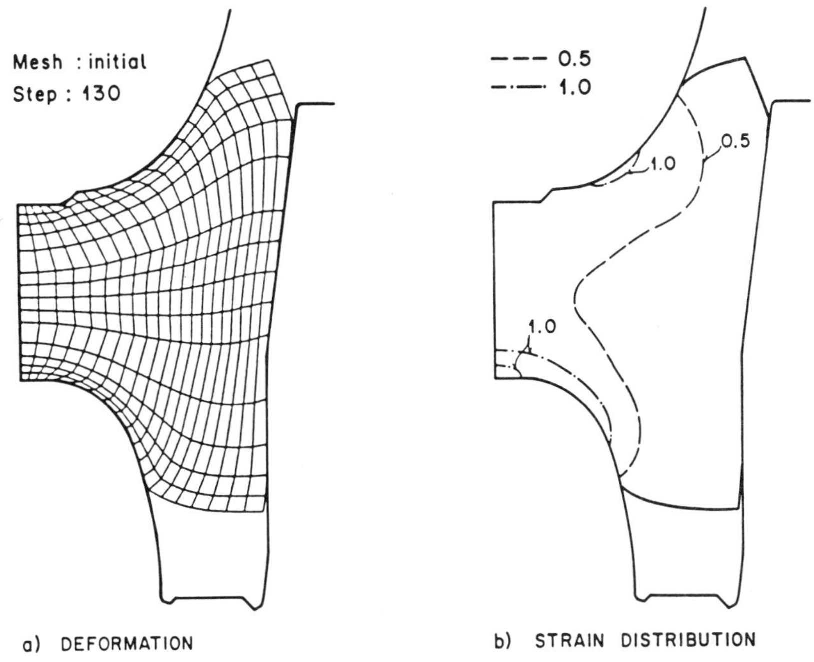

Figure 4.77: Deformation and strain distribution for axisymmetric forging.

From the above results, it can be deduced that the position and
the thickness of the flash gave the desired die filling.

4.5.2 Strains

The strain distribution also provides very important information
for the designer. Figures 4.77 (b) to 4.80 (b) show the strain
distribution at the same deformation steps as the corresponding
Figures 4.77 (a) to 4.80 (a). The values of the strain contours
are kept constant throughout the whole sequence and therefore
there is a concentration of lines at the end of the process and
a sparse distribution at the beginning. A first appearance of
contour lines is in Figure 4.77 (b), where a value of logarithmic
(von Mises) strain larger than 1 is present at the contact surface of the lower die and the billet along the radius of the

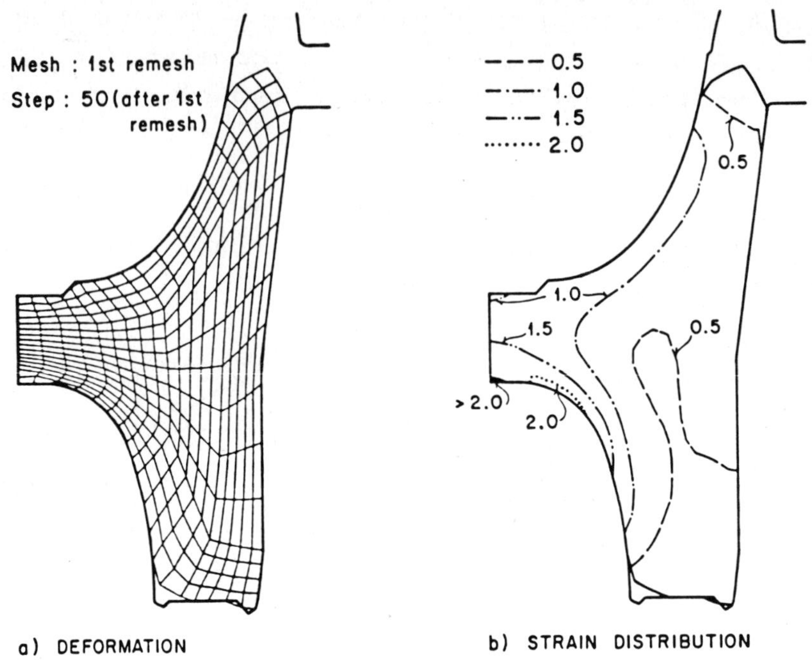

Figure 4.78: Deformation and strain distribution for axisymmetric forging.

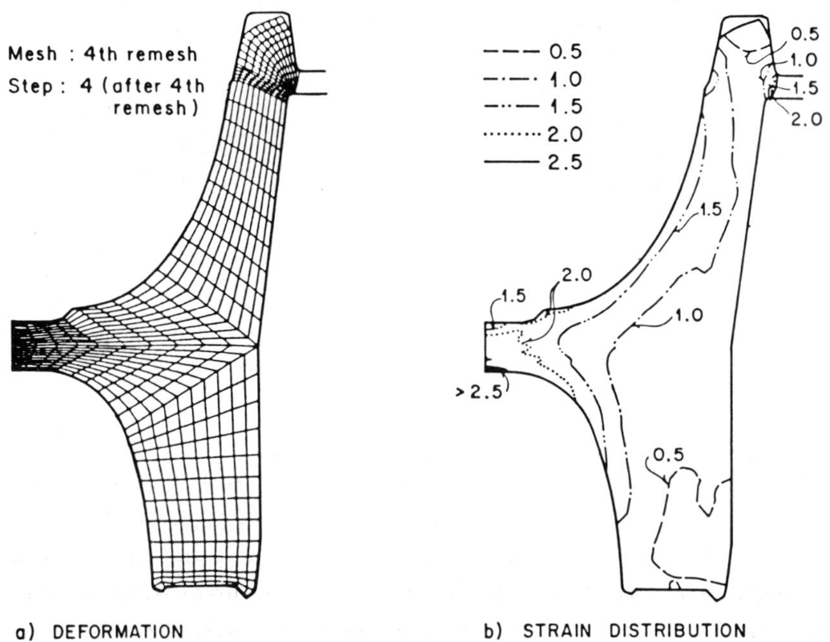

Figure 4.79: Deformation and strain distribution for axisymmetric forging.

entrance of the rib. The strain reaches a value of 2 in Figure
4.78 (b), again in the same place as described above. In fact,
in this region, the material is considerably stretched. Even when
the material reaches the flash region, the zone that is most
strained is the central portion of the workpiece, while other
portions are at as low a strain as the lower portion of the
workpiece that has filled the lower die and the upper portion
with the still free surface (see Figure 4.79 (b)).

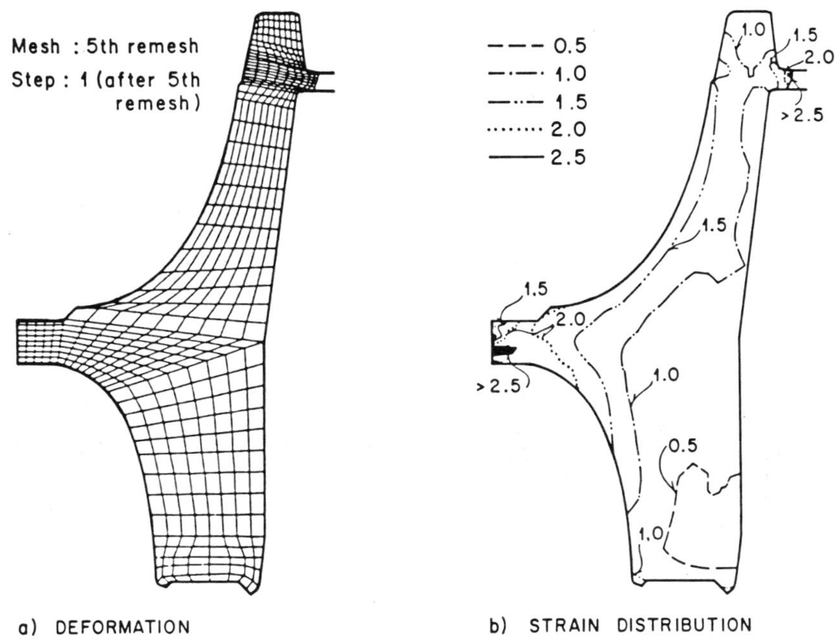

Figure 4.80: Deformation and strain distribution for axisymme-
              tric forging.

The modelling of the material flow and thereby the strain and
stress distribution gives the designer the possibility to predict the material properties in the final forging. A comparison
with experimental results is shown in Figure 4.81 where a part
with very similar geometry has been forged. The material, the
nickel-base alloy Waspaloy was forged in a condition where a
certain cold-deformation-hardening took place. Figure 4.81a shows

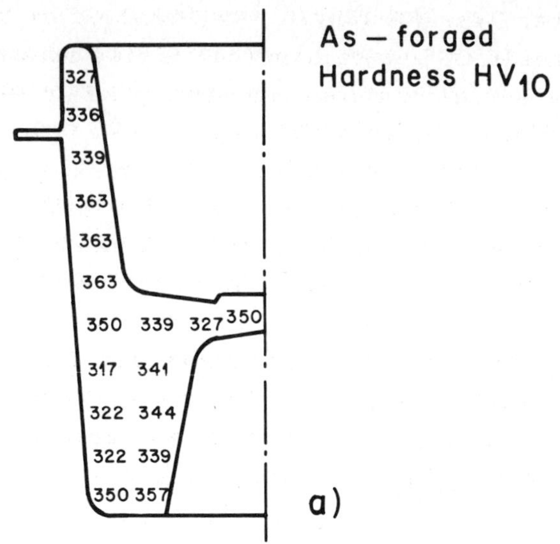

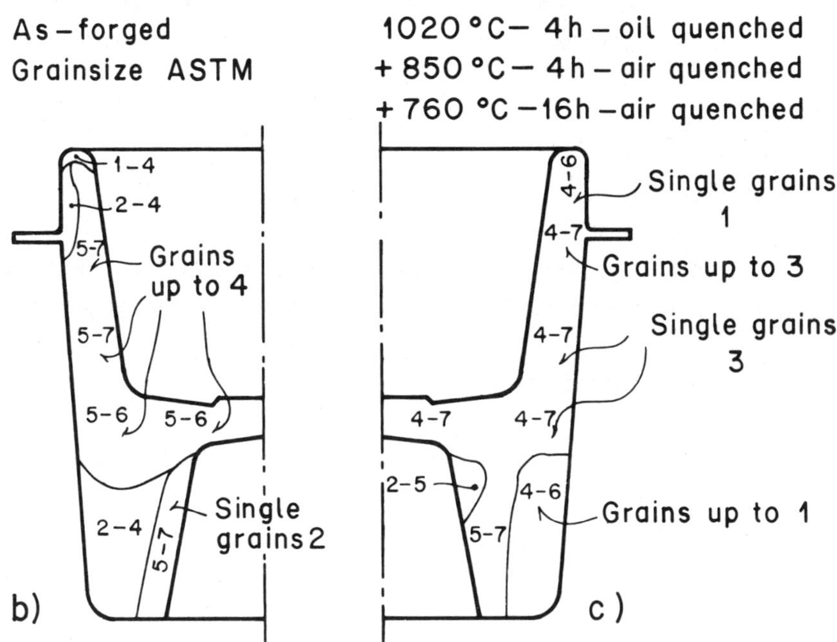

Figure 4.81: a) HV$_{10}$ hardness distribution in the as-forged condition.
b) Grain size distribution in the as-forged condition.
c) Grain size distribution in the fully heat treated condition.

the HV_{10} hardness distribution in the part after forging. There is a clear connection between the calculated strain distribution and the hardness distribution. In Figure 4.81b and c the grain size distribution (ASTM-standard) is given in the as-forged and in the fully heat-treated condition. The initial grain size was coarse with ASTM 0-1 surrounded by grains with ASTM 5-6. Also here a clear dependance between the predicted strain distribution and the grain size distribution can be seen.

In Chapter 3 model material experiments are described where the forging of a similar shape as the one in Figures 4.77 - 4.79 was simulated. The aim of that study was mainly to see how the material flow was during the forging operation. This material flow could be satisfactorily simulated. The strain distribution in the model material part was similar to the ones predicted by FEM-calculation. However, the technique used was not accurate enough to get a conclusive picture of the strain distribution.

4.5.3 Die Stress

There are three major causes for die failure [4.18]:

1. Die cavity wear due to billet material flow or reaction between the die material with lubricants or the atmosphere

2. Plastic die deformation

3. Crack initiation and propagation due to high local stress concentration.

The plastic die deformation can either lead to loss of tolerance in the forged part due to local geometrical changes or to a complete die failure due to too high overall stresses. An example of the first is the forging of O.D.S. (oxide dispersion strengthened) Ni-base alloys, which have high flow stresses under the condition at which they have to be forged to retain their desired properties, and therefore give rise to high stresses [4.21].

Figure 4.82 shows an example where a complete die failure was

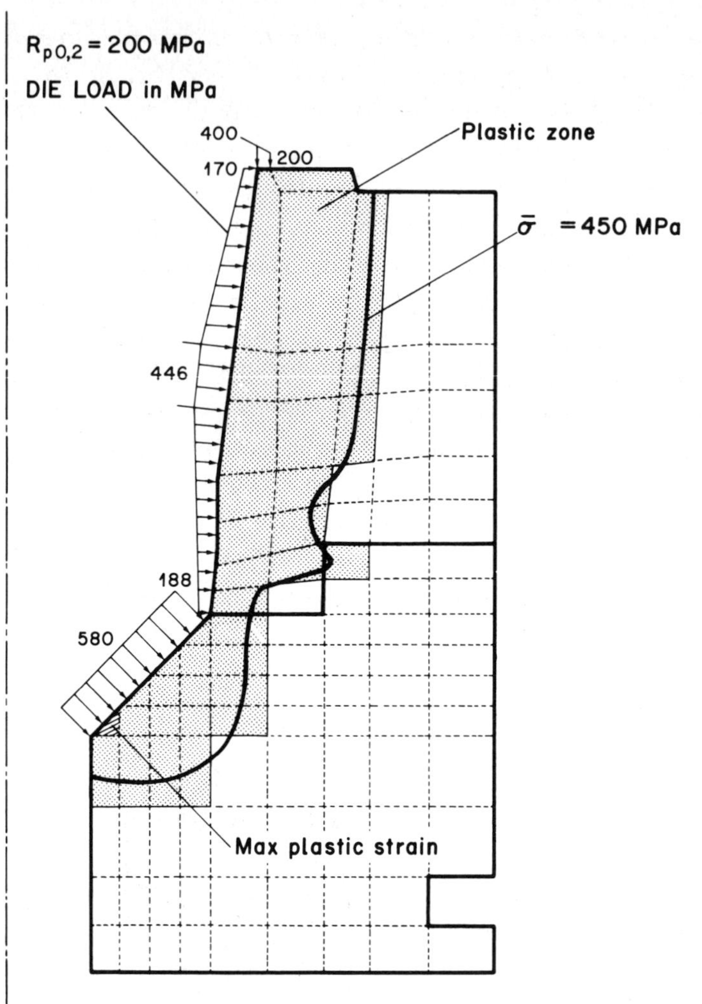

Figure 4.82: Cross-section of a die with the stress distribution on the die cavity and the plastic zone in the die marked. The die cavity stress distribution was calculated with the slab method and the stress distribution in the die with elastic-plastic FEM.

a possibility. The die load was calculated using the slab method with an assumed flow stress of 200 MPa. Then, using a FEM package

the die stress distribution in the circular-symmetric die was
calculated using an elastic-plastic material model for the die
material. The calculations show in what part of the die equivalent stresses higher than the flow stresses of the material
occur. The case for which the calculations were done was an
isothermal forging case where the yield stress of the die material was assumed to be 450 MPa. The calculations show how large
is the part of the die in the plastic region. It was also shown
where the maximum plastic strain occurs. When the die diameter
is too small, catastrophic die failure occurs, Figure 4.83. By
being able to simulate the die load and then calculate the die
stresses, the die designer can change the die geometry to avoid
plastic die deformations.

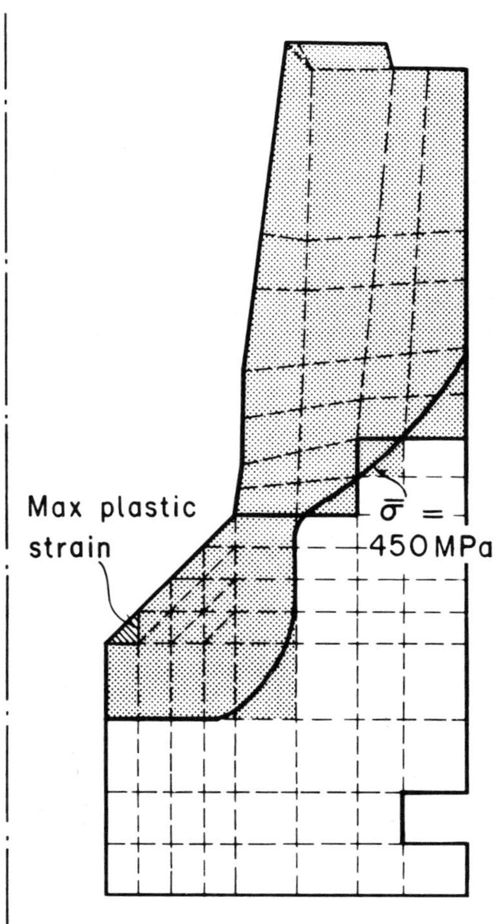

Figure 4.83: Example of an under-dimensional die in which the
die would plastically deform.

References Chapter 4

[4.1]   Boër C.R., Jovane F., Computer-Aided Design of Metal Forming Systems, Keynote paper of the Forming Session, Annals CIRP 33/2 (1984) 433-449.

[4.2]   Boër C.R., Rydstad H., Schröder G., Choosing Optimum Forging Conditions in Isothermal and Hot-Die Forging, J. of Applied Metalworking, Vol.3, No. 4 (1984)421-431.

[4.3]   Boër C.R., Schröder G., Heat Transfer during Hot Upsetting in Heated Dies, Proceedings 21th MTDR Conference (1980) 209-215.

[4.4]   Boër C.R., Schröder G., Temperature in the Die-Billet Zone in Forging, Annals CIRP 30/1 (1981) 153-157.

[4.5]   Boër C.R., Rydstad H., Schröder G., Process Modelling for Upsetting of Cylindrical Billets: Experimental Verification of the Heat Transfer Calculations, Proceedings 1981 ASM Metal Congress, ed. C.C. Chen, (1983) 375.

[4.6]   Schröder G., Fracture Mechanics Methods for Life Prediction of Forging Tools (in German), wt Zeitschrift für Industrielle Fertigung, 74 (1984) 207-210.

[4.7]   Altan T., et al., Forging Equipment, Materials, and Practices, Air Force Materials Laboratory, MCIC-HB-03 (1973).

[4.8]   Schröder G., Rebelo N., Umformverhalten induktiv erwärmter Rohteile beim Schmieden, wt-Z ind. Fertig. 73, No. 9 (1983) 565-568.

[4.9]   Decker E., Stahl mit elektrischer Energie rationell erwärmen, Ind. Anz. 104, No. 55 (1982) 14-18.

[4.10]   Jürgens H., Anwendung der induktiven Erwärmung in der Schmiede, Elektrowärme Internat. 37, No. 1 (1979) 33-36.

[4.11] Grulke N., Induktive Erwärmung in der leicht- und schwer-
        metallverarbeitenden Industrie, Metall 29, No. 1 (1975)
        22-27.

[4.12] Roll K., Tekkaya A.E., Prozesssimulation in der Umform-
        technik mit der Methode der Finiten Elemente - Stand und
        Entwicklungstendenzen, Umformtechnik '84, K. Lange, Inst.
        f. Umformtechnik Univ. Stuttgart (1984).

[4.13] Kulkarni K.M., Isothermal Forging - from Research to a
        Promising New Manufacturing Technology, Proc. SME-Conf.
        Detroit, MF 77-299 (May 1977).

[4.14] Corti C.W., Gessinger G.H., Shabaik A.H., Superplastic
        Isothermal Forging: a Model Metal Flow Study, J. Mech.
        Work. Tech., 1 (1977) 35-51.

[4.15] Wüthrich C., Schröder G., Use of Fracture Mechanics Me-
        thods for Lifetime Improvement of Forging Tools (in Ger-
        man), Z. Werkstofftech., 11 (1980) 417-422.

[4.16] Unksow E.P., An Engineering Theory of Plasticity, London,
        Butterworths (1961).

[4.17] Rebelo N., Boër C.R., A Process Modelling Study of the
        Influence of Friction During Rib Forging, Proceeding
        NAMRC (1984) 146-150.

[4.18] Rydstad H., Hoffelner W., PM Molybdenum Isothermal
        Forging Dies - Property Requirements and Service Life,
        Proc. 11th Plansee Seminar, Reutte, Austria (1985).

[4.19] Rydstad H., Boër C.R., Process Modelling of Hot-Die
        Forging - Experimental Verification using a Computer-
        Controlled Press, Proc. 11th North American Manufac-
        turing Research Conference, NAMRC (May 1983) 211-217.

[4.20] Rebelo N., Rydstad H., Schröder G., Simulation of Mate-
        rial Flow in Closed Die Forging by Model Techniques and
        Rigid-Plastic FEM, in Numerical Methods in Industrial

Forming Processes, edited by J.F.T. Pittman, R.D. Wood, J.M. Alexander and O.C. Zienkicwicz, Pineridge Press, Swansea, U.K. (1982) 237-246.

[4.21] Schröder G., Rydstad H., Singer R.F., Nazmy M., Forging and Thermomechanical Treatment of ODS-Nickel-Base Alloys, Proc. First Int. Conf. on Technology of Plasticity, Tokyo, Japan, Vol. 1 (1984) 39-44.

[4.22] Geisel H., Der Temperaturverlauf beim induktiven Erwärmen. Masch. u. Werkz. 24 (1970) 32-37.

5 Modelling of Rolling

<u>List of Symbols</u>
(other symbols are defined in the text)

A0   Area of the final profile

A1   Area of displacement

A2   Area of lateral spread

C   constant of the flow stress law

D   Maximum allowable deformability

$D_L(i)$ Deformation at X(i)

F   Lateral spread ratio

F1, F2   Y-coordinates of the upper and lower part of the
              starting profile

G1, G2   X and Y coordinate of the center of gravity

H0   Height of the starting profile

H9   Percentage increase from the final to starting profile

$H_M(\ell,i)$ Height of the pass ℓ at X(i)

H_P Minimum height of the final profile in the range
              L1 - L2

k   Index of the center of gravity X-coordinate

L1, L2   X-coordinates of left-hand and right-hand sides of the
              starting profile

n   Number of coordinates (total)
              or exponent of the flow stress law

P   Number of passes

P1(ℓ,i)   Y-coordinates of the upper and lower roll of pass ℓ
P2(ℓ,i)   at X(i)

W0   Width of starting profile

Y1(i)   Y-coordinates of the lower roll (final profile)

Y2(i)   Y-coordinates of the upper roll (final profile)

X(i)   X-coordinates of the final profile

Δx Interval of x-coordinates (equal spacing)

$\bar{\sigma}$ flow stress (effective)

ε strain (logarithmic)

5.1 Introduction

Warm or cold roll forming is a deformation process used for economic production of profiles, not only for structural purposes (e.g. angles, T-bars, channel sections etc.) but also for precision work, such as for turbine blades. For simple standard profiles, there are guidelines available and experimental findings for roll pass design and for working out the roll pass schedule from the starting pass to the finished section [5.1]. Flat rolling, too, has also been the subject of detailed investigation [5.2]. However, for designing complicated special profiles, the rolling tools are, as a rule, designed empirically. There are two reasons for this:

1. Roll forming is an experience-based technology, whereby improvements and developments have come about principally through a pragmatic approach by trial-and-error methods.

2. Although numerous publications deal with lateral spread of the material within the roll gap, pre-determination of spread behavior during roll forming gives rise to considerable problems and is possible to a sufficient degree of accuracy only for special cases by the application of lateral spread formulae [5.3-5.6] or calculations on the basis of plasticity theory [5.7-5.9]. As lateral spread is influenced by many parameters [5.10] - geometry, friction, temperature, microstructure, yield stress - for new materials, there is no alternative but to determine lateral spread by experiment.

For rolling steels, the designer can fall back on empirical values which define the spread behavior of certain materials and on permissible deformations, which, however, are frequently not published, as they represent company-specific know-how.

It is precisely these interrelationships which enable computer-aided roll pass design [5.14], as will be described in the following. The method presented makes use of experimentally-de-

---

Note: For this chapter the contribution of H. Riegger (ISOPROFIL GmbH, Mannheim, W.Germany) is gratefully acknowledged.

termined spread values, which are stored in the computer as a
function of various parameters. The principal advantage is that
time-consuming tasks, such as planimetry computation, measurement, modifications and drafting are performed by the computer,
whereby - since the computer program is interactive - the
designer has the possibility, by appropriate entries, of direct-
ing the roll pass design process to achieve the desired results.
Other advantages are: direct storage of the coordinates of all
rolls and passes for NC-production, and simultaneous compilation
of and call-up from a data file. By application of computer-aided
roll pass design, it is a realistic objective to execute a correct roll pass design from receipt of drawings to issuing a tool
production order by means of technical drawings within a few
hours [5.11-5.14].

5.2 Measurements of Lateral Spread

The basis of the computer program here presented is the result of
tests on the lateral spread during flat rolling tests [5.10] as
a function of various parameters. The cross-sectional areas of
the various bars were projected prior to and after flat rolling
and the area categories below defined (see also list of symbols):

1. Remaining cross-sectional area A0:

    The remaining core cross-sectional area A0 is that part of
    the profile which <u>apparently</u> remains undeformed. This area
    is a function of the roll gap and the initial width of the
    rod (Figure 5.1).

2. The area of displacement A1 is the difference between the
    initial cross-sectional area of the bar and of the area A0
    in the remaining cross-sectional area. This area A1 is also
    shown in Figure 5.1.

3. The area of lateral spread A2 is limited by the width of
    the profile as-rolled and the initial width of the bar (Fi-
    gure 5.1). It is the difference between the cross-sectional
    area of the rolled bar and the remaining cross-sectional
    area A0.

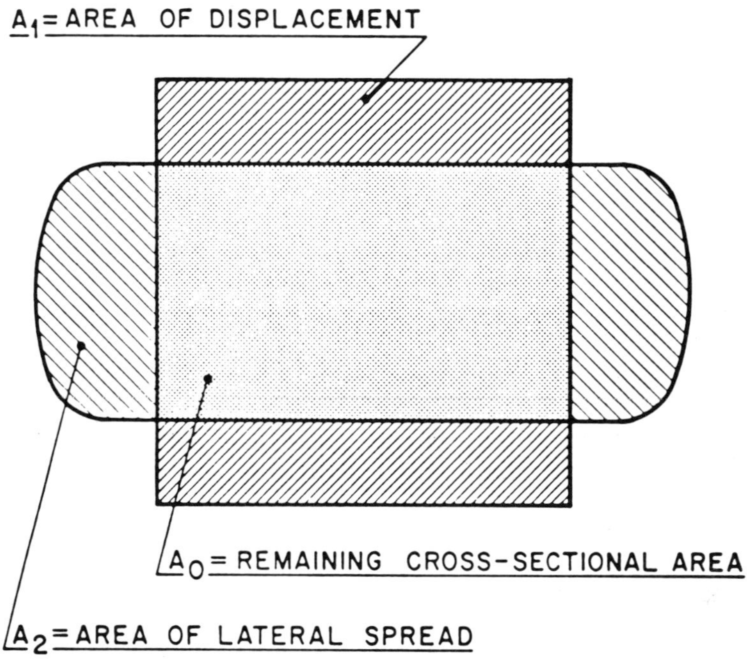

Figure 5.1: Definition of the areas for the lateral spread.

The measured areas do not agree with predictions based on the fundamentals of metal deformation of changes in area or volume during rolling, but are particularly suitable for the method of approximate prediction of lateral spread during the pass which is here presented. The function of lateral spread F = A2/A1, results in quantitative representation of lateral spread. If F = 0, there is no lateral spread during rolling. For F = 1, there is no elongation of the bar during rolling and purely lateral material flow occurs. The influence of the following parameters on the lateral spread function F = A2/A1 was investigated:

1. Material: copper (2.0060*), low carbon steel St 37 (1.0120*), Ti6Al4V (3.7164*).

---

* FRG - German materials number (DIN)

2. Width to thickness ratio, W0/H0; during profile rolling, an average height H0 was fed in, which resulted in an average width-to-thickness ratio.

3. Reduction in area.

4. Rolling temperature.

5. Friction coefficient.

6. Velocity of the rolls (strain rate).

Although the lateral spread behavior of materials in the roll gap depends on many other parameters (roll radius, bar width to length of the arc of contact ratio, anisotropy of the material etc.), attention was restricted during the experiments to the 6 parameters mentioned. For the purposes of computer-aided roll pass design, further simplifying assumptions were made in order to limit attention to the important parameters.

Thus friction coefficient, surface roughness, the velocity of the rolls, roll radius and rolling temperature were kept constant for the purposes of calculation. Further, it was assumed that material behavior is isotropic and that texture or strain hardening have no influence on lateral spread. Moreover, it could be demonstrated [5.10] that for large reductions in area (over about 30 %), the value of the function of lateral spread remains constant.

Under these conditions, the values of the function of lateral spread were stored in the computer depending on the two most important parameters: the influence of the material, which can be specified by the designer, and the influence of the average width-to-thickness ratio, which is automatically determined by the computer for the specified profile and with which the entered values can be made to agree. The entered values are shown in Figure 5.2, whereby the rolling temperature is 20°C for copper, 800°C for the steel St 37 and 950°C for TiAl6V4.

In order to check the computer program, lateral spread as predicted by the program was compared with a sequence of experimental roll passes whereby the results showed good agreements. Barrelling occurring during rolling at the sides of the profiles was ignored in the first approximation. Thus conditions for compilation of a drafting and computing program were created, which determines material flow on the basis of experimental data and which performs roll pass design for each individual pass with the aid of entries made by the designer.

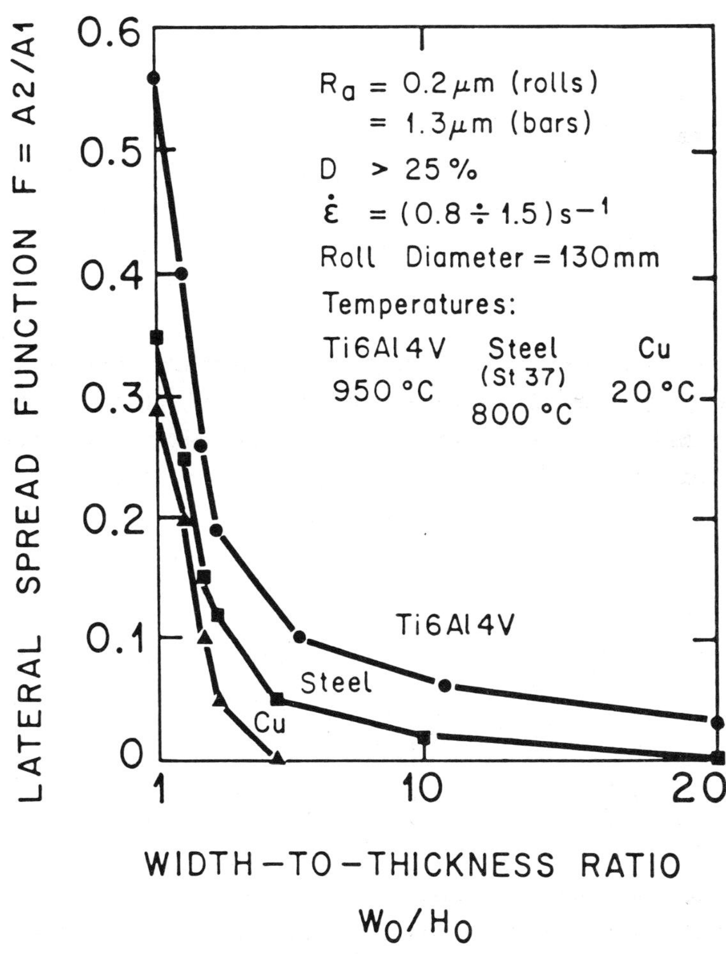

Figure 5.2:   Lateral spread function versus width-to-thickness ratio for three different materials.

5.3 Computer-Aided Roll Pass Design

The program was divided into two sections (see Figure 5.3) - the
process of profile input or modification, and the actual execu-
tion of roll pass design, which also includes issuing of data
and drawings as well as transferring the intermediate passes to
the memory.

5.3.1 Input of the Geometry of the Profile

The profiles can be entered into the computer on the basis of a
dimensional drawing. In so far as coordinates are required for
the profile to be rolled, these can be entered either explicitly
or via a digitizing board. For computation purposes, the computer
generates coordinates with equidistant abscissa values, which
divide the profile into strips. A program for smoothing coordinates thus calculates new equidistant coordinates and at the
same time smooths the digitized profile with a second-degree
polynomial through all digitized points. The individual coordinates are thereby redefined. Thereafter, the digitized profile
and the newly-calculated smoothed profile with equidistant coordinates are stored in a data bank, together with further profile
data (name, material, customer, date, form factor etc.). From
here it can be called up at any time for roll pass design and
drawing or for further modifications.

5.3.2 Modifications to the Profile

After completion of smoothing of the profile contours, there
still remains a possibility of modifying individual points of
the profile "by hand". This makes it possible for the designer
to change a geometry to make it suitable for rolling, e.g. by
avoiding recesses and 90° angles. Thus the basic shape of a profile (e.g. rectangle) is modified (e.g. rectangle with groove) or
vice-versa.

A further possibility for manual profile modification is the
introduction of a flash (Figure 5.4). The flash influences roll
pass design, as then, in the main program, calculation for over-
filling of the pass is made. However, the coordinates additio-

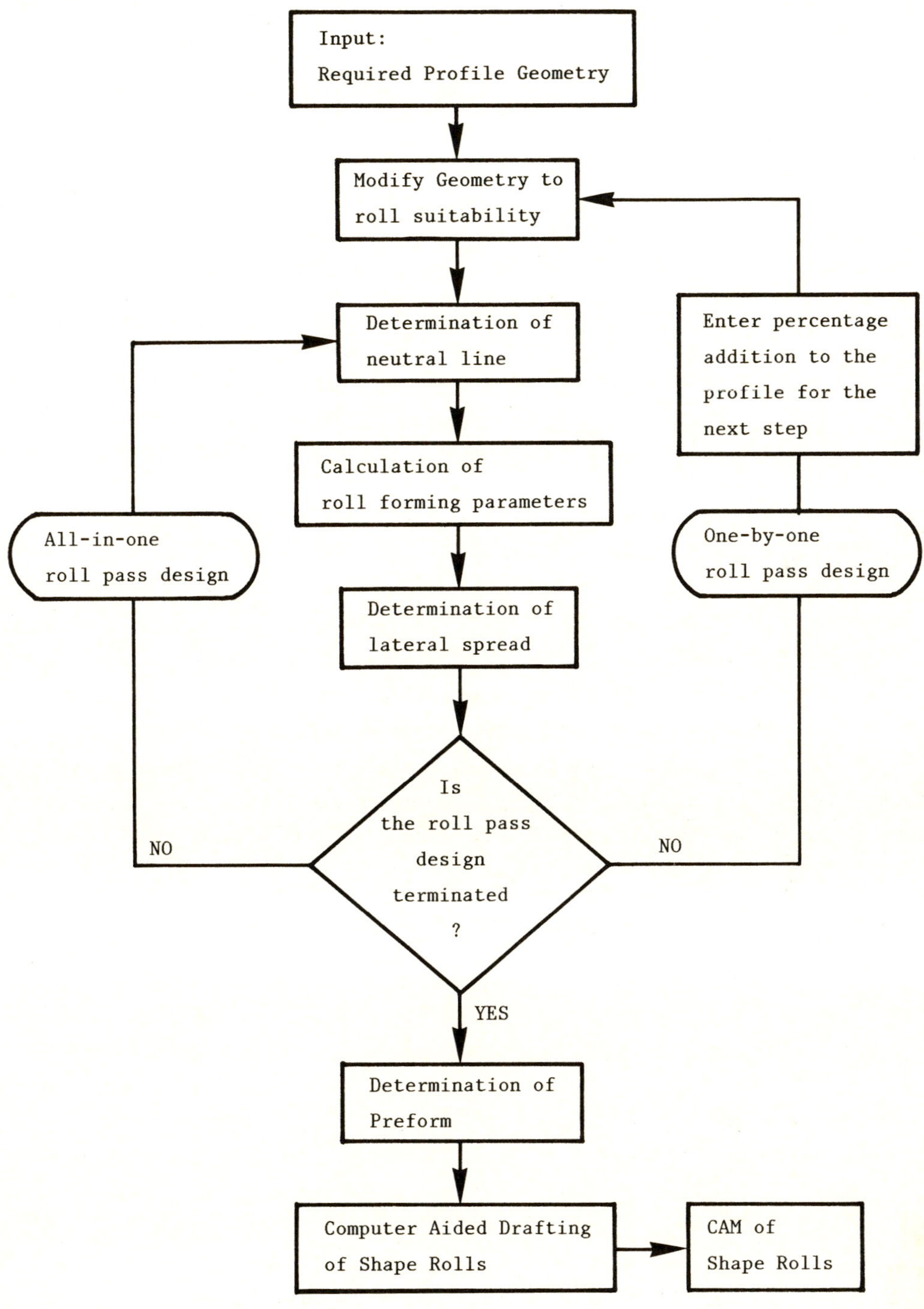

Figure 5.3:   General logical flow chart for the Roll Pass Design.

297

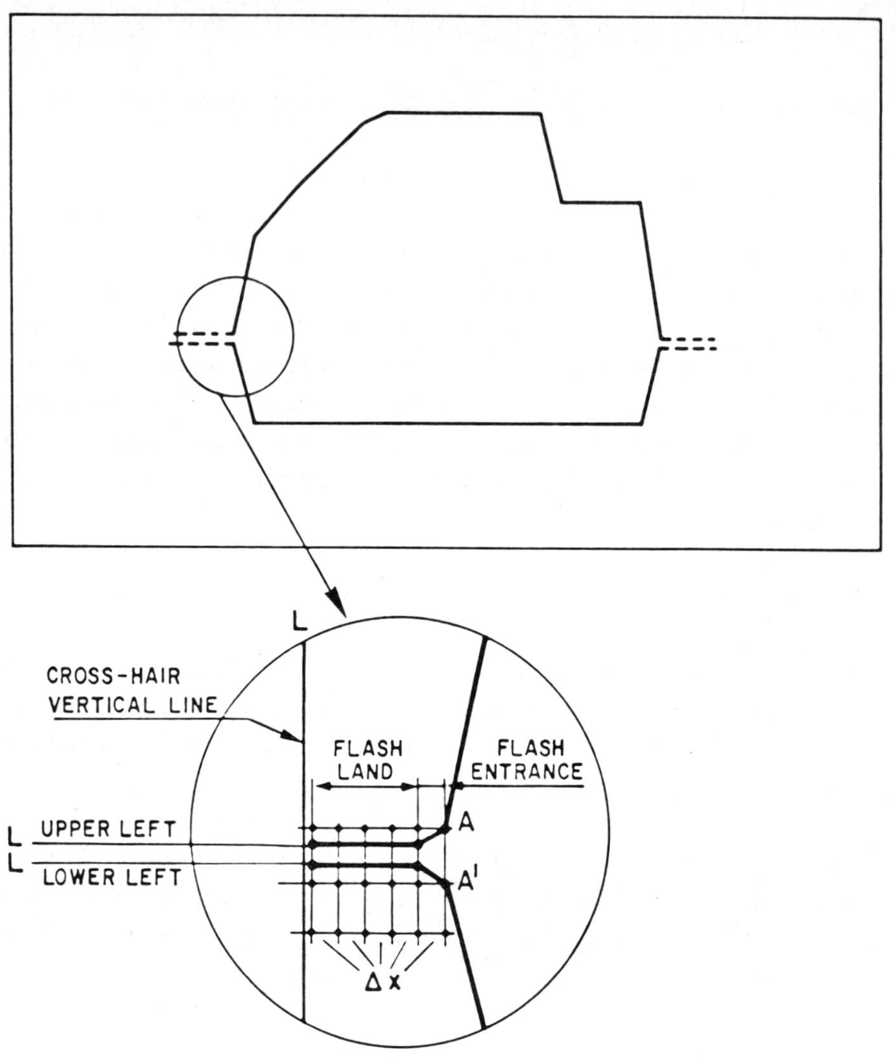

Figure 5.4: Introduction of flash.

nally introduced for the flash (Figure 5.4) are not stored in
the data bank. As for the modification of individual points, the
designer specifies the extent of the modification. The interactive program makes it possible to specify the required thickness and width of flash and to change this if it is not suitable.

298

5.3.3 Calculation and Modification of the Neutral Line

The neutral line position can be determined in two ways:

a) Center of gravity

For each profile section, the computer program determines the
center of gravity, inserts it into the drawing together with its
coordinates and uses it as a first approximation of the neutral
line for the material flow. For profiles with a high degree of
asymmetry, such as some turbine blades, this assumption does
not hold. In such a case, the roll pass designer can shift the
neutral line in the required direction on the basis of his
experience.

b) Slab method

The computer program calculates the stress distribution across
the profile section using the slab method of analysis. The position of the neutral line will coincide with the maximum stress
value.

For calculation of lateral spread, the neutral line is very
important, as the program calculates lateral spread at both
sides of the profile, thus greatly influencing the modified
width-to-thickness ratio (Figure 5.5).

5.3.4 Determination of the Starting Profile (Preform)

The area of the final profile is first determined according to
Figure 5.5 by:

$$AO = \sum_{i=1}^{n-1} \frac{1}{2} \left[Y2(i) - Y1(i) + Y2(i+1) - Y1(i+1) \right] \cdot$$

$$\cdot \left[X(i+1) - X(i) \right] \tag{5.1}$$

The height (HO) of the starting profile, which is usually a rectangle, is also fixed by the designer after the program has specified a minimum starting height using the entered coordinates

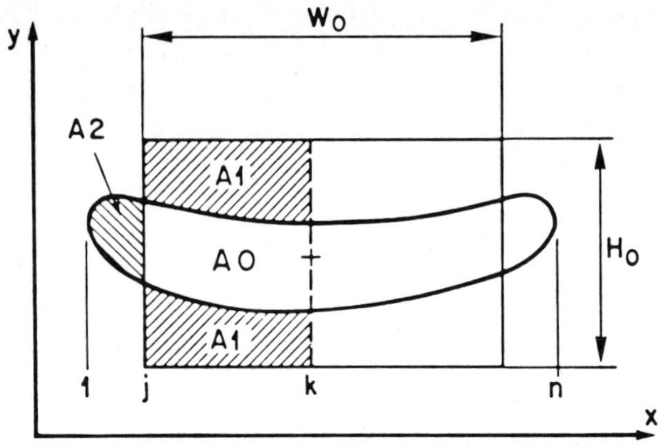

<u>Figure 5.5:</u>   Schematic drawing of the calculated areas (see also
              Figure 5.1).

as follows:

$$HO = [\underset{i=1}{\overset{n}{MAX}} Y2(i) - \underset{i=1}{\overset{n}{MIN}} Y1(i)] \cdot (\frac{H9}{100} + 1) \qquad (5.2)$$

On the basis of this minimum, the designer can propose a starting
height either on the basis of his own experience or according to
those bar dimensions stocked. The program now calculates the lateral material spread for the given profile with the aid of the
material data which are also prescribed, and thus determines the
first approximation for the width of the initial profile. The
determination of the starting profile is done using the spread
function defined as:

$$F = \frac{A2}{A1} \qquad (5.3)$$

where A1 and A2 are the areas given by the following equations
(see Figure 5.5):

$$A2 = \sum_{i=1}^{j-1} \frac{1}{2} [Y2(i) - Y1(i) + Y2(i+1) - Y1(i+1)] \cdot$$

$$\cdot [X(i+1) - X(i)] \qquad (5.4)$$

300

$$A1 = \sum_{i=j-1}^{k-1} \frac{1}{2} [H0 - (Y2(i) - Y1(i) + Y2(i+1) - Y1(i+1))] \cdot$$

$$\cdot [X(i+1) - X(i)] \tag{5.5}$$

with the index i taking the following values:

left-hand side	$i = 1 \dots k-1$
right-hand side	$i = k \dots n-1$

The value F obtained by equation (5.3) is compared with the value of the function given in Figure 5.2 for the specific WO/HO-ratio (width-to-thickness ratio). The width WO is given by the position of the index j (see Figure 5.5). The procedure is iteratively repeated until the value of F from equation (5.3) and the value F of the experimentally-determined lateral spread coincide. The accuracy is given by the Δx interval.

The first proposal must not necessarily be the correct width of the initial bar. For this reason, the program executes a built-in self-correction, which, after calculation of the number of passes, checks the spread once again and corrects the initial width of the bar.

Three other forms of the preforms are possible: round, oval and square. Each form can be chosen interactively by the designer and then the computer program determines the right dimensions.

5.3.5 Roll Forming Parameters Calculations

Once the form of the pass has been determined, the plot of the deformability (reduction in height) is shown on the screen to allow the designer to make modifications if he thinks that the deformation is excessive in some portion of the pass.

The neutral line is also displayed and the designer has the possibility of modifying its position interactively with the cursor. At present the possibility of a sticking zone in the rolling gap is ignored.

The lateral spread is then calculated and displayed on the screen.

At this stage, it is possible to calculate several parameters concerning the metal forming. If the designer decides to do so, he is requested to read in the following data:

    roll radius
    roll speed
    material properties: constant   C
                          hardening exponent n

The material flow stress is expressed according to the equation $\bar{\sigma} = C \cdot \varepsilon^n$.

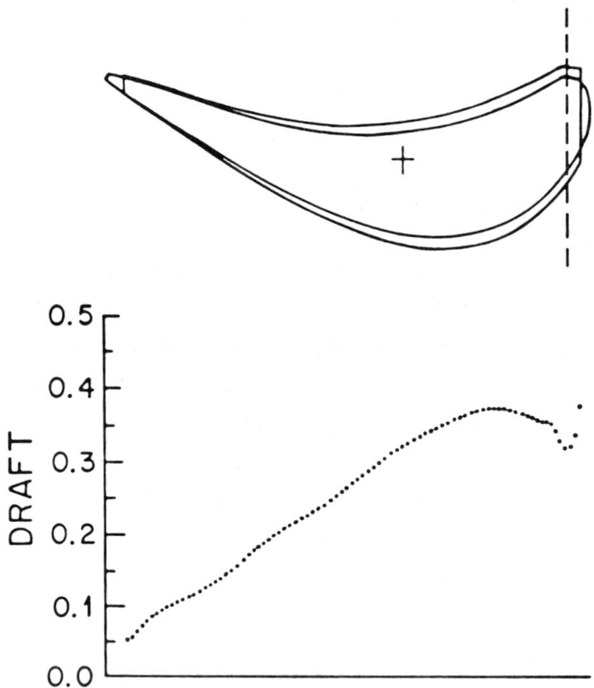

<u>Figure 5.6</u>: Calculation and presentation of various rolling parameters (in this case: the draft).

These data, together with the geometry of the pass, are used to

302

calculate the following parameters:

    draft (or reduction in height in mm) as in Figure 5.6
    area of the section of the profile
    reduction in height (as a percentage)
    deformed roll radius
    mean values of strain rate, yield stress, shear yield stress
    equivalent values of strain rate and yield stress
    total roll force (with or without roll flattening)
    total roll torque (roll flattening)

The designer can examine the results of the calculations above
and decide to go to the next design step or modify the present
pass design and go back to a previous step.

5.3.6 Roll Pass Schedule and Roll Geometry

Two methods are available for designing the roll pass schedule
and roll geometry.

a) All-in-one roll pass design

After entering the maximum allowable deformation of the material
$D = 1 - (H0 - H)/H0$, the number of passes is determined by the
computer. The maximum deformation describes the rollability of
the material, i.e. for no cracks to appear in the material. Usually, the designer knows these values from his own experience,
and, additionally, sufficient roll pass schedules are given in
the literature for the various materials [5.13-5.16], these providing data concerning rollability and maximum deformation. For
the titanium alloy Ti6Al4V, maximum deformations for conventional
hot rolling [5.10] and for isothermal rolling [5.12, 5.13] have
also been determined. For unlimited deformations, such as for
isothermal rolling, the entered value is limited by the capacity
or geometry of the rolling tools (bite angle, maximum roll
force). The roll pass program calculates the number of passes in
which it divides the interval from the starting to the final
profile by the entered height reduction of the rolling process at
the thinnest point of the finished profile, which, however, must
be within the pre-calculated width of the starting profile. The

program finds the minimum height (H_P) of the final profile in the range L1 - L2, on the left-hand and right-hand sides of the starting profile:

$$H_P = \min_{i=L1}^{L2} \; [Y2(i) - Y1(i)] \tag{5.6}$$

The number of passes P is then calculated with the equation:

$$(1 - D) = \sqrt[P]{\frac{H_P}{H0}} \tag{5.7}$$

The same process also takes place at all further points (coordinate pairs) of the profile, whereby the reduction in height is calculated as:

$$D_L(i) = 1 - \exp\left[\frac{\ln\,(Y2(i) - Y1(i)) - \ln\,(H0)}{P}\right] \tag{5.8}$$

and drawn out over the width of the profile by the program. The designer assesses the deformation graph and can correct the entered reduction in height. The height H_M and the coordinates of the upper P1 and lower P2 profile of each pass are then computed:

$$H_M(\ell+1,i) = H_M(\ell,i) \cdot (1 - D_L(i)) \tag{5.9}$$

$$P1(\ell+1,i) = \frac{H_M(\ell+1) \cdot (Y1(i)-F1)+Y2(i) \cdot F1-Y1(i) \cdot F2}{Y2(i)-F2-Y1(i)+F1} \tag{5.10}$$

$$P2(\ell+1,i) = H_M(\ell+1) + P1(\ell+1,i) \tag{5.11}$$

for: $i = 1 \ldots n$
         $\ell = 1 \ldots P$
         F1 and F2 (see Figure 5.7)

The lateral spread for each pass is calculated with a procedure similar to the one described in the previous paragraph, but with the equations:

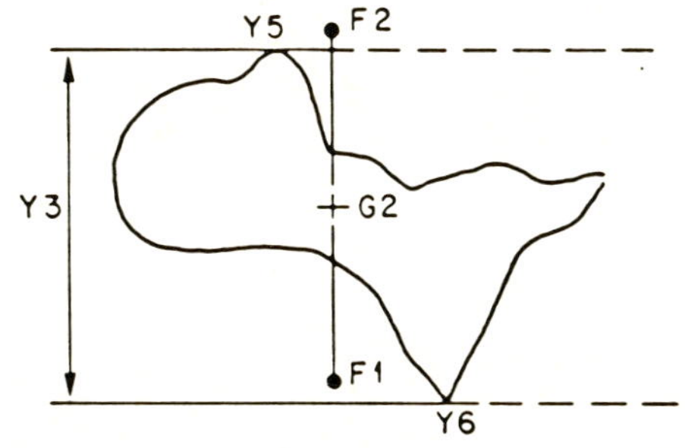

$$Y5 = \overset{N}{\underset{I=1}{\text{MAX}}} \left[Y1(I) \right] \qquad\qquad F1 = G2 - HO/2$$

$$Y6 = \overset{N}{\underset{I=1}{\text{MIN}}} \left[Y2(I) \right] \qquad\qquad F2 = G2 + HO/2$$

$$Y3 = Y5 - Y6 \qquad\qquad S = Y5 - F2$$

$$H2 = \frac{H9}{100} + 1 \qquad\qquad R = Y6 - F1$$

$$HO = H2 * Y3$$

Y5 < F2 AND Y6 > F1	Y5 > F2	Y6 < F1	SYMMETRY
F1 = F1 F2 = F2	F1 = F1 − S F2 = F2 + S	F1 = F1 − R F2 = F2 + R	YES
F1 = F1 F2 = F2	F1 = Y6 F2 = F2 + S	F1 = F1 − R F2 = Y5	NO

Figure 5.7: Determination of the starting profile, decision table.

$$A2 = \sum_{i=1}^{j-1} \frac{1}{2} \left[P2(\ell,1) - P1(\ell,1) + P2(\ell,i+1) - P1(\ell,i+1) \right] \cdot$$

$$\cdot \left[X(i+1) - X(i) \right] \qquad\qquad (5.12)$$

$$A1 = \sum_{i=j}^{k-1} \frac{1}{2} [P2(\ell-1,i) - P1(\ell-1,i) - (P2(\ell,1) - P1(\ell,1)) +$$

$$+ (P2(\ell-1,i+1) - P1(\ell-1,i+1) - (P2(\ell,i+1) - \qquad (5.13)$$

$$- P1(\ell,i+1))] \cdot [X(i+1) - X(i)]$$

where: $i = 1 \ldots k-1$ for the left-hand side
         $i = k \ldots n-1$ for the right-hand side
and: $\ell = 1 \ldots P$.

The coordinates of the intermediate passes are transferred to memory after the program has once again calculated the lateral spread of the individual passes. At this point, the program performs the self-correction mentioned. As the width-to-thickness ratio changes with each pass, the lateral spread also assumes other values. The amount of lateral spread is reduced as the number of passes increases, i.e. the values of lateral spread are not additive.

Figure 5.8 shows this result for a thin turbine blade which is rolled in 4 passes. Even at this juncture, the designer can reconsider his decision and return to the starting point of the program, as the time-consuming activities (planimetry, plotting the deformation graph and calculations) are performed by the computer. Finally, a technical drawing of all passes is prepared by a high-resolution precision plotter which is connected directly to the computer.

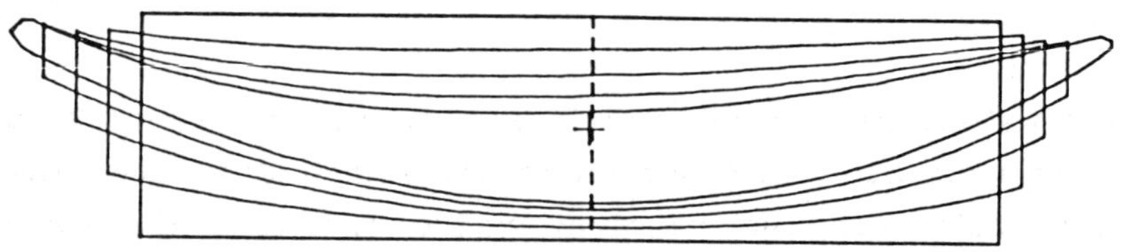

Figure 5.8 a):   Roll pass design of a thin turbine blade made of Ti6Al4V.

306

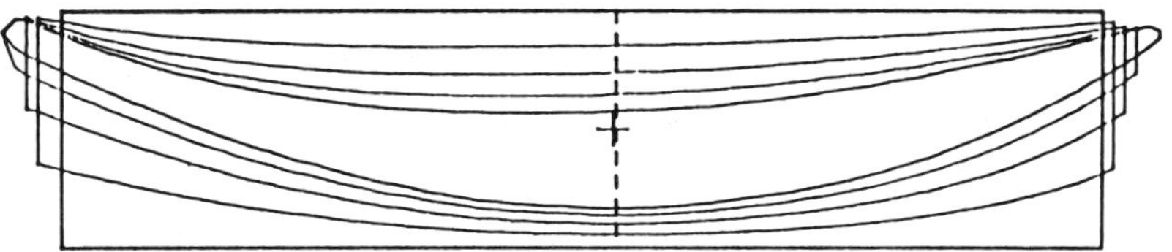

<u>Figure 5.8 b)</u>: Roll pass design of a thin turbine blade made of
                     steel.

b) <u>One-by-one design method</u>

Each pass can be designed according to one of the methods described in the following pages. These various methods were implemented in order to give more freedom to the designer to modify
and to try several solutions for the roll pass design. In each
method, the previous geometry is displayed and the designer
enters the necessary data in order to calculate the pass geometry.

<u>Variable percentage addition</u>

The y coordinates are given by entering an addition to the previous y coordinates expressed as a percentage decrease. This is
accomplished by simply positioning the cursor (a cross) on one
of the squares drawn above the profiles, Figure 5.9, and by
entering the addition required. If the calculated height is not
correct, it can be immediately changed and a new value can be
reentered. The smoothing and the editing of the profile is then
done as in the method above.

<u>Constant percentage addition</u>

When a constant percentage addition over the whole length of the
profile is possible, this method is the simplest to use. The program asks only for the required value and then the upper and
lower profiles for the next pass are automatically calculated

and drawn.

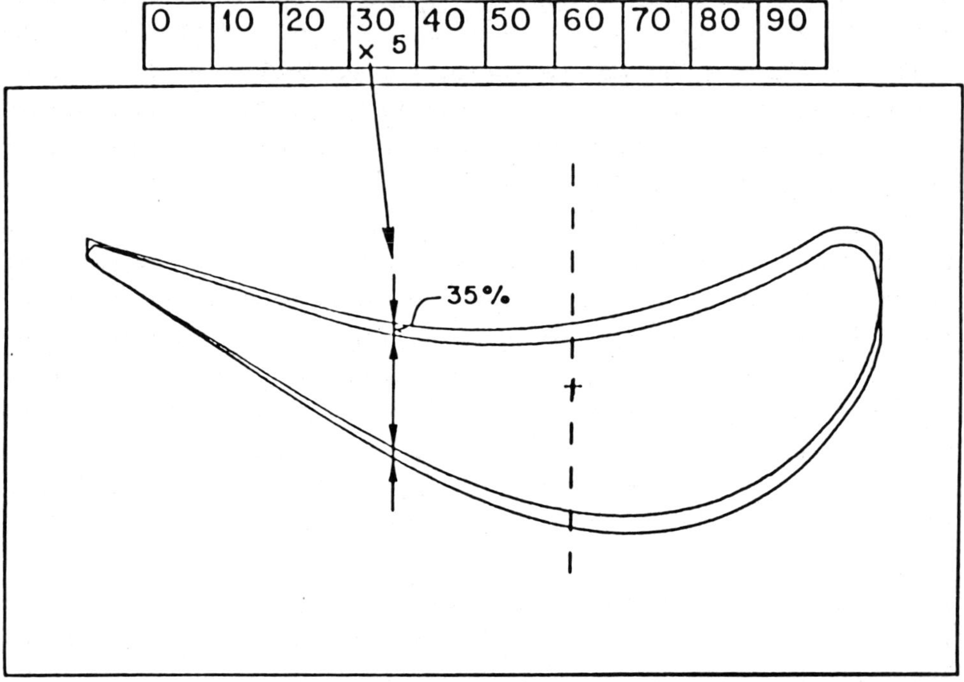

Figure 5.9:   Interactive generation of a roll pass.

5.4 Results

The result in Figure 5.8 (a) shows the roll pass design for a
slightly asymmetrical turbine blade profile made from the titanium alloy Ti6Al4V. As a comparison, Figure 5.8 (b) shows the
roll pass design for the same profile of steel. The differences
demonstrate the versatility of the program. The spread values
for steel are lower than for titanium; the number of passes is
less than for titanium.

The geometry of the sides of the sequences of passes and profiles
shown in Figure 5.8, does not yet correspond to the design necessary in the roll pass. This side geometry must subsequently be
introduced by the designer according to factors depending on
rolling techniques. This process requires little time, as most
CAD systems offer software for graphic operations. By suitable
entry of parameters, the modification of the side geometries can
also be performed with the aid of a variant program.

A complete roll pass design is shown in Figure 5.10 to show the
results that can be attained by a designer with the help of the
described program including side modifications.

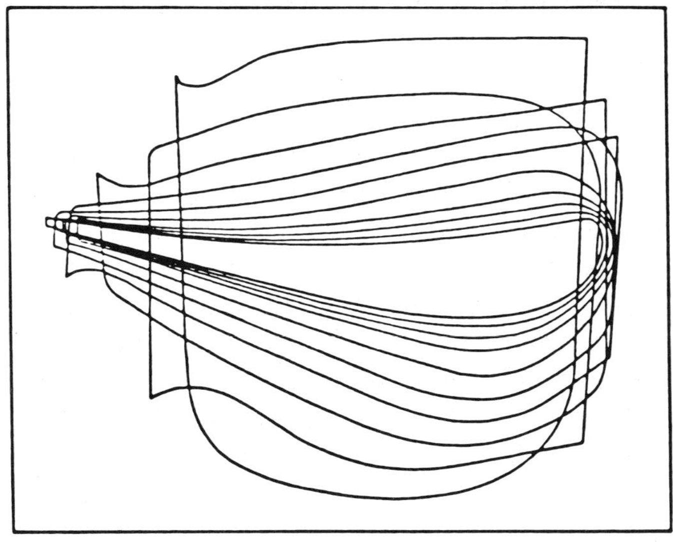

Figure 5.10: Example of roll pass schedule.

309

5.5 CAD/CAM in Steel Profile Production

The process modelling of rolling described previously has been
implemented in industry for the production of complex steel profiles. An industrial application requires that all the steps of
commissioning, technical drawing, roll and draw pass schedule,
die manufacturing, etc. be integrated in CAD/CAM technology. In
this chapter such a successful application is described [5.15].

5.5.1 CAD of Shape Rolls

For loading the geometry of the profile, the geometrical data
supplied by the customer are read in (in the form of punched
tape, magnetic tape etc.) or, if they are in the form of a drawing, they are entered into the computer interactively, where the
geometrical elements line, circle, arc, ellipse, hyperbola and
spline (for turbine blade profile) are required.

A dimensioned production drawing to a magnified scale is sub-
mitted to the customer for confirmation and approval, whereby
any incorrect dimensions specified in the customer's drawing are
immediately identified on the basis of the dimensions generated
by the computer. The same drawing to a scale of 1:1 is entered
into the profile book (Figure 5.11) for purposes of internal
documentation.

For the manufacturing of the profile, the designer specifies the
number of operations and the tools required. The process of
"roll and draw pass design" determines the production steps
from the raw material to the finished product, whereby economic
aspects and boundary conditions associated with the technology
of metal forming (see previous paragraphs) influence the final
decision to a very high degree, as in most cases there are several possibilities for manufacturing a profile by the combination
of hot rolling, cold rolling and drawing. Interactive work at
the graphic display workplace is supported by input menus. Figure
5.12 shows a menu for the rolling of profiles and for variants
of several shapes.

Specialized program modules and variant programs which contain

company-internal empirical information and data on materials
support the designer, for instance for calculating drafts and
for taking into account the law of constant volume for different-
ly-formed parts of the profile.

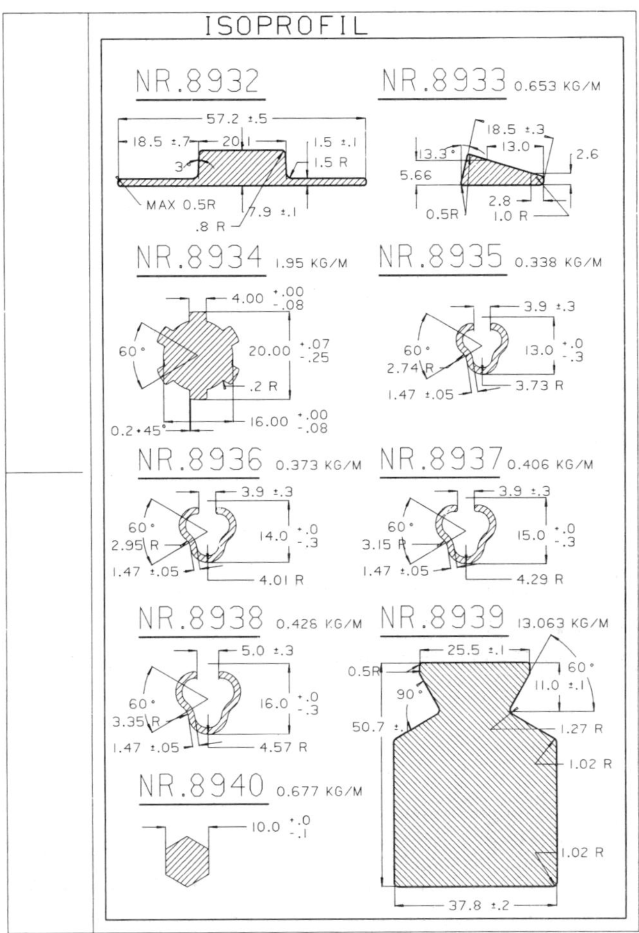

Figure 5.11: Excerpt from the Isoprofil GmbH profile book for
company internal documentation.

5.5.2 CAM of Shape Rolls at the CAD Workstation

Manufacturing drawings and additional information, such as parts
lists, are the most important information media between the de-

sign office and the production department. The drawings and parts list prepared by the design office form the basis for all further programming activities in production planning.

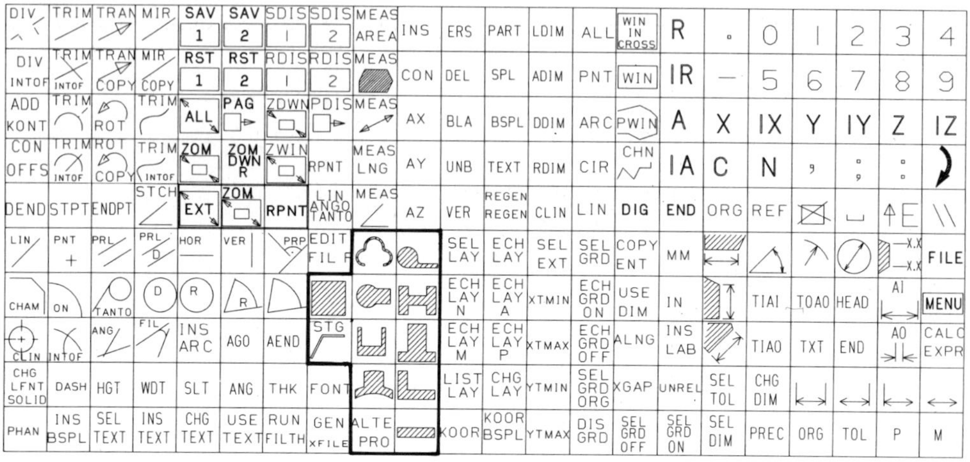

Figure 5.12: Command menu for producing layouts and designs of profiles.

The aim of integrating CAD with CAM is that the data that are compiled in the design office are directly available to production planning in the required form. Thus, for instance, the representation of a drawing die or of a shape roll produced by the designer at the computer can be used directly for NC programming. As the work for geometrical description is eliminated, the usual clarification processes, as well as programming and transcription errors at the interface of the two processing phases, will also disappear.

The results are shorter processing times for the product, enhanced quality (with reduced scrap rate) and higher efficiency due to the better form of "data exchange".

From a designer's viewpoint, a transfer of data actually no longer takes place, as he converts his design into punched tape at the same work station. Within the computer, however, the graphical data are transferred to an NC programming system and translated into the corresponding machine code.

In addition to the usual software for graphic operations, the CAD system also has a program module for NC programming. By means of graphic-interactive dialogue, this is used to generate tool paths with the aid of the layout shown on the graphic display, Figure 5.13.

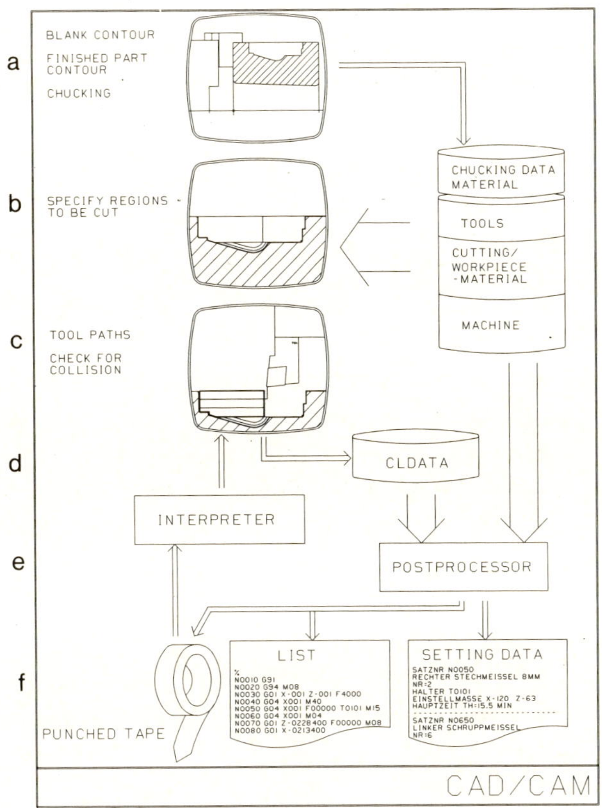

Figure 5.13: Internal data flow through computer during interactive NC programming in the graphic mode for the turning operation.

From a file, a suitable chuck, Figure 5.13 (a) is selected and placed in position. The required profile is obtained by a copying command from the pass design drawing (see Chapter 5.3). The final roll geometry is completed by interactive graphic work.

After specifying the sequence of operations (grooving, roughing, finishing), the material to be cut away is divided up into regions interactively (Figure 5.13 (b)).

For the grooving process, the designer selects an appropriate tool from a tool file, identifies the region to be machined and digitizes the starting point of the tool path (Figure 5.13 (c)).

The computer automatically determines the segmentation and all technological data, such as feed rate, cutting speed, cutting force, power and torque, taking into account the combination of workpiece material and tool material, the tool geometry angles and radius and the method of chucking.

After all required tool paths have been generated and if the optical collision check was negative, the computer generates the CLDATA file (Figure 5.13 (d)). The postprocessor checks the technological data for their compatibility with the stored machine characteristics, automatically determines the approach paths for changing the tool, generates the machine code and issues this in the form of a list and as a punched tape, Figure 5.13 (f). In addition, a list of the required tools is printed, giving settings and the actual machining time for each tool. The punched tape can be reread into the computer to check for punching errors. An interpretation program displays all tool paths on the graphic screen to allow an optical check for collision of the workpiece and the tool in machining or positioning mode.

References Chapter 5

[5.1]   Kennedy K.F., Altan T., Lahoti G.D., Computer-Aided Analysis of Metal Flow Stresses and Roll Pass Design in Rod Rolling, Iron and Steel Engineer (June 1983) 50.

[5.2]   Li G.J., Kobayashi S., Spread Analysis in Rolling by the Rigid-Plastic Finite Element Method, Proceedings of Conf. Numerical Methods in Industrial Forming Processes, Pineridge Press (1982).

[5.3]   Wusatowski Z., A Study of Draught, Spread and Elongation, Iron and Steel (1955) 49.

[5.4]   Helmi A., Alexander J.M., Geometric Factors Affecting Spread in Hot Flat Rolling of Steel, J. Iron Steel Inst., 206 (1968) 1110.

[5.5]   Sparling L.G.M., Formula for Spread in Hot Flat Rolling, Proc. IME, 175 (1961) 604.

[5.6]   Shinokura T., Takai K., A New Method for Calculating Spread in Rod Rolling, J. Applied Metalworking, 2/2 (January 1982).

[5.7]   Lahoti G.D., Kobayashi S., On Hill's General Method of Analysis for Metal-Working Processes, Int. J. Mech. Sci. 16 (1974) 521.

[5.8]   Oh S.I., Kobayashi S., An Approximate Method for a Three-Dimensional Analysis of Rolling, Int. J. Mech. Sci. 17 (1975) 293.

[5.9]   Wilson W., Kennedy K.F., An Analytical Model for Side Spread in Rolling of Flat Products, Proc. 6th NAMRC, (1978) 119.

[5.10] Riegger H., Lateral Spread in Rolling of Ti-6Al-4V, Proc. 8th NAMRC Conf., Rolla, Missouri (1980) 187.

[5.11]   Kozono H., CAD/CAM System for Roll Pass Design, 19th MTDR
         Conference (1978) 163.

[5.12]   Lahoti G.D., Akgerman N., Altan T., Computer-Aided Ana-
         lysis and Design of the Shape Rolling Process for Pro-
         ducing Turbine Engine Airfoils, NASA Report CR-159445
         (1978).

[5.13]   Akgerman N. et al., Computer-Aided Roll Pass Design in
         Rolling of Airfoil Shapes, J. of Applied Metalworking,
         1/3 (July 1980) 30.

[5.14]   Riegger H., Boër C.R., Roll Pass Design by Interactive
         Graphic Computing, J. Appl. Metalworking, Vol. 3, No. 4
         (January 1985) 432-438.

[5.15]   Boër C.R., Riegger H., CAD/CAM for Rolling Complex Shape
         Profiles, Annals of the CIRP, Vol. 34/1/1985.

[5.16]   Boër C.R., Riegger H., Pattern Recognition in Steel Pro-
         file Production, to be published on the Int. J. Machine
         Tool Design and Research.

6 Modelling of Drawing

<u>List of Symbols</u>

(other symbols are defined in the text)

m	constant shear factor at the die-workpiece interface
M	die surface area
n	geometrical parameter ($2 \leq n \leq \infty$)
s_{ij}	deviatoric stress tensors
Si	inlet cross-sectional area
So	outlet cross-sectional area
v_1, v_2, v_3 or $\dot{u}_1, \dot{u}_2, \dot{u}_3$	velocity field
V	volume under deformation
Δv_t	tangential velocity difference
α_s, α_c	die angles
ε	reduction of area
$\dot{\varepsilon}_{ik}$	strain rate components
$\bar{\sigma}$	effective stress
σ_f	flow stress of the material
σ_z	axial drawing stress

6.1 Introduction

In an attempt to reduce costs, increasing attention is being paid
to the drawing of section rods from round bar in a minimum possible number of steps. However, a direct theoretical approach to
this problem has been only partially attempted in the past [6.1]
due to the difficulty in applying simple techniques to represent
the deformation pattern even in the relatively simple section
analyzed in this chapter. The analysis of the deformation process
from a round bar to a square rod with different corner radii was
chosen as a first step towards the study of more complicated
shape drawing [6.2, 6.3].

6.2 Upper-Bound Analysis of Round-to-Square Drawing

An upper-bound solution for a rigid perfectly plastic solid is
given. A kinematically admissible velocity field (in the deforming region) is constructed, allowing the upper-bound power to be
calculated. The drawing stress is predicted as a function of
area, friction and form of exit section. The frictional stress is
assumed constant between the surface of die and the workpiece.

6.2.1 The Upper-Bound Solution

The theory of the upper-bound technique [6.4, 6.5] is expressed
by:

$$J^* = \dot{W}_i + \dot{W}_f + \dot{W}_s \qquad\qquad (6.1)$$

where:

J^* is the actual external supplied power,

$\dot{W}_i$ is the internal power of deformation,

$\dot{W}_f$ is the power absorbed by frictional losses,

$\dot{W}_s$ is the power dissipated in shearing the material
            at both the inlet and exit interface between de-
            formed and rigid material.

The upper-bound solution is obtained (see Chapter 2.5) when, assuming a velocity field, the actual kinematically admissible velocity field is the one which minimizes equation (6.1).

The internal power of deformation is expressed by:

$$\dot{W}_i = \frac{2}{\sqrt{3}} \sigma_f \int_V \sqrt{\frac{1}{2} \dot{\varepsilon}_{ik} \dot{\varepsilon}_{ik}} \; dV \tag{6.2}$$

where:

σ_f is the flow stress of the material,

$\dot{\varepsilon}_{ik}$ are the strain rate components, and

V is the volume that is actually under deformation.

The power loss in friction at the die-workpiece interface is:

$$\dot{W}_f = m \frac{\sigma_f}{\sqrt{3}} \int_M [\Delta v_t] \; dS \tag{6.3}$$

where:

m is the constant shear factor at the die-workpiece interface,

M is the die surface area, and

$[\Delta v_t]$ is the velocity of the material at the die surface

dS is a surface element.

The shear losses are given by:

$$\dot{W}_s = \frac{\sigma_f}{\sqrt{3}} \{ \int_{Si} [\Delta v_t] \; dS + \int_{So} [\Delta v_t] \; dS \} \tag{6.4}$$

where:

Si   is the inlet cross-sectional area,

So   is the outlet cross-sectional area, and

$[\Delta v_t]$ is the tangential velocity on the inlet or
        outlet surface.

6.2.2   The Geometrical Model

The geometry of the die and the undeformed and deformed section
rod is given in Figure 6.1. It is assumed that the volume under
deformation, V, is included between the surface Si at z = o and
So at z = h and the die surface M.

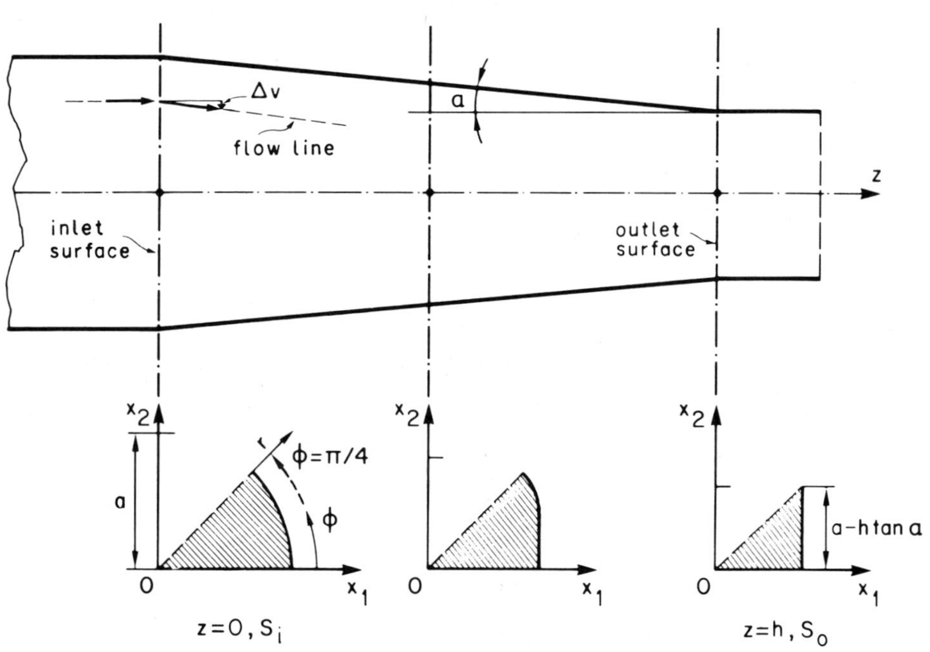

Figure 6.1:   The geometry of the die and the undeformed and de-
              formed section rod.

The basic equation defining the geometry of the billet at some
section Z is given by:

$$\frac{x_1}{ka}^n + \frac{x_2}{ka}^n = 1 \tag{6.5}$$

where:

  a   is the radius of the initial billet,

  k   is a parameter between 0 and 1, and

  n   is a parameter ranging from 2 (circle) to ∞ (square).

Assuming a billet of initial radius equal to 1 and $0 \leq r \leq 1$, then equation (6.5) simplifies as:

$$\frac{x_1}{r}^n + \frac{x_2}{r}^n = 1 \tag{6.6}$$

Introducing the coordinates of Figure 6.1, it is possible to express equation (6.6) in Cartesian or spherical coordinates. It follows that:

$$x_1 = f(z) \; r \; g(z, f, r) \tag{6.7}$$

$$x_2 = f(z) \; r \; g(z, f, r) \; t(\phi) \tag{6.8}$$

$$x_3 = z \tag{6.9}$$

with:

$$f(z) = a - z \tan \alpha_c \tag{6.10}$$

$$g(z, r, \phi) = [1 + t(\phi)^{n\,(z,\,r)}]^{-1/n\,(z,\,r)} \tag{6.11}$$

$$t(\phi) = \tan \phi \tag{6.12}$$

Due to symmetry, the domain of variation is:

$$o \leq r \leq 1, \quad o \leq \phi \leq \frac{\pi}{4}, \quad \text{and} \quad o \leq z \leq h \tag{6.13}$$

6.2.3 The Velocity Fields

The Jacobian of equations (6.7), (6.8) and (6.9) is given by:

$$A = \begin{bmatrix} fb & frg_\phi & r(fg)_z \\ fbt & fr(gt)_\phi & r(fg)_z\,t \\ o & o & 1 \end{bmatrix} \tag{6.14}$$

where: $b = (rg)_r$

and the subscript indicates partial differentiation (i.e. $g_\phi = \frac{\partial g}{\partial \phi}$).

The determinant of A is given by:

$$D = \det(A) = f^2\,rgb\,\frac{1}{\cos^2\phi} \tag{6.15}$$

The velocity components in Cartesian coordinates are:

$$v_1 = \frac{D^o}{D}\,(frg)_z$$

$$v_2 = \frac{D^o}{D}\,(frgt)_z \tag{6.16}$$

$$v_3 = \frac{D^o}{D}$$

where:

$$D^o = D(r,\phi,o) \tag{6.17}$$

For:

$$n(r,o) = 2 \tag{6.18}$$

it follows that:

$$D^o = a^2 r \tag{6.19}$$

Hence, (6.16) can be written as:

$$v_1 = \frac{a^2}{f^2} \frac{\cos^2\phi}{gb} r \, (fg)_z$$

$$v_2 = v_1 t \tag{6.20}$$

$$v_3 = \frac{a^2}{f^2} \frac{\cos^2\phi}{gb}$$

Calculating the divergence of (6.20):

$$\operatorname{div} v_k = \sum_{k=1}^{3} \frac{\partial v_k}{\partial x_k} = \operatorname{tr} \frac{\partial v}{\partial x} = o \tag{6.21}$$

i.e. the velocity field given by (6.20) is divergence free.

The velocity field at the entrance Si is obtained from (6.20) for z = o and on the exit for z = h.

The flow lines are kinematically admissible from the entrance throughout the deformation region up to the exit, but not at the exit. The form of velocity discontinuity surface at the exit, which gives a kinematically admissible velocity field, is given in [6.3] using a different die geometry.

6.2.4 <u>Reduction of Area</u>

The area of a section S at z is given by (see Figure 6.1):

$$A(z) = 8 \int_{o}^{\pi/4} d\phi \int_{o}^{1} dr \, D(r, \phi, z) \tag{6.22}$$

where $D(r, \phi, z)$ is the determinant of the Jacobian giving the change of coordinate system $x_1 x_2$ to r, ϕ (see section 6.2.3).

The solution of equation (6.22) yields:

$$A(z) = 4f^2 \, F(N) \tag{6.23}$$

where:

$$f = f(z) = a - z \tan \alpha \tag{6.10 rep}$$

$$F(N) = \int_0^1 dt \, (1 + t^N)^{-2/N} \tag{6.24}$$

and:

$$N = N(z) = n(1, z) \tag{6.25}$$

Equation (6.24) can be evaluated analytically for a few special cases, but, in general, the evaluation has to be done numerically.

i.e. for $N = 0$, $F(N) = 0$ and for
$N = \infty$, $F(N) = 1$

For $z = 0$ and assuming $n(1, 0) = 2$, i.e. round rod, equation (6.23) yields:

$$A_i = \Pi a^2 \tag{6.26}$$

and for $z = h$ and $n = n(1, h)$:

$$A_f = 4 \, (a - h \tan \alpha)^2 \, F(N) \tag{6.27}$$

Setting:

$$\varepsilon = \frac{A_i - A_f}{A_i} \tag{6.28}$$

yields, from (6.26) and (6.27):

$$\varepsilon = 1 - \frac{4}{\pi} \left(1 - \frac{h}{a} \tan \alpha\right)^2 F(N) \tag{6.29}$$

or equivalently:

$$\frac{h}{a} = \left(1 - \sqrt{\frac{1}{F(N)} \frac{\pi}{4}(1 - \varepsilon)}\right) \cotan \alpha \qquad (6.30)$$

6.2.5 Flow Lines

The parameter n is a function of the radius r and the axis z. The selection of n-function is critical to obtain a good upper-bound solution. To compare the results generated with a selected n-function and the experimental results, a computer program was written to visualize the flow lines within the deforming region.

From equation (6.6) and Figure 6.2 it is possible to obtain:

$$x_1 = \frac{r}{[1 + (\tan \phi)^n]^{1/n}} \qquad (6.31)$$

$$x_2 = x_1 \tan \phi$$

for:

$$\phi = \frac{\pi}{4}, \quad x_1 = x_2 = \frac{r}{2^{1/n}} \qquad (6.32)$$

and if:

$$n = 2, \quad x_1 = x_2 = \frac{r}{\sqrt{2}} \qquad (6.33)$$

if:

$$n = \infty, \quad x_1 = x_2 = r \qquad (6.34)$$

At the same intermediate length z, a point on the line O'O" (see Figure 6.2 c) has the coordinates ($\phi = \frac{\pi}{2}$):

$$x_1 = o$$

$$x_2 = 1 - z \tan \alpha_c \qquad (6.35)$$

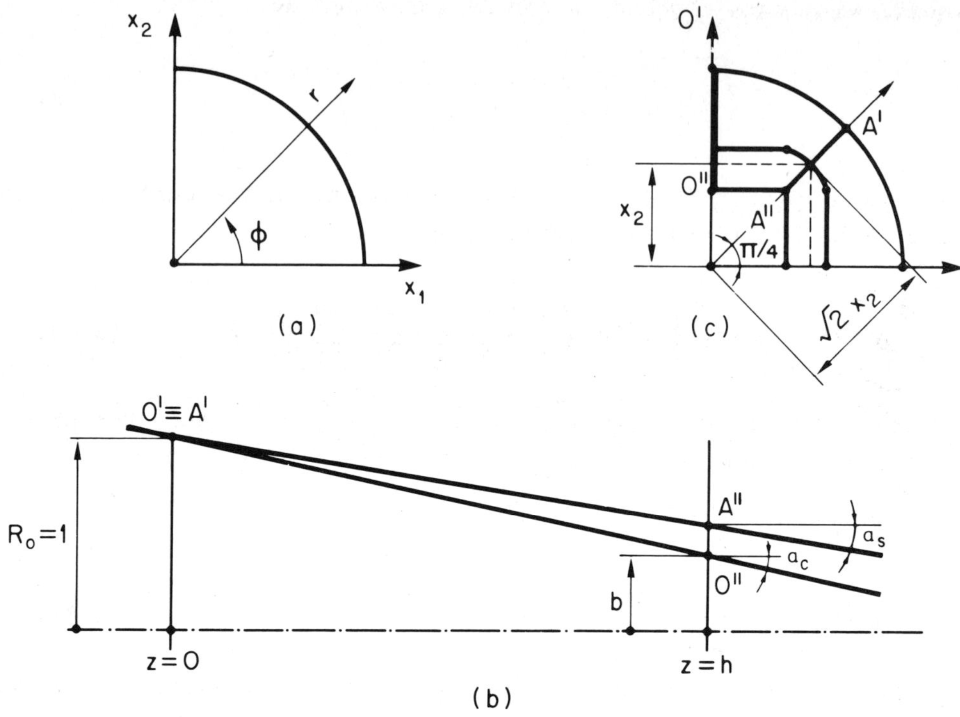

Figure 6.2: Geometry and variables for equation (6.40).

and from equation (6.32) with $r = 1 - z \tan \alpha_c$:

$$x_2 = \frac{1 - z \tan \alpha_c}{2^{1/n}} \tag{6.36}$$

But it is also:

$$x_2 = \frac{1 - z \tan \alpha_s}{\sqrt{2}} \tag{6.37}$$

Therefore from (6.36) and (6.37):

$$\frac{1 - z \tan \alpha_c}{2^{1/n}} = \frac{1 - z \tan \alpha_s}{\sqrt{2}} \tag{6.38}$$

is obtained.

326

Next, with some algebra, n can be expressed as:

$$n = \frac{1}{2} + \frac{1}{\log 2} \log \left(\frac{1 - z \tan \alpha_c}{1 - z \tan \alpha_s} \right)^{-1} \tag{6.39}$$

The relationship (6.39) is valid for points on the surface of the die. Equation (6.39) is rewritten as:

$$n = \frac{1}{2} + \frac{1}{\log 2} \log \left(\frac{1 - \phi(r, z) \tan \alpha_c}{1 - \phi(r, z) \tan \alpha_s} \right)^{-1} \tag{6.40}$$

where $\phi(r, z)$ is a function which closely represents the physical deformation of the rod.

It was found that:

$$\phi(r, z) = r^{\beta} z \tag{6.41}$$

gives the best results. The exponent β is an optimization factor which minimizes the upper-bound solution for different input parameters.

The angle α_s can be obtained from Figure 6.2 and has the expression:

$$\alpha_s = \arctan \; \frac{1}{h} - 2^{\frac{n_f - 2}{2 n_f}} \left(\frac{1}{h} - \tan \alpha_c \right) \tag{6.42}$$

when n_f is the value of n for $r = 1$ and $z = h$.

Figure 6.3 is a plot of the flow lines described by equation (6.40).

6.2.6 <u>Optimization of the Theoretical Flow Field</u>

The plot of the relative drawing stress versus the reduction of area for different n-function is given in Figure 6.4. The lines for function (6.40) are also represented in more detail in Figure 6.5. It can be shown that for different angles, α_c, there

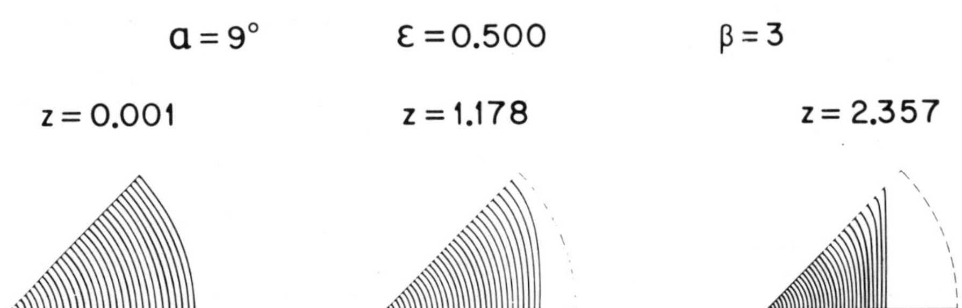

Figure 6.3: Flow lines for drawing from round-to-square ($n_f = 100$).

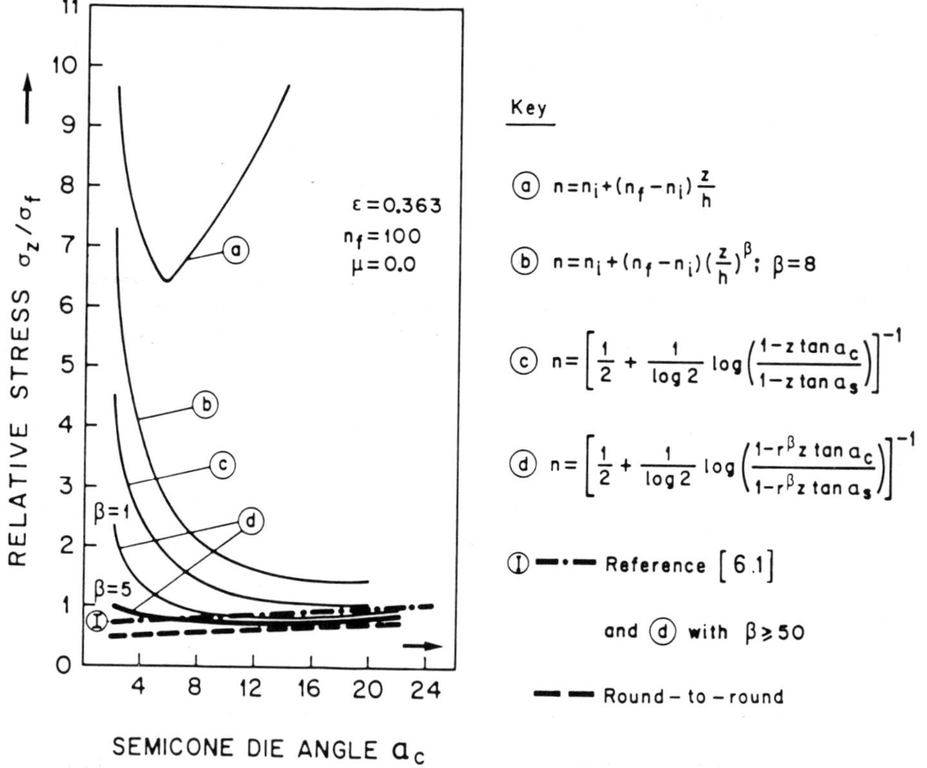

Figure 6.4: Relative drawing stress versus reduction of area for different n-functions.

is an optimum value of the parameter β.

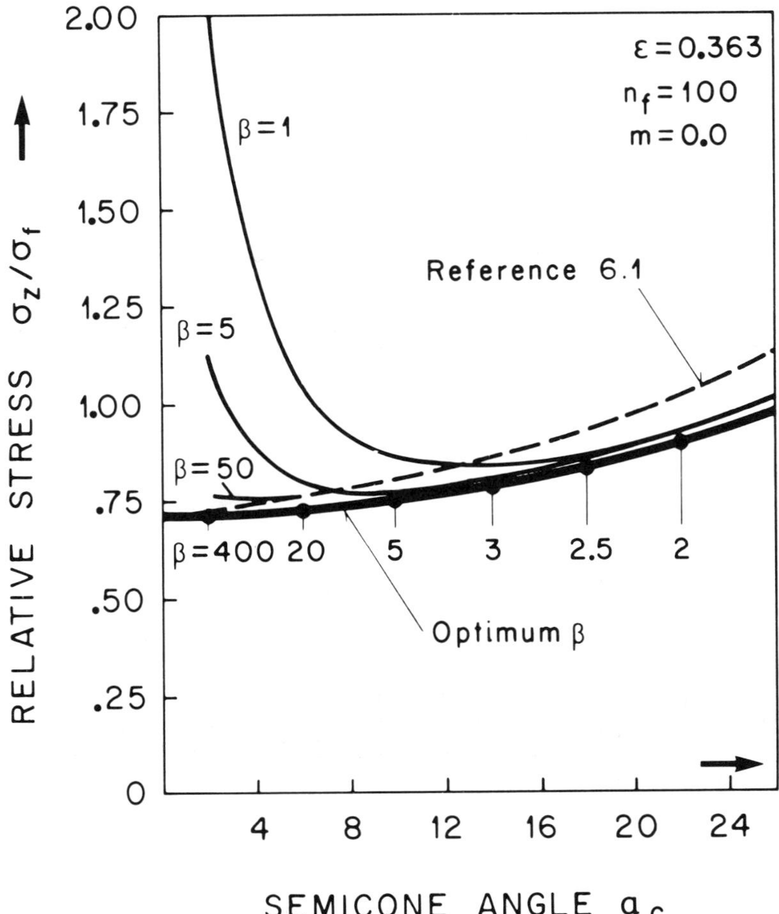

Figure 6.5:   Relative drawing stress caculated by upper-bound with equation (6.40).

The strain rate components are obtained as [6.4]:

$$\dot{\varepsilon}_{ik} = \frac{1}{2} \left(\frac{\partial v_i}{\partial x_k} + \frac{\partial v_k}{\partial x_i} \right) \tag{6.43}$$

The partial derivatives of the above equation are obtained with the aid of coordinate transformation as:

$$\frac{\partial v_i}{\partial x_k} = \sum_{j=1}^{3} \frac{\partial v_i}{\partial u_j}\frac{\partial u_j}{\partial x_k} = \sum_{j=1}^{3} \frac{\partial v_i}{\partial u_j} b_{j,k} \qquad\qquad (6.44)$$

where $b_{j,k}$ are the coefficients of the inverse of the Jacobian given in paragraph 6.2.3.

6.2.7 Analysis of Process Parameters

Knowing the velocity field, it is possible to calculate the three power terms of equation (6.1). A numerical integration has to be performed in order to obtain the solutions of equations (6.2), (6.3) and (6.4). The non-dimensional variable σ_z/σ_f or relative drawing stress is calculated.

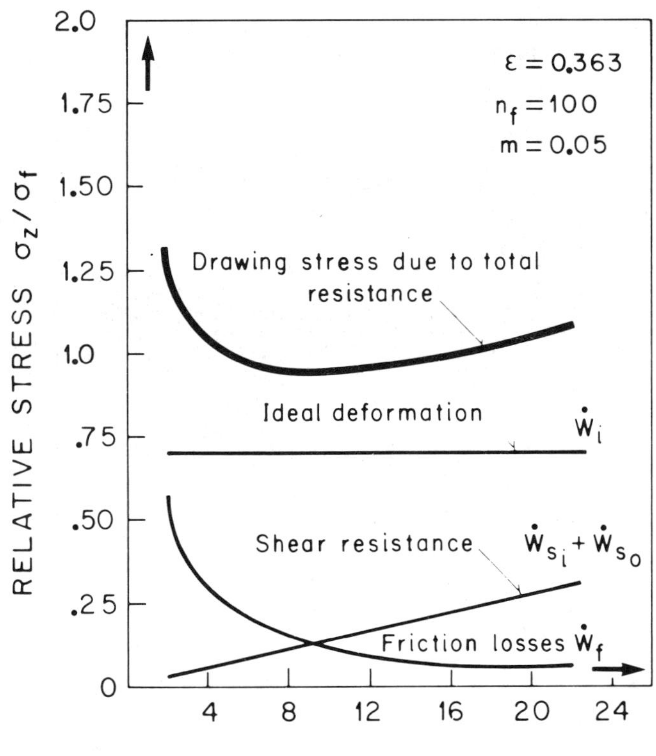

Figure 6.6: Partial contributions of the drawing stress to overcome resistance in drawing from round-to-square cross-section.

330

Figure 6.6 represents the partial contributions of the drawing stress to overcome resistance in drawing from a round-to-square cross-section. The total relative drawing stress exhibits a minimum when the friction losses and the shear resistance are equal. Therefore it is possible to have an optimum die angle for a different coefficient of friction, Figure 6.7, or a different reduction of area, Figure 6.8.

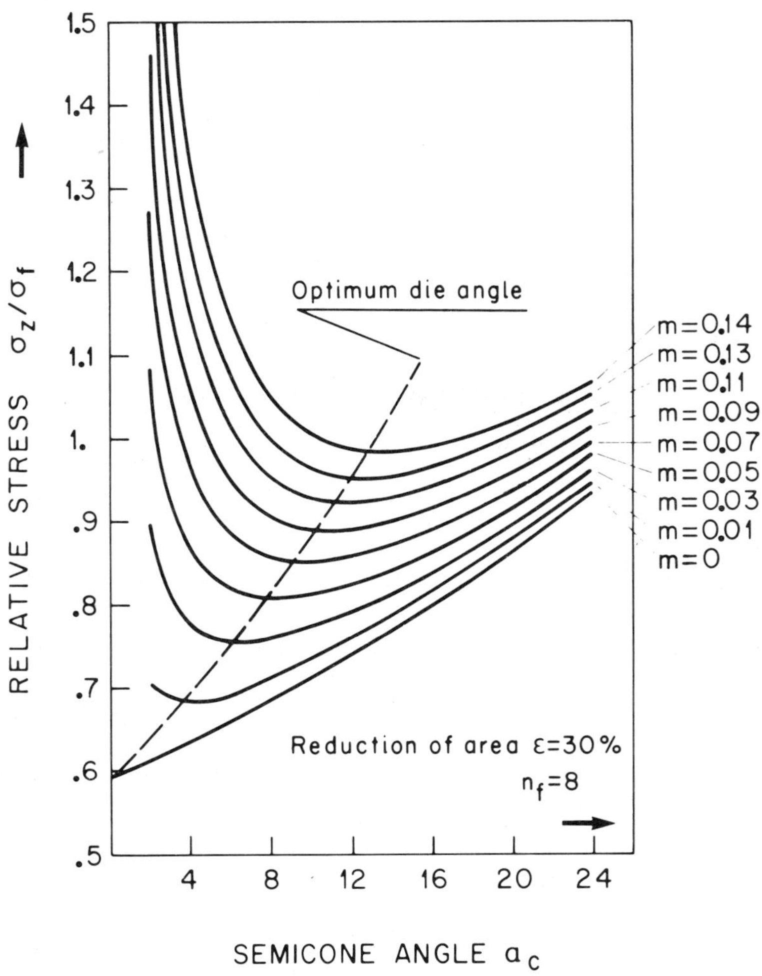

Figure 6.7: Optimum die angle for different coefficient of friction (round-to-square section).

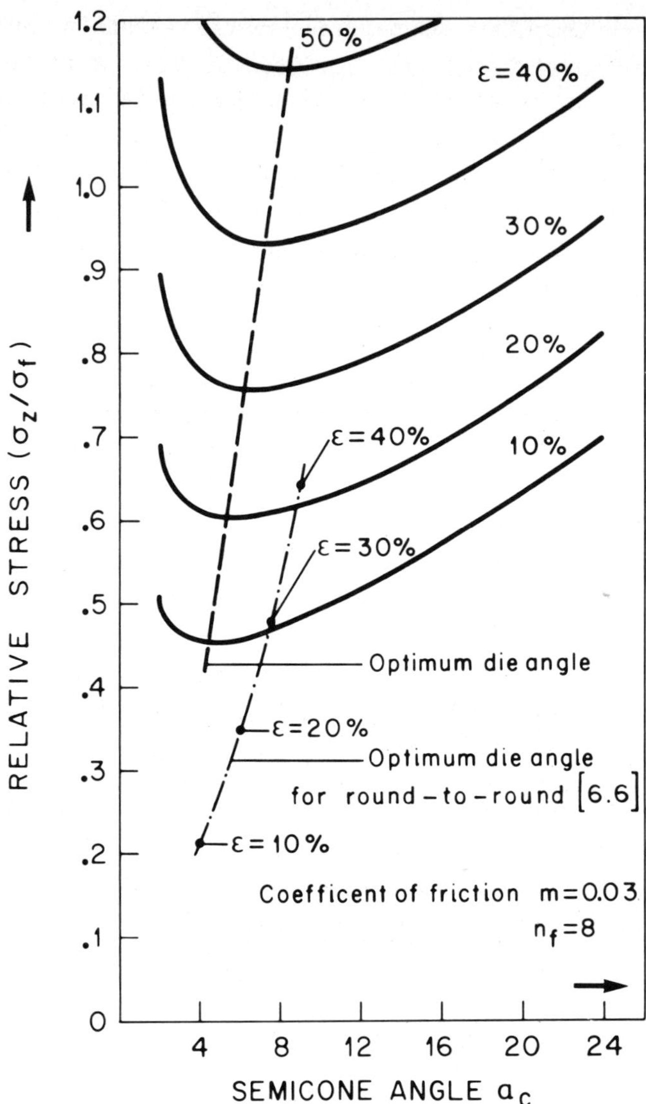

<u>Figure 6.8</u>: Optimum die angle for different reduction of area.

In Figure 6.8 the experimental results of drawing from round-to-round [6.6] dashed-dotted line are also presented for comparative purposes. The optimum angle shifts towards lower values for drawing non-round sections.

The influence of the form of the exit section is shown by plotting the drawing stress against the coefficient of form n_f as in Figure 6.9. The coefficient of form in the present analytical treatment is an indication of the degree of deformation from round-to-square. A radius of the corner of the die may also be related to the coefficient of form n_f. In Figure 6.10, the flow lines plotted by the computer are given for different coefficients of form. The radius R as a function of the side length l of the square is also given.

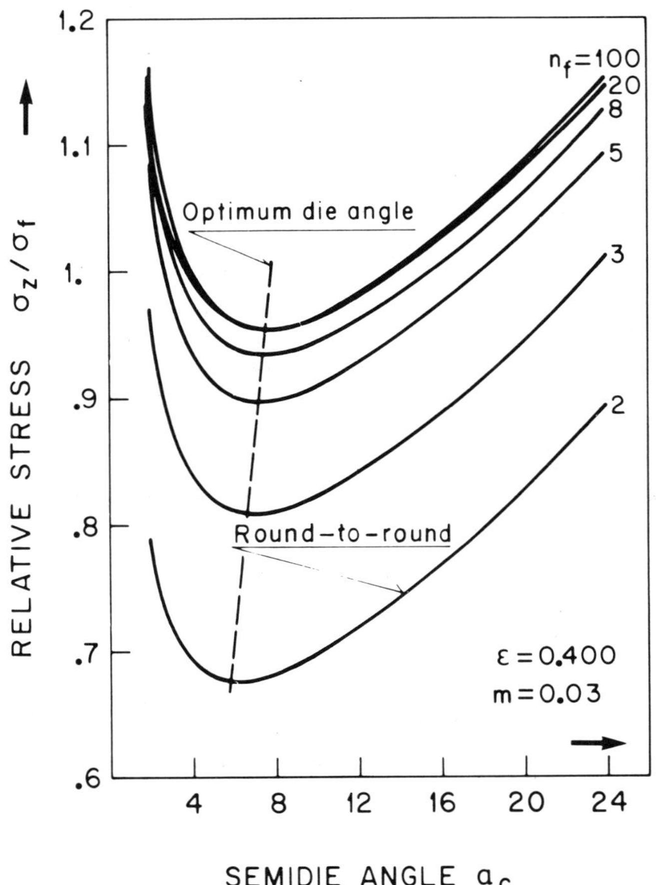

Figure 6.9: Optimum die angle for different form factor.

An important result appears from Figure 6.9. With the same reduction of area and coefficient of friction, different optimum

die angles are obtained for different coefficients of form.
Although this difference is not too great, it is, however, enough
to imply that for more complicated shapes, the optimum die angle
should be better defined than using the quasi-empirical methods
generally adopted in industry.

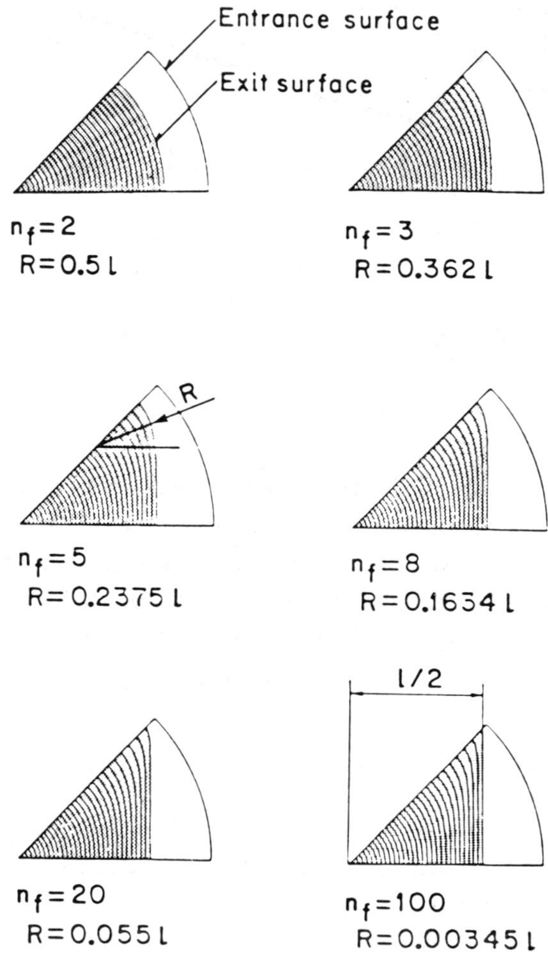

Figure 6.10: Flow lines, geometric variables for different coefficient of form n_f ($\varepsilon = 0.400$).

Comparison was also made with experiments. In Figure 6.11, some
of the results are presented. The die angle α_c is 9° (total 18°).

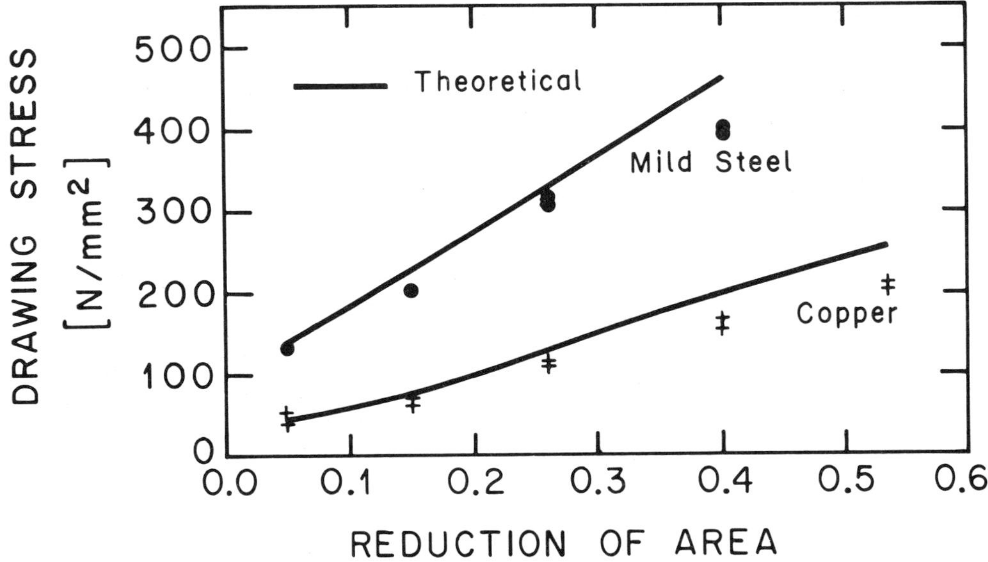

<u>Figure 6.11</u>:   Comparison experiment-analytical of drawing round-
to-square in steel and copper.

Different reductions of area were experimented with and the
drawing stress was measured. Two materials were tested (mild
steel and copper). In the figure, the experimental points are
shown, together with the theoretical calculated values. The
correlation is good and, as is expected, the theoretical values
are always higher than the experimental ones.

6.2.8   Conclusions

An upper-bound solution is presented for the drawing process from
a round-to-square cross-section rod with different corner radii.
The results are expressed in terms of relative drawing stress as
a function of the semicone angle of the die, reduction of area,
friction and form of exit section. With the aid of a computer,
different mathematical functions of the flow lines in the deforming region were examined. An optimizing factor was found which
gives the minimum value of the upper-bound solution.

The drawing process can also be modelled by other methods.

The FEM seems to be very promising, but in the case presented in this chapter, the deformation is three-dimensional and the computer time (and cost), at the present state of the art, is very high and still confined to research. One successful application of the FEM to the round-to-square drawing (and extrusion) is given in Chapter 6.3. In this case, the upper-bound theorem is used in the description of the Finite Element and an optimization technique is used to find the velocity field which gives the minimum power.

6.3 Finite Element Analysis of Bar Drawing

A growing interest in the modelling of metal forming processes in
recent years has brought the development of different analytical
and/or numerical techniques. Each technique has its own advantages and disadvantages and rarely are they compared directly to
test their performances. In this chapter two methods are compared
which have the same basic approach (upper-bound-theorem) but
fundamental differences in the solution methods.

The first requires a knowledge of a kinematically admissible
velocity field in the deforming region and then uses a direct
solution for the minimum power.

The second method uses the Finite Element approach. The deforming
region is subdivided into small elements and the velocities are
calculated at each node with the condition to minimize the power
required [6.7]. This is a slightly different-approach from other
FEM-programs described in this book.

The results obtained with the two methods are compared, together
with the advantages and disadvantages.

6.3.1 The Finite Element Approach

Equation (2.134) may be viewed as the variation of a functional
$\dot{w}$ representing the total power dissipated in a system:

$$\dot{w} = \int_V \sigma_{ij}\dot{\varepsilon}_{ij}dV - \int_{S_f} f_i v_i dS_f = \int_V \bar{\sigma}\dot{\bar{\varepsilon}}dV - \int_{S_f} f_i v_i dS_f \quad (6.45)$$

Provided incompressibility is imposed, the minimum of such a
functional provides the solution to the solid mechanics boundary
value problem.

Note: For this chapter, the contribution of W.D. Webster, Jr.,
      GMI Engineering & Management, Flint, Michigan, U.S.A., is
      gratefully acknowledged.

The first term in eq. (6.45) represents the power dissipated in internal deformation, and the second the power dissipated by the external forces applied. The search for the minimum of such a functional is represented exactly by equation (2.93), derived as the basis for the upper-bound method. Upon Finite Element discretization, the parameters with respect to which the over-all velocity field is minimized are the nodal velocities. The approach presented here minimizes the functional resulting from the discretization of the functional referred to above, subjected to the external constraint of incompressibility, and stated as:

$$\dot{\varepsilon}_{ii} = 0 \qquad (6.46)$$

6.3.2 Finite Element Formulation

A three-dimensional, subparametric, Finite Element was used [6.8] (Figure 6.12). The hexahedron-shaped element was described with twenty geometry nodes of which the eight corner nodes had 12 generalized displacements (3 for velocity, 9 for velocity derivatives), giving a total of 96 degrees of freedom.

The definition for strain rates is given by:

$$\dot{\varepsilon}_{ij} = 1/2 \left(\frac{\partial \dot{u}_i}{\partial x_j} + \frac{\partial \dot{u}_j}{\partial x_i}\right) \qquad (6.47)$$

where $\partial \dot{u}_i/\partial x_j$ are the derivatives of the velocities ($\dot{u}_i$) in Cartesian coordinates. The relationship of the derivatives of the velocities in Cartesian coordinates (x_i) to those in local coordinates (α_i) is given by:

$$\frac{\partial \dot{u}_i}{\partial x_j} = \sum_{m=1}^{3} J_{jm}^{-1} \frac{\partial \dot{u}_i}{\partial \alpha_m} \qquad (6.48)$$

where J_{jm}^{-1} is the inverse of the Jacobian matrix for the transformation from local coordinates to Cartesian coordinates and $\partial \dot{u}_i/\partial \alpha_m$ are the derivatives of the velocities in local coordinates.

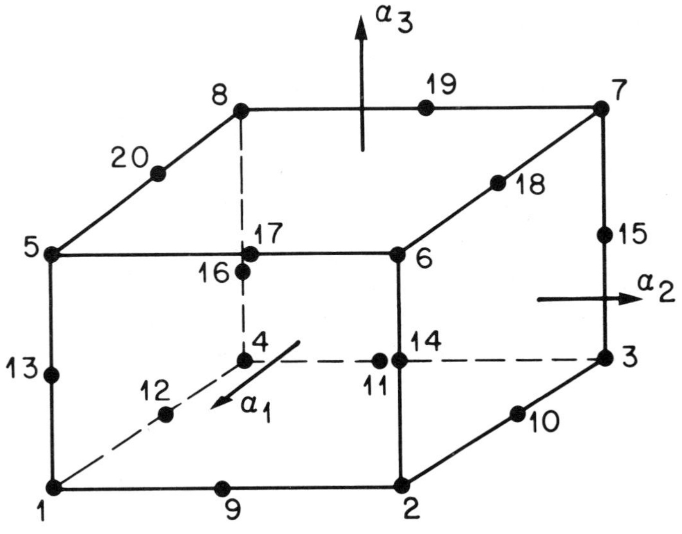

<u>Figure 6.12</u>: A three-dimensional subparametric element.

The velocity field in local coordinates is defined by:

$$\dot{u}_1 \;=\; \sum_{k=1}^{32} a_k f_k \qquad\qquad (6.49\ a)$$

$$\dot{u}_2 \;=\; \sum_{k=33}^{64} a_k f_k \qquad\qquad (6.49\ b)$$

and:

$$\dot{u}_3 \;=\; \sum_{k=65}^{96} a_k f_k \qquad\qquad (6.49\ c)$$

where a_k represents a vector of 96 constants and f_k represents a vector of 32 distinct polynomial terms expressed in local co-ordinates.

The derivatives of the velocity field are:

$$\frac{\partial \dot{u}_1}{\partial \alpha_j} \;=\; \sum_{k=1}^{32} a_k \frac{\partial f_k}{\partial \alpha_j} \qquad\qquad j = 1,\ 2,\ 3 \qquad\qquad (6.50\ a)$$

$$\frac{\partial \dot{u}_2}{\partial \alpha_j} = \sum_{k=33}^{64} a_k \frac{\partial f_k}{\partial \alpha_j} \qquad\qquad j = 1,\ 2,\ 3 \qquad\qquad (6.50\ b)$$

$$\frac{\partial \dot{u}_3}{\partial \alpha_j} = \sum_{k=65}^{96} a_k \frac{\partial f_k}{\partial \alpha_j} \qquad\qquad j = 1,\ 2,\ 3 \qquad\qquad (6.50\ c)$$

Taking the velocities and the derivatives of the velocities together at the corner nodes gives 96 equations and 96 unknown constants a_k:

$$\dot{u}_1 = \sum_{k=1}^{32} a_k f_k \qquad\qquad\qquad (6.51\ a)$$

$$\dot{u}_2 = \sum_{k=33}^{64} a_k f_k \qquad\qquad\qquad (6.51\ b)$$

$$\dot{u}_3 = \sum_{k=65}^{96} a_k f_k \qquad\qquad\qquad (6.51\ c)$$

$$\frac{\partial \dot{u}_1}{\partial x_j} = \sum_{m=1}^{3} J_{jm}^{-1} \frac{\partial \dot{u}_1}{\partial \alpha_m} = \sum_{m=1}^{3} J_{jm}^{-1} \sum_{k=1}^{32} a_k \frac{\partial f_k}{\partial \alpha_m} \qquad\qquad (6.52\ a)$$

$$\frac{\partial \dot{u}_2}{\partial x_j} = \sum_{m=1}^{3} J_{jm}^{-1} \frac{\partial \dot{u}_2}{\partial \alpha_m} = \sum_{m=1}^{3} J_{jm}^{-1} \sum_{k=33}^{64} a_k \frac{\partial f_k}{\partial \alpha_m} \qquad\qquad (6.52\ b)$$

and

$$\frac{\partial \dot{u}_3}{\partial x_j} = \sum_{m=1}^{3} J_{jm}^{-1} \frac{\partial \dot{u}_3}{\partial \alpha_m} = \sum_{m=1}^{3} J_{jm}^{-1} \sum_{k=65}^{96} a_k \frac{\partial f_k}{\partial \alpha_m} \qquad\qquad (6.52\ c)$$

The strain rates are:

$$\dot{\varepsilon}_{ij} = 1/2 \left[\sum_{m=1}^{3} J_{jm}^{-1} \sum_{k=K_{oj}}^{K_{fj}} a_k \frac{\partial f_k}{\partial \alpha_m} + \sum_{m=1}^{3} J_{im}^{-1} \sum_{k=K_{oi}}^{K_{fi}} a_k \frac{\partial f_k}{\partial \alpha_m} \right] \qquad (6.53)$$

where the indices of the summation of k vary according to the values of i and j.

$$K_{o1} = 1 \qquad K_{o2} = 33 \qquad K_{o3} = 65$$

$$K_{f1} = 32 \qquad K_{f2} = 64 \qquad K_{f3} = 96$$

The Jacobian matrix is calculated using the shape functions of the 20 geometry nodes. The transformation equations are:

$$X = \sum_{k=1}^{20} N_k X_k \qquad\qquad (6.54\ a)$$

$$Y = \sum_{k=1}^{20} N_k Y_k \qquad\qquad (6.54\ b)$$

and:

$$Z = \sum_{k=1}^{20} N_k Z_k \qquad\qquad (6.54\ c)$$

where X_k, Y_k, and Z_k are the coordinate values of the element nodes. The definition for the Jacobian matrix is:

$$J_{1j} = \frac{\partial X}{\partial \alpha_j}$$

$$J_{2j} = \frac{\partial Y}{\partial \alpha_j}$$

and:

$$J_{3j} = \frac{\partial Z}{\partial \alpha_j}$$

Substituting gives:

$$J_{1j} = \sum_{k=1}^{20} \frac{\partial N_k}{\partial \alpha_j} X_k \qquad\qquad (6.55\ a)$$

$$J_{2j} = \sum_{k=1}^{20} \frac{\partial N_k}{\partial \alpha_j} Y_k \qquad (6.55\ b)$$

and:

$$J_{3j} = \sum_{k=1}^{20} \frac{\partial N_k}{\partial \alpha_j} Z_k \qquad (6.55\ c)$$

Equations (6.55) are substituted into the power of deformation, equation (6.45), and the integration is performed using Gaussian Quadrature. Equation (6.45) becomes:

$$\dot{W} = \sum_{k=1}^{N} \sum_{l=1}^{N} \sum_{m=1}^{N} \bar{\sigma}\ (\tfrac{2}{3}\ \dot{\varepsilon}_{ij}\dot{\varepsilon}_{ij})^{1/2}\ [J]\ w_k w_l w_m \qquad (6.56)$$

where N denotes the order of integration, w_k, w_l, and w_m are the factors of integration, [J] is the determinant of the Jacobian matrix, and $\dot{\varepsilon}_{ij}$ is evaluated from equation (6.47) at the pre-scribed integration points.

Equation (6.56) permitted the integration for the power of deformation over a cubic volume. For the same order of integration, the factors of integration would retain the same values from one element integration to the next.

The approximation at any iteration for the equation of incompressibility $\dot{\varepsilon}_{ii}^{*} = 0$ becomes for each element:

$$\int_V \dot{\varepsilon}_{ii}^{*} dV = \sum_{k=1}^{N} \sum_{l=1}^{N} \sum_{m=1}^{N} \dot{\varepsilon}_{ii}^{*}\ [J]\ w_k w_l w_m = 0 \qquad (6.57)$$

6.3.3 Optimization

A different solution technique from Chapter 2 is used here to find the actual velocity field. A direct optimization is made subject to the constraints previously specified. A full description of the method is given in Ref. [6.7].

Many methods of minimization are possible and several have been

342

tried, [6.15, 6.16, 6.17, 6.18, 6.19].

The method found most suitable was a simplification of Rosen's Gradient Projection method [6.20]. This method, originally intended for inequality constraints, was simplified for equality constraints only.

It is based on the fact that the gradient direction does not ensure that the velocities at the next iteration will satisfy the constraint conditions; therefore the gradient must be modified in order for the new velocities to satisfy the constraint conditions.

6.3.4 <u>Finite Element Model</u>

A one-quarter symmetrical section of the billet was modelled by 12 hexahedron-shaped elements (Figure 6.13). Transition zones were defined by smaller elements. At the points where the flat surface begins at the die entrance and ends at the die exit, triangular faces are required. Since the program at this time does not allow triangular faces, model distortion in the form of almost triangular faced elements was used (Figure 6.14).

In the deformation zone the elements are sized according to boundary velocities. If a velocity discontinuity occurs, then the element thickness is 10 % of the axial distance in the deformation zone. The remaining elements are then proportioned equally.

The length of the entrance region was set equal to the entrance radius. The length of the exit region was set equal to the entrance radius divided by the square root of the ratio of initial billet area to final billet area. The radius was set equal to unity.

A computer program generates models based on the requested die angles.

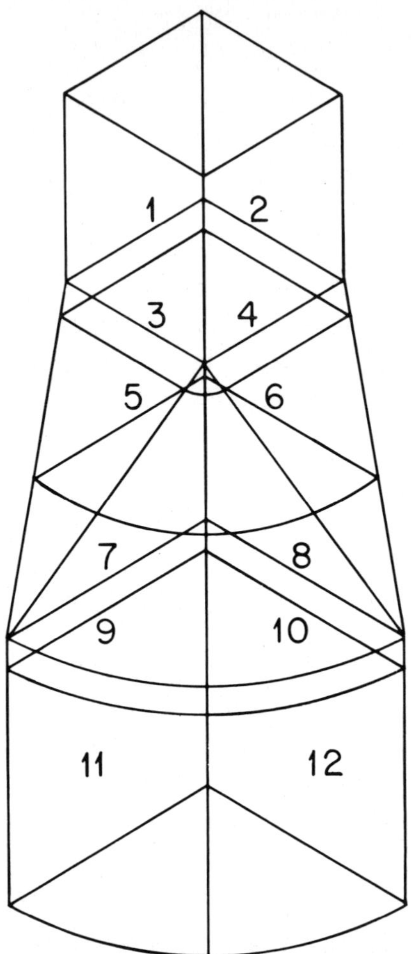

Figure 6.13:   Round-to-square FE model.

6.3.5   Restraints, Die Boundaries and Initial Conditions

Since the Finite Element incorporates nodal derivatives, it is possible to restrain nodal derivatives of velocity with respect to geometry as well as nodal velocities.

The velocities of those nodal points on the die boundary but not on the xz- or yz-plane were restrained in the direction normal to the die surfaces. ($\dot{V}_n$ = 0) This was accomplished using the

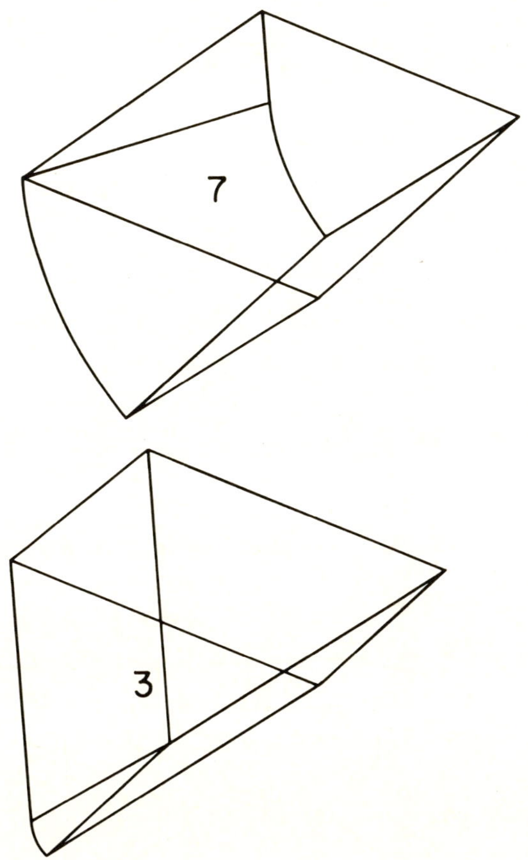

<u>Figure 6.14</u>:   Distortion of the Finite Element.

relationship:

$$\alpha\dot{U} + \beta\dot{V} + \gamma\dot{W} = 0$$

where α, β, and γ were direction cosines [6.21] and:

$$\alpha = \frac{j_1}{da}, \quad \beta = \frac{j_2}{da}, \quad \text{and} \quad \gamma = \frac{j_3}{da}$$

where:

$$j_1 = \begin{array}{cc} \dfrac{\partial y}{\partial r} & \dfrac{\partial z}{\partial r} \\[2mm] \dfrac{\partial y}{\partial s} & \dfrac{\partial z}{\partial s} \end{array} \qquad j_2 = \begin{array}{cc} \dfrac{\partial x}{\partial r} & \dfrac{\partial z}{\partial r} \\[2mm] \dfrac{\partial x}{\partial s} & \dfrac{\partial z}{\partial s} \end{array} \qquad j_3 = \begin{array}{cc} \dfrac{\partial x}{\partial r} & \dfrac{\partial y}{\partial r} \\[2mm] \dfrac{\partial x}{\partial s} & \dfrac{\partial y}{\partial s} \end{array}$$

and:

$$da = (j_1^{\ 2} + j_2^{\ 2} + j_3^{\ 2})^{1/2}$$

Because of the shape of the round-to-square die, a boundary equation generator was incorporated such that α, β, γ were automatically calculated at any given node requested. The location of the boundary nodal points occurred only in the die forming area since the entrance and exit regions could be handled using restraints.

Note that these boundary equations did <u>not</u> include the nodal derivatives.

A starting feasible solution must be reasonable. It has been found that a successful approach is to set all velocities and derivatives to zero except the axial velocities ($\dot{W}$). The axial velocities would begin with the known exit velocity and decrease in an amount inversely proportional to the increase in area of the cross-section of the die.

The initial values used were:

NODES			$\dot{W}$
1	-	12	1.5708
13	-	18	1.5000
19	-	24	1.2500
25	-	42	1.0000

The program has the capability of plotting the model in its entirety or in sections from any view desired. The figures of the model shown in this report were all generated by the program.

6.3.6 <u>Finite Element Results</u>

The program was allowed to iterate until the following criterion was satisfied:

$$\frac{\text{Power}_{i+1} - \text{Power}_i}{\text{Power}_i} < \varepsilon$$

where $\varepsilon = .001$. The results shown were determined in 4 to 8 iterations. A rigid-perfectly plastic effective stress-effective strain relationship was used. The results are listed in Table 6.1, and shown in Figure 6.15.

TABLE 6.1

Die Half Angle (degrees)	Power (in-lb/sec)	σ_z/σ_f
4	.4973	.6332
8	.5310	.6761
12	.5505	.7009
16	.5777	.7355
20	.6096	.7762
24	.6458	.8222

6.3.7 Conclusions

The results from the direct upper-bound solution and the Finite Element approach are compared in Figure 6.15.

The relative drawing stress is plotted versus the semicone die angle. The results from Reference 6.1 are also included. It can be said that the agreement is fairly good. The results of the Finite Element approach are reasonable considering the number of elements and the use of distorted element shapes.

The advantages of the direct upper-bound solution are the possibility to incorporate friction and to calculate a whole range of die angles (6 different values) in a very short computer time. It is also easy to obtain results for any form intermediate between the circle and the square by changing the form coefficient n_f. The disadvantages are in the inflexibility of the solution. For any other shape, the velocity field has to be calculated or to be found by some empirical method and converted into mathematical expressions.

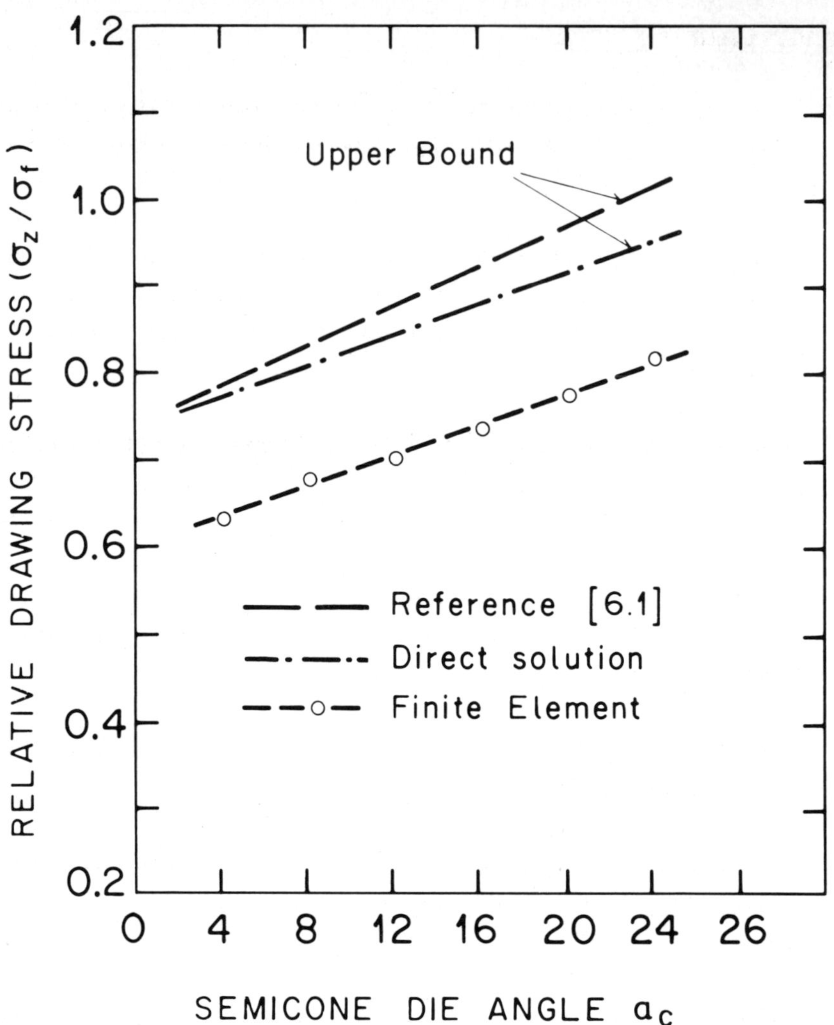

Figure 6.15: Comparison of the direct upper-bound solution and the Finite Element approach.

The advantages of the Finite Element approach are mainly in its great flexibility to adapt to almost any shape. But for every shape, a new mesh has to be generated. This very time-consuming step can be considerably reduced with the use of a well-designed preprocessor for automatic mesh generation. Also, the computing time is considerable.

References Chapter 6

[6.1]   Basily B.B., Sansome D.H., Some Theoretical Considera-
        tions for the Direct Drawing of Section Rod from Round
        Bar, Int. J. mech. Sci., 18 (1976) 201.

[6.2]   Boër C.R., Avitzur W.R., Schneider W.R., Eliasson B., An
        Upper-Bound Approach for the Direct Drawing of Square Rod
        from Round Bar, Proceedings 20th MTDR Conference (1979)
        149-156.

[6.3]   Hoshino S., Gunasekera J.S., An Upper-Bound Solution for
        the Extrusion of Square Section from Round Bar Through
        Converging Dies, Proceedings 21st MTDR Conference (1980)
        97-105.

[6.4]   Avitzur B., Metal Forming: Processes and Analysis,
        McGraw-Hill, New York (1968).

[6.5]   Johnson W., Mellor P.B., Engineering Plasticity, Van No-
        strand Reinhold, London (1973).

[6.6]   Wistreich J.G., Investigation of the Mechanics of Wire
        Drawing, Proc. Inst. Mech. Engrs. (London), 169 (1955)
        659.

[6.7]   Webster W., Davis R., Finite Element Analysis of Round-
        to-Square Extrusion Processes, Manufacturing Engineering
        Transactions, Proceeding NAMRC (1978).

[6.8]   Zienkiewicz O.C., The Finite Element Method in Engine-
        ering Science, London: McGraw-Hill (1971).

[6.9]   Drucker D.C., Prager W., Greenberg H.J., Quarterly of
        Applied Mathematics, 9 (1952) 381-389.

[6.10] Chacour S., Transactions of the ASME, Journal of Basic
        Engineering (1972) 71-77.

[6.11] Kudo H., International Journal of Mechanical Science, 1 (1960) 57-83.

[6.12] Kobayashi S., Transactions of the ASME, Journal of Engineering for Industry, 86, Series B, 2 (May 1964) 122-126.

[6.13] Lee C.H., Altan T., American Society of Mechanical Engineers (1971).

[6.14] Nagpal V., Transactions of the ASME, Journal of Engineering for Industry, 96 (1974) 1197-1201.

[6.15] Wilde Douglass J., Beightler Charles S., Foundations of Optimization, Englewood Cliffs, New Jersey: Prentice Hall, Inc. (1967).

[6.16] Pun L., New York: Introduction to Optimization Practice, John Wilex & Sons, Inc. (1969).

[6.17] Wilde Douglass J., Optimum Seeking Methods, Englewood Cliffs, New Jersey: Prentice Hall, Inc. (1964).

[6.18] Fletcher R., Optimization, New York: Academic Press (1969).

[6.19] Cooper L., Steinber D., Method of Optimization, Philadelphia, Penn.,: W.B. Sounders Company (1970).

[6.20] Rosen J.B., Journal of the Society for Industrial and Applied Mathematics, 8 (1960) 181-217.

[6.21] Olmsted John M.H., Advanced Calculus, New York: Appleton-Century-Crofts, Inc. (1961).

7 Modelling of Thermomechanical Treatment

<u>List of Symbols</u>

a	thermal diffusivity $[m^2/s]$	

a $\qquad$ thermal diffusivity $[m^2/s]$

C_p $\qquad$ specific heat $[J/kg\ ^\circ C]$

h $\qquad$ heat transfer coefficient $[W/m^2 {}^\circ C]$

k $\qquad$ thermal conductivity $[W/m^\circ C]$

Q $\qquad$ heat production per volume and time unit $[W/m^3]$

$\dot{Q}$ $\qquad$ heat flow rate $[W]$

$\dot{q}$ $\qquad$ heat flow rate per unit area (W/m^2)

T_e $\qquad$ environment temperature $[^\circ C]$

T_s $\qquad$ surface temperature $[^\circ C]$

$\dfrac{\partial T}{\partial x}$ $\qquad$ temperature gradient $[^\circ C/m]$

α $\qquad$ thermal expansion $[^\circ C^{-1}]$

ρ $\qquad$ density $[kg/m^3]$

7.1 Introduction

According to the traditional meaning, thermomechanical treatments are combinations of heat treatments and plastic deformations which may or may not be simultaneous and which aim at controlling material properties that could not be obtained otherwise. To a certain extent, the deformation aspects of this concept have been dealt with in Chapter 4.3.4, when deformation and heat transfer analyses were made in parallel. It was shown then how a deformation can be modelled at any temperature with a variety of heat transfer boundary conditions. That chapter therefore covered the mechanical aspects during a thermomechanical treatment.

This chapter will be dedicated to the modelling of the heat treatment aspects of a thermomechanical process. It is not within the scope of this book to focus on the metallurgical aspects of a thermomechanical process nor on the choice of the proper sequence of actions that would produce some desired final material properties. Studies of this type are usually performed with simple specimens and great caution has to be taken in generalizing results to complicated industrial parts.

As the examples have shown so far, and will continue to show in this chapter, histories of deformation and temperature cycles can vary immensely from point to point in a part. Therefore, to model an industrial thermomechanical process, it is necessary to perform sequences of modellings of deformations and heat treatments, the results being very much dependent on the geometry chosen. It is not uncommon to see extensive experimental verifications being made in the developmental stages of new products after certain thermomechanical treatments are chosen in the laboratory.

Note: For the chapter 7.5 the contribution of J. Albrecht (Brown Boveri Research Center, Baden, Switzerland) is gratefully acknowledged.

As a result, most of the published literature is "specimen" dependent and generalizations are dangerous. This applies, of course, to the examples presented in this chapter. Attention will be focussed on the procedures and methods of analysis rather than on the specific results obtained.

It is felt that the quenching process is probably the heat treatment that is most difficult to describe. Problems found in modelling other heat treatments are a subset of problems found in the quenching process. The fact that quenching problems are being addressed specifically does not minimize the generality of the approaches.

Chapter 7.2 deals with the most elusive parameter necessary to model a heat treatment - the heat transfer coefficient between the material and the quenching medium.

Chapter 7.3 discusses heat transfer when cooling plate through water spraying, using as an example quenching of a steel plate. A method for analyzing the cooling process is described and experimental results presented and discussed.

Chapter 7.4 considers the modelling of the quenching of an impeller, a reasonably complex part used in turbomachinery. Temperature histories are obtained first; with them, the mechanical consequences in terms of distortions and residual stresses.

In Chapter 7.5 an example is given in which cooling curves at the center of large forged parts such as rotors are obtained. Such curves are not obtainable experimentally and are of prime importance for the understanding of the material transformation that takes place during quenching.

7.2 Determination of Heat Transfer Coefficients in Quenching

Heat treatment is a process that has as input the temperature histories at various material points. To obtain such histories we will deal with a problem of heat conduction in a body as described in section 2.4.4 which is driven particularly by boundary conditions with the environment. These conditions are especially severe in the quenching process. In order to perform a simulation, several material properties are needed, such as heat conductivity and specific heat, which are relatively easy to obtain, as well as a description of the heat flow to the quenching medium.

This is a complex convection boundary condition, usually described by means of a heat transfer coefficient h defined as:

$$\dot{q} = h \, (T_e - T_s) \tag{7.1}$$

with
$\quad \dot{q}$ rate of heat flow per unit area
$\quad T_s$ surface temperature
$\quad T_e$ environment (quenching medium) temperature

Very detailed analyses of the heat transfer problem between a hot surface and the environment (gas, liquid, or solid) exist in the literature. The reader is referred to texts such as [7.1-7.3] which show how complex the phenomena involved are and further, how for specific conditions it is possible to elaborate on (7.1) in order to obtain expressions that are both predictable and reproducible. Such procedures are equivalent to expanding h into functions that take into account the specifics of the problem being analyzed.

Different functions are applied to model influences of parameters such as surface temperature, surface finish (and oxidation), flow at the interface, boundary layers, boiling (nucleate, transition or film) of the contact liquid, surface inclination, pressure, turbulence, and so on.

It is obvious that only under very well-controlled circumstances and for well-defined (simple) geometries found in heat exchangers

and boilers in the power industry, for instance, can one hope
to describe the problem accurately. Unfortunately, this is not
the case in a quenching process because it is transient, has
variable geometries, and there are too many variable parameters.
What is sought, usually, is h as a function of surface temperature, given a set of operating conditions. Even so, in laboratory experiments the scatter in values obtained is very large
[7.4, 7.5].

This section describes a method of obtaining heat transfer coefficients during quenching and shows results obtained in a variety
of conditions.

7.2.1 Approach to the Determination of the Heat Transfer Coeffi-
cient

The heat transfer coefficients to be determined are those between
a hot metal surface and a quenching medium. The range of surface
temperatures when quenching aluminum is between about 500°C and
room temperature. Experiments such as those described in the next
section show that the cooling process often changes regime during
one quench in that temperature range. Therefore, large variations
of the heat transfer coefficient with temperature are expected.

The methods commonly used to determine the heat transfer coefficient h fall into three categories:

1) Simulating a simple case for which an exact solution is
   known, measuring some variable, and matching h in such a way
   that the exact solution produces equal values for such a
   variable.
   The variable mentioned can be a temperature at the surface,
   a temperature gradient, or any other measurable quantity.

2) Measuring the temperature gradient as close as possible to
   the surface, and using the equation:

$$\dot{q} = k \frac{\partial T}{\partial x} = h (T_e - T_s) \qquad\qquad (7.2)$$

   with: k   heat conductivity

$$\frac{\partial T}{\partial x}$$ temperature gradient

3) Creating a stable situation in which, on one side of an experimental apparatus, heat is supplied in the same amount as the heat that is convected under the desired conditions on another side. Knowing $\dot{q}$ and T, h is easily calculated.

It was decided to do experiments in such a way that a one-dimensional situation could be approximated. The full transient cooling problem could then be numerically simulated in small time increments and the heat transfer coefficient fitted at every step in order to follow the experimental measurements. In this way, h is determined as a function of surface temperature. It should be pointed out that exact solutions for a one-dimensional heat transfer problem with convection boundary conditions consist of complex functions from which it is not easy to extract h. Moreover, these solutions exist for constant h. If h is to be calculated as a function of temperature at the surface, matters become more complicated.

The experiment presented in section 7.2.2 simulates an infinite plate insulated on one surface, with convection on the other surface. It becomes a one-dimensional problem defined by constant starting temperature: no heat flow at one end and convection heat flow at the other.

An elementary one-dimensional Finite Element program for transient heat transfer analysis was written. The theoretical background for it was presented in section 2.4.5.2 and performing this program corresponds to uncoupling the heat transfer analysis from the deformation and considering no heat generation.

Both heat conductivity and specific heat are functions of temperature. In cases where there are phase changes, these two parameters can also vary with temperature gradient. Such changes have not been included in our calculations.

7.2.2 <u>Experimental Setup</u>

One aim was to find heat transfer coefficients under different

quenching conditions. Therefore, it was necessary to pre-heat
the setup and then immerse it in the quenching medium. At the
same time, measurements were made of the effect of the inclination of the surfaces losing heat. With boiling water especially,
the entrapment of vapor bubbles depends upon whether the surface
is upward or downward. Therefore, it was not possible to make use
of the symmetry of a plate. One surface had to be insulated. With
having to withstand heating and cooling cycles, the easy application of any insulator is not possible. The final geometry used
is represented in Figure 7.1. Two plates are used, connected by
a thin short ring. The whole is welded and air-tight so that the
air between plates acts as an insulator. The material of the
plates is stainless steel, which provides an oxidationfree metal
surface and has a low heat conductivity.

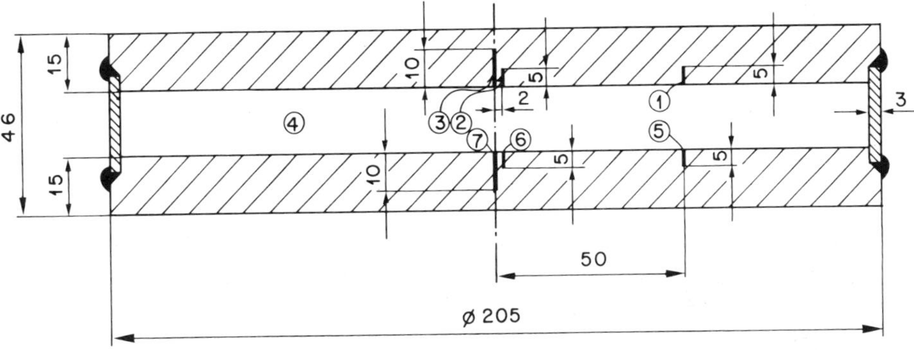

Figure 7.1: Experimental setup for determination of h on flat
surfaces.

Seven thermocouples were placed as indicated in the figure. Thermocouple 4 serves as a check for water entering the cavity. Its
readings were always higher than, or at the same level as, the
innermost thermocouples. Thermocouples 1 and 5 were placed originally to check on the one-dimensional approximation. Although
their readings show temperatures considerably lower than thermocouples 2 and 6, some simple calculations with the temperature
gradients indicate that the radial heat flow is only a few percent of the axial heat flow.

The heat transfer coefficient evaluations are made with thermocouples 3 and 7, i.e., the heat transfer coefficient is changed
until the calculated temperature at the thermoelement coincides
with the measured one. Measured temperatures in these thermocouples are within 3°C. Calculations could have been done for
thermocouples 2 and 6, but they were too insensitive to variations and therefore the accuracy in evaluating < h > is poor.
They were used mostly as a check on the procedure by comparing
the calculated values with the measured values at such locations.
The plate system is placed in a steel frame built in such a way
that it can hang horizontally, vertically, at 30°, at 60° or at
other inclinations. The frame does not interfere with the large
surface of the plates. In each experiment, the system is heated
in a furnace up to 530°C and then immersed into the bath object
of the study. Initially, two experiments in each case were made.
Reproducibility in heat transfer coefficients was quite good.
During the experiment, thermocouple temperatures were directly
read through a data acquisition system. The time increment in
the acquisition system was between 2 and 5 seconds, according to
the experiment.

7.2.3 <u>Results</u>

Data for the following quenching media have been obtained:

- boiling water in a "large" container (200 dm^3) (Fig. 7.2)
- boiling water in a "small" container (50 dm^3) (Fig. 7.3)
- water at 95°C, 200 dm^3 container   (Fig. 7.4)
- water at 90°C, 200 dm^3 container   (Fig. 7.5)
- water at 60°C, 200 dm^3 container   (Fig. 7.6)
- oil at room temperature, 200 dm^3 container   (Fig. 7.7)
- polymer quenchant[*] (25 % concentration) at room temperature,
  200 dm^3 container.   (Fig. 7.8)

Previous experiments had shown the influence of constraining
bubble flow when quenching near boiling water conditions. This
led to measurements in both a large container (200 litres) and

———

[*] Aquaquench 251

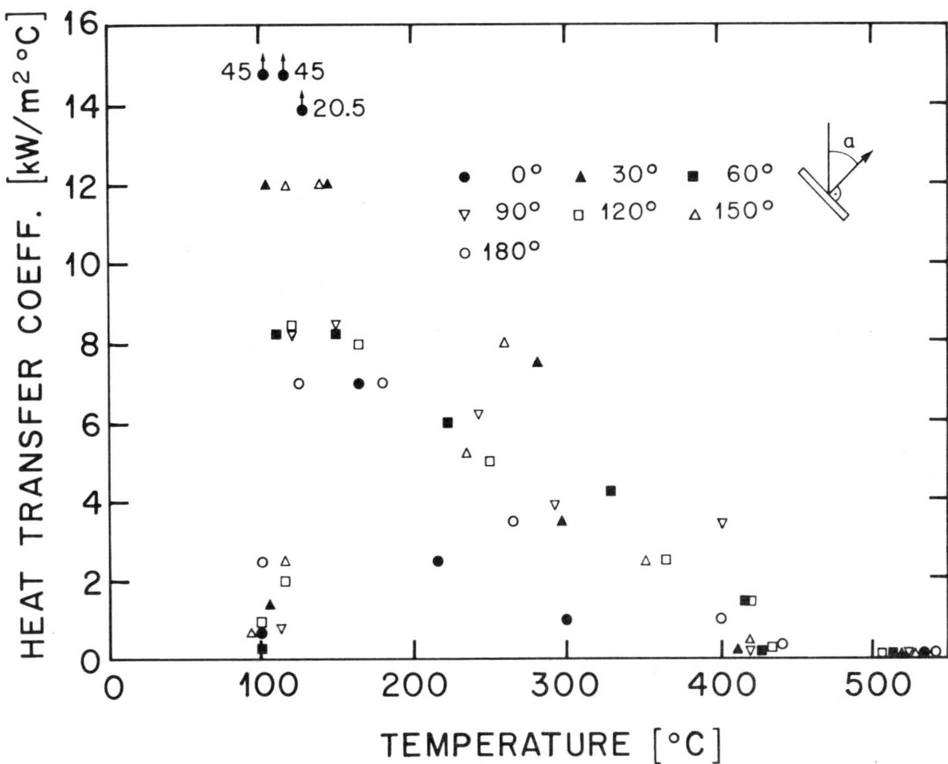

Figure 7.2: h vs. T when quenching in water at 100°C. 200 dm^3 container.

in a smaller one (50 litres). In fact, the results show some differences.

Typical values for the heat transfer coefficients obtained are presented in Figures 7.2 to 7.8. Generally the reproducibility is better for upper surfaces than for lower ones because bubble entrapment is an unstable process. It can be observed that the surface inclination is most important when there are vapor bubbles and vapor film present. In both oil quench and polymer quench where no evaporation takes place, the variations are within the scatter of the experiments.

In boiling water (Figs. 7.2 and 7.3), the peaks in heat transfer coefficient appear close to 100°C because at higher temperatures,

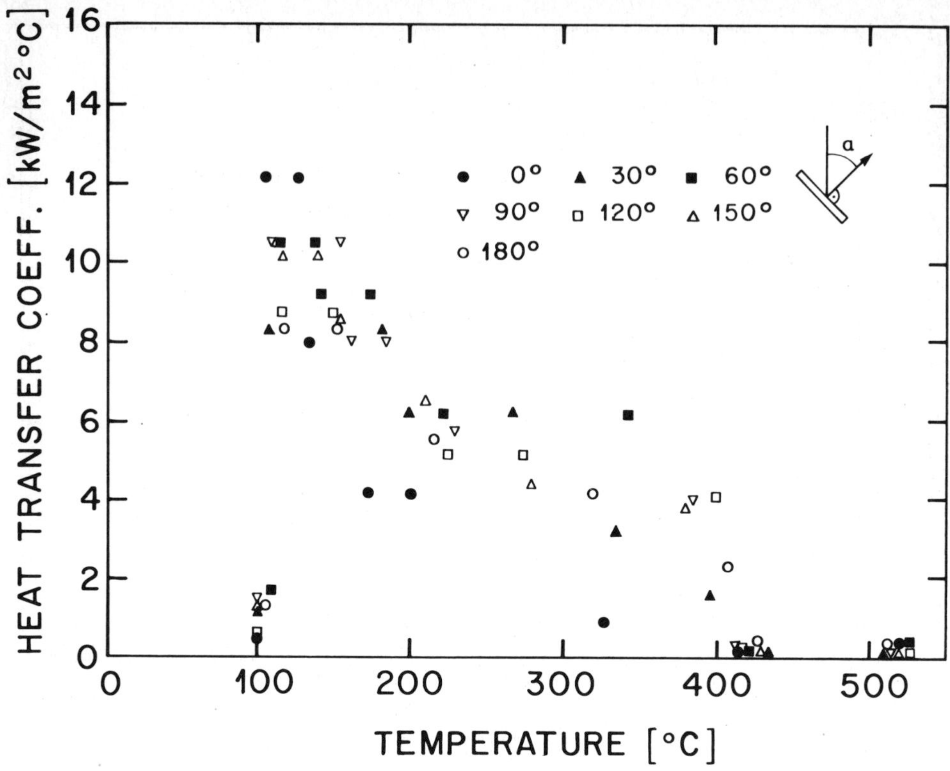

Figure 7.3: h vs. T when quenching in water at 100°C. 50 dm³ container.

a vapor film decreases the heat transfer. Values only rise when these layers break up and bubbling starts. This shift is more pronounced when the plate is directed downwards because it is more difficult for the vapor to float away.

These peaks decrease with the water temperature (Figs. 7.4, 7.5 and 7.6). In water at 60°C, h is much more uniform, probably because there is no complete formation of vapor layers and some kind of bubbling is always present.

In both oil and polymer quench (Figs. 7.7 and 7.8), the values of the heat transfer coefficient are much lower than in water and have their peaks at the highest temperatures. In oil quench, temperatures are not high enough to produce boiling; however, some kind of change in regime is observed around 350°C. In poly-

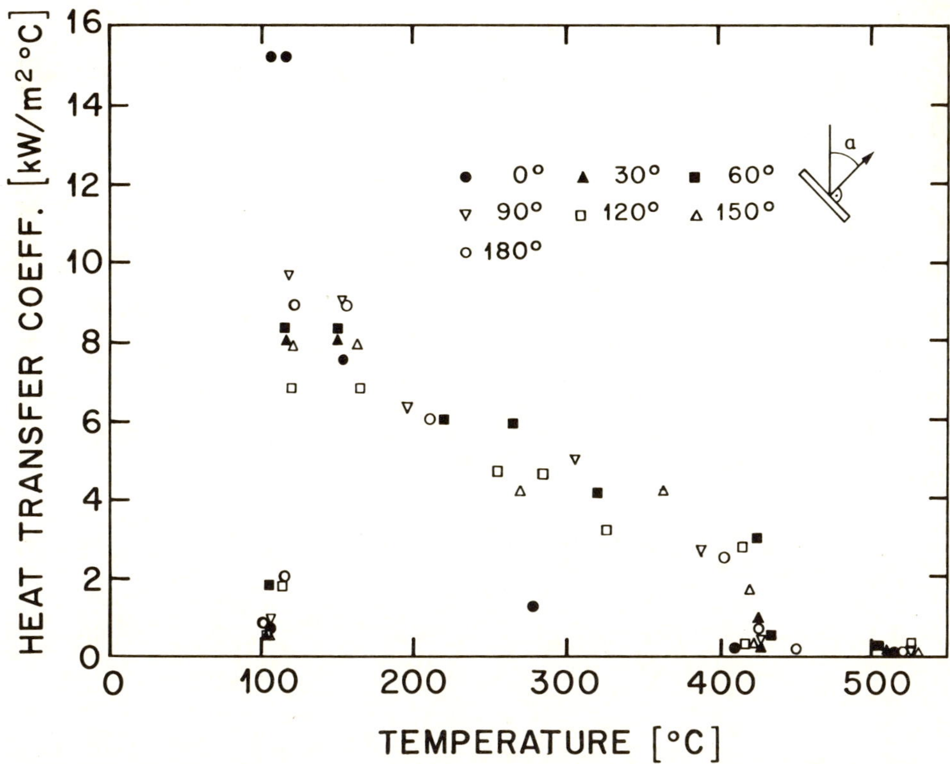

<u>Figure 7.4</u>: h vs. T when quenching in water at 95°C. 200 dm^3 container.

mer quench, at most, minute bubbles form. Therefore, the dominant regime is pure convection.

Figure 7.3 shows the heat transfer coefficient in boiling water, but with a small amount of water. The curves are more similar to the ones for 90°C than the ones for boiling water.

Due to the odd and variable shape of these curves, it is hard to define parameters for them. On the other hand, the widespread use of numerical computations makes it of lesser interest to find expressions to fit the data obtained. Therefore, it is preferred to present the data as they are. In this way, the introduction of further errors in an approximation is avoided and other users can manipulate the data at their convenience. The value h = 0.1 which invariably appears at the highest temperature is the

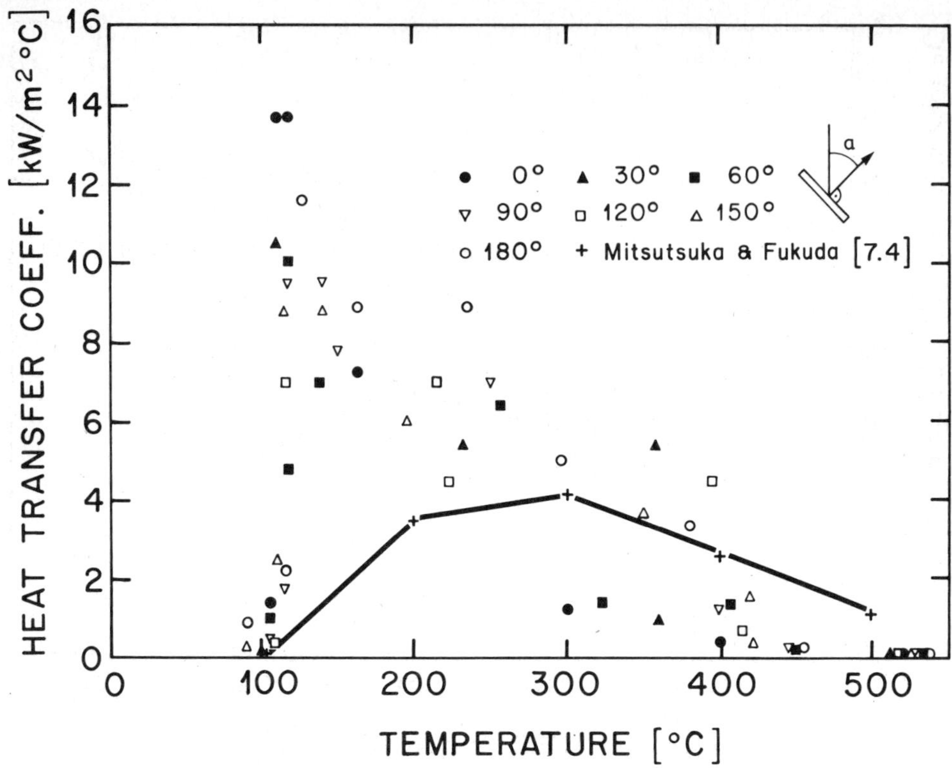

Figure 7.5: h vs. T when quenching in water at 90°C. 200 dm^3 container.

start-up value supplied to the computer program. It is obvious from the data in which cases this value should be discarded.

It may be interesting to compare the values obtained with values found in the literature. Mitsutsuka and Fukuda [7.4] have determined an experimental function for quenching in water that averages many tests and produces h as a function of water temperature between 30°C and 90°C and surface temperature:

$$h = 10^{(A+B \cdot T_s)} [1 - K (T_e - 26)] \quad (Kcal/m^2 \cdot h \cdot °C) \qquad (7.3)$$

with A, B, K constants, which change with the surface temperature T_s (°C).

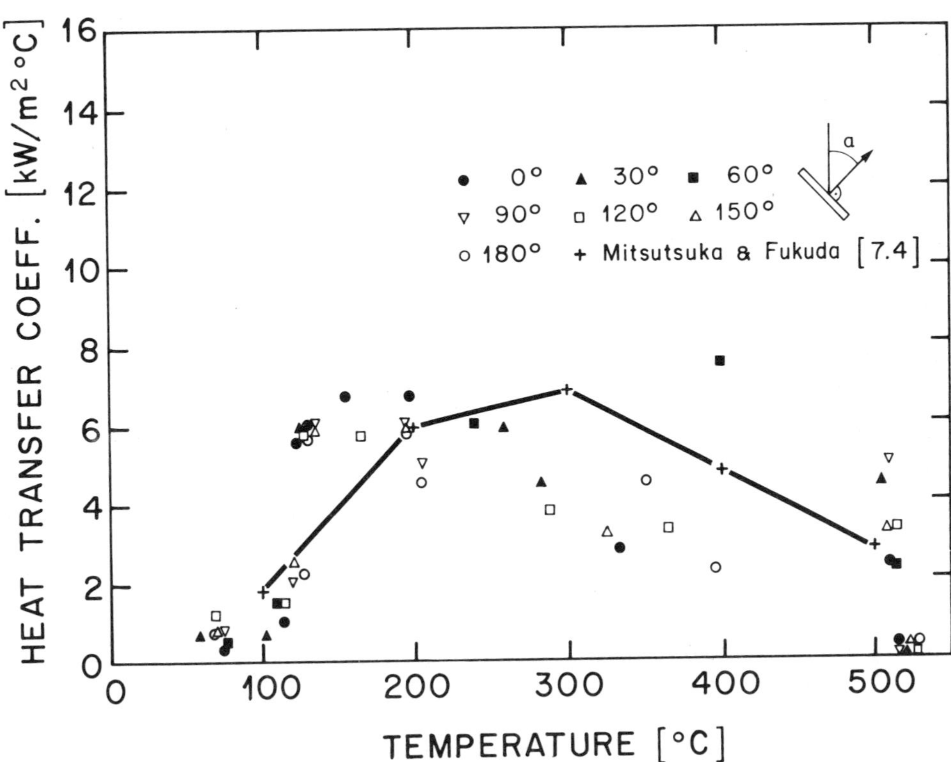

Figure 7.6:   h vs. T when quenching in water at 60°C. 200 dm³ container.

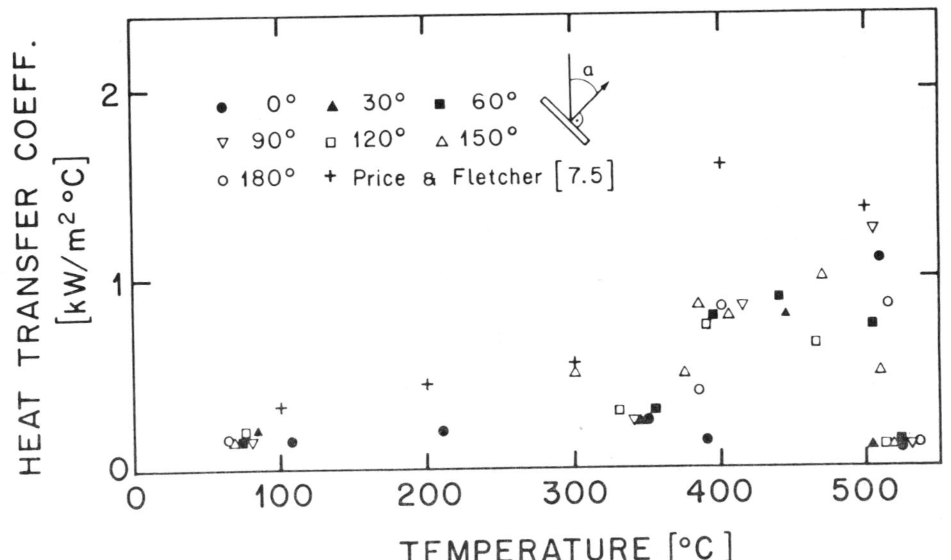

Figure 7.7:   h vs. T when quenching in oil at 25°C. 200 dm³ container.

Figure 7.9 shows the values calculated with such a function for a quenching water temperature of 30°C. Calculated values according to [7.4] were superimposed in our plots for water at 60°C and 90°C. Whilst the 60°C calculations show reasonable agreement, the 90°C calculations are quite different. Values measured by Price and Fletcher [7.5] are also presented in Figure 7.9. These values were obtained from experiments done using both a thin strip and a thick strip of stainless steel. It can be seen that only those values calculated using the thin strip agree with those of Mitsutsuka and Fukuda. Using the thick strip, Price and Fletcher have measured heat transfer coefficients for an oil quench and a polymer quench which have been superimposed on our results. They differ from ours by a factor of up to 2. The calculations in reference [7.5] are made using a Finite Difference solution to the heat transfer problem, while in reference [7.4], the method is not specified.

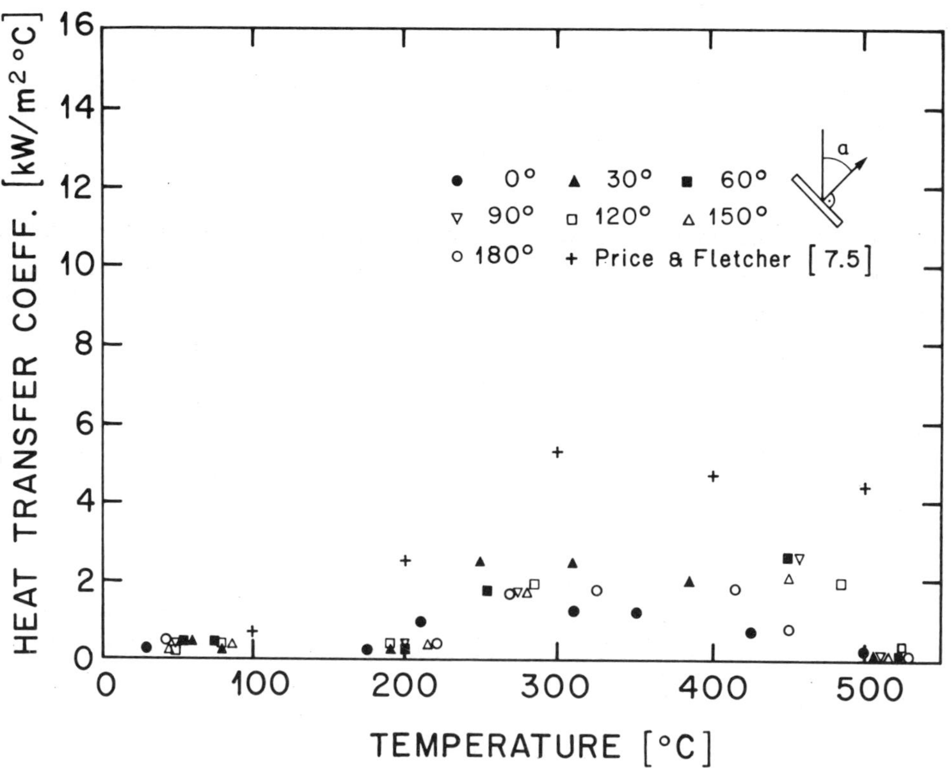

Figure 7.8: h vs. T when quenching in a polymer solution at 25°C. 200 dm³ container.

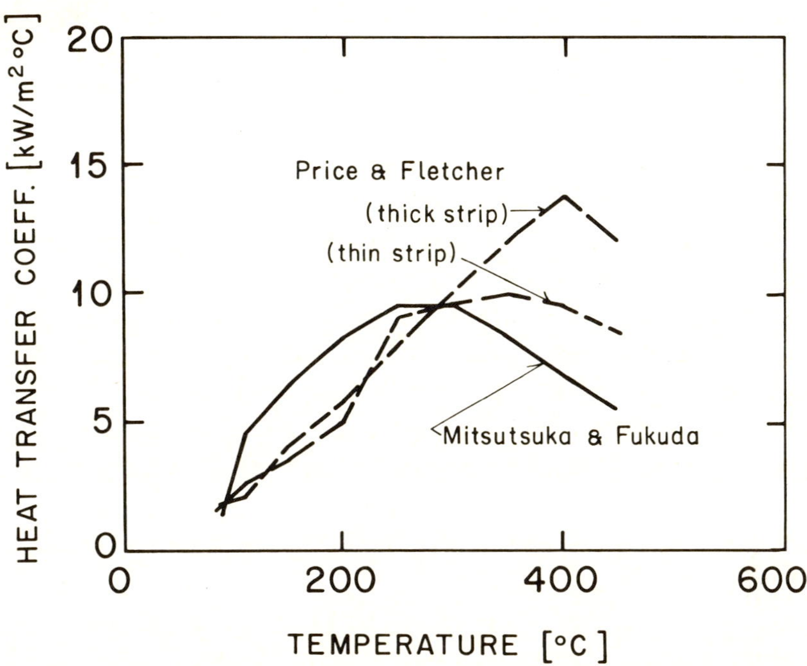

Figure 7.9:   Literature values of U vs. T when quenching in water
at 30°C.

It can be concluded from a comparison of these results that
extreme care has to be taken when interpreting results. Not only
can the method of calculation of the heat transfer coefficient
introduce errors but the results can also be especially sensitive to operating and experimental conditions. Specimen geometry,
because it interferes with flows around itself, can influence
the results and, in unstable conditions, variations can be too
large.

If greater accuracy is desired, it is almost essential to determine heat transfer coefficients using conditions that simulate
the surfaces and conditions of the process being studied as
closely as possible.

7.3 Quenching of Steel Plate through Water Flushing

7.3.1 <u>Introduction</u>

When quenching steel plate, an even and fast quenching rate is
sought. Uneven cooling leads to waviness and can also lead to
microstructural differences in the plate. The case depth for a
given steel is mainly dependent on the achievable heat transfer
rate. The most important part of the quenching operation is its
first part, where the gradients are high and where the quenching
is most difficult to analyse.

The heat transfer from the quenched surface consists of four
parts: conduction, convection, radiation and vaporization (Fig.
7.10), where vaporization is not a truly separate process, but
more a consequence of the three others [7.6]. During water-
quenching of steel, four steps can be differentiated: a high-
temperature process, film boiling, nucleative boiling and convection.

Type of heat transfer	Cooling Step	Convection	Nucleative Boiling	Film Boiling	High Temperature Process
Radiation		-	-	(x)	-
Convection		x	x	-	-
Conduction		-	x	x	-
Vaporization		-	x	x	x

<u>Figure 7.10</u>

The high-temperature process is characterized by an intensive
heat transfer between the water and the steel surface. This process is only active until a vapor film has been created on the
surface, which then acts as an insulating layer and drastically
reduces the heat transfer rate.

During the film boiling stage the heat is transferred through conduction in the film and through radiation. This leads to an increase of the film thickness.

When using water as a quenching medium, the major part of the energy is used to vaporize the water, since the heat of vaporization is seven times that needed to heat the same amount of water between 20°C and 100°C.

In a plate immersed in water, the film thickness increases until the net lifting force is high enough to overcome the surface tension forces that keep the film in place. When this happens, the film breaks and rises in the fluid and a new film is created at the surface and so on. This process, which is called unstable film boiling, continues until the surface has reached a temperature where stable nucleate boiling starts. The vapor bubbles are nucleated at rough spots on the hot surface. Only a small proportion of the heat is transferred to the vapor bubbles through radiation and conduction; the largest part is transferred through convection to the bubbles. In the temperature range below that where nucleate bubbling takes place, the heat is transferred by convection. Figure 7.11 shows the typical relationship between heat transfer rate and surface temperature.

However, the heat transfer in a given fluid is not only a function of the surface temperature; it is also dependent on a number of other factors such as surface roughness, geometry, hydromechanical variables such as pressure, water velocity at the surface and turbulence, as well as temperature of the cooling medium. It is also well-known that additions of polymers or salts to the water change the heat transfer conditions dramatically. When salt is used, the temperature at which stable nucleate boiling starts (the Leidenfrost temperature) is higher [7.7]. This is attributed to lower surface tension. When quenching by flushing the surface with water, the breakdown of the vapor film is accelerated. Here the velocity at which the water hits the surface, which can be translated into pressure in the quenching apparatus and the flow rate of water, is of importance. The high pressure water jets break up the vapor film and bring water in direct contact with the surface to be cooled. The water flow

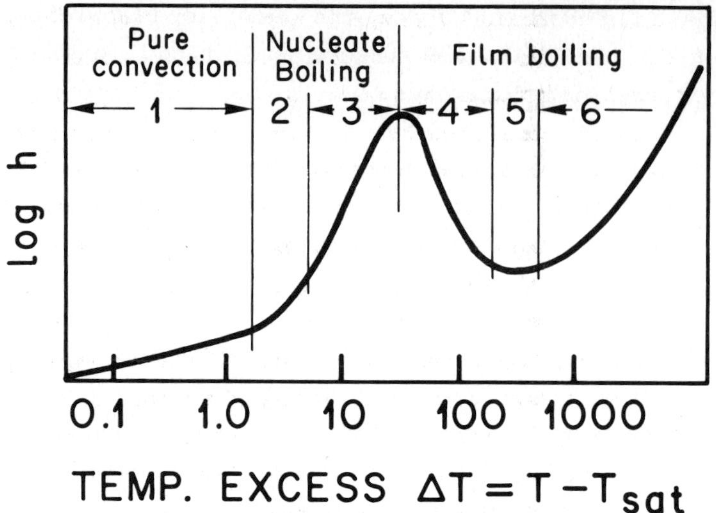

Figure 7.11: Graph of logarithm of heat-transfer coefficient, h, vs. temperature excess, ΔT, for an electrically heated platinum wire submerged in water. Region 1: Pure convection heat transfer by superheated liquid rising to the liquid-vapor interface where evaporation takes place. Region 2: Bubbles condense in superheated liquid rising to the liquid-vapor interface. Region 3: Bubbles rise to the liquid-vapor interface. Region 4: Partial nucleate boiling and unstable film boiling. Region 5: Stable film boiling. Region 6: Radiation coming into play.

hinders the growth of the vapor film and since the heat transfer coefficient in the film is proportional to the film thickness, this influences the heat transfer, equation (7.4):

$$\frac{1}{h} = \frac{t_v}{k_v} + \frac{1}{h_r} \tag{7.4}$$

where t_v is the film thickness, k_v the vapor conductivity and h_r the heat transfer coefficient due to radiation.

The influence of temperature on the quenching rate when using

368

water flushing or spraying has been investigated by several
authors [7.5, 7.8, 7.9]. However, the water temperature influences the cooling rate more in immersion cooling [7.4].

In the following, it is demonstrated how a heat transfer problem
in quenching can be analyzed using an example where the quenching
of a heavy steel plate is investigated.

Spray quenching is a technically important process since most
rolled products that are water quenched either directly after
rolling or after solution heat treatment are quenched with water
sprays. In spray quenching, it is also possible to regulate the
heat transfer rate in different parts of a profile, thereby en-
suring an even quench-rate throughout the part.

7.3.2 Analysis of the Heat Transfer

Based on a production setup for water quenching of heavy steel
plate, a method for the analysis of the heat transfer has been
described [7.10]. Figure 7.12 shows schematically how the plate
is quenched. The parameters that can be varied are:

- the pressure in the nozzle slit, p
- the slit width, t_n
- the angle of attack, β
- the water temperature, T_w
- the initial temperature of the plate, T_o
- the plate velocity, v

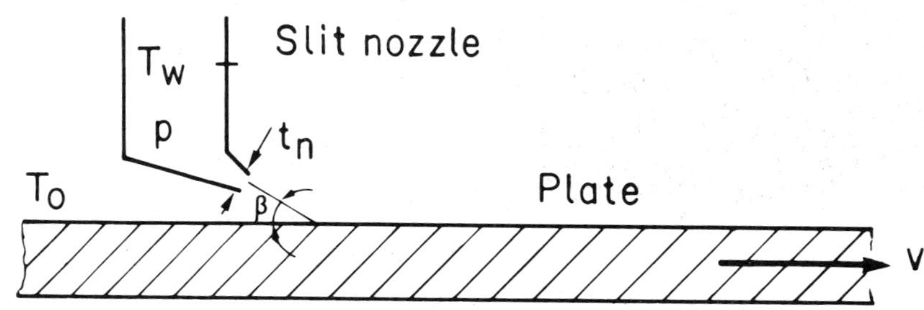

<u>Figure 7.12</u>: Schematic of heavy steel plate water quench.

This gives us the following calculated variables:

- the velocity of the water jet, v_w
- the velocity of the water perpendicular to the steel surface, $v_w \sin \beta$
- the water flow rate, $\dot{q}_w$

The temperature distribution in the plate can be described by:

$$C_p \cdot \rho \cdot \frac{\partial T}{\partial t} = \nabla (k \cdot \nabla T) + Q \qquad (7.5)$$

where:

$$C_p = f_1 \left(T, \frac{\partial T}{\partial t}\right) \qquad \text{specific heat}$$

$$\rho = f_2 \left(T, \frac{\partial T}{\partial t}\right) \qquad \text{density}$$

$$k = f_3 \left(T, \frac{\partial T}{\partial t}\right) \qquad \text{heat conductivity}$$

$$T = f_4 (x, y, z, t) \qquad \text{temperature}$$

$$Q = f_5 (x, y, z, t) \qquad \text{heat production per volume and time unit.}$$

The material properties are dependent on temperature and cooling rate. For the analysis of the problem, a number of assumptions that lead to simplifications can be made:

- The released energy during the austenite-martensite transformation can be ignored in comparision with the energy content in the plate.

- The plate width is large in comparison with the thickness. The corner effects are therefore only local and the problem is reduced to a two-dimensional one.

Study the heat flow along the direction of the plate movement, Figure 7.13. For a typical parameter combination of $\beta = 15°$ and

$t_n = 1$ mm, Δx equals 3 mm. The plate moves typically with a velocity of 50 mm/s. The water jet therefore hits the plate during 60 ms. Assume a temperature difference:

$$\Delta T = T_1 - T_0 = 150°C$$

where T_1 is the temperature 0.5 mm away from the surface which corresponds to $h = 20000$ [W/m^2°C], $T_0 = 850$ [°C] and $T_e = 20$ [°C] where T_0 is the initial temperature of the plate.

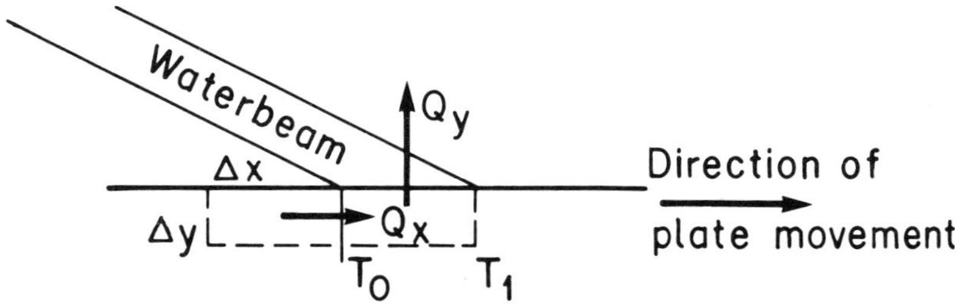

Figure 7.13: Heat flow during the plate quench.

This leads to:

$$\frac{\partial T}{\partial x} \sim \frac{\Delta T}{\Delta x} \sim \frac{150}{3 \cdot 10^{-3}} = 5 \cdot 10^4 \; [°C/m] \qquad (7.6)$$

and the heat flux:

$$Q_x = k \cdot \frac{\partial T}{\partial x} \sim 1.2 \cdot 10^6 \; [W/m^2] \qquad (7.7)$$

with k as in Figure 7.15.

The heat transfer from the plate to the cooling medium is:

$$Q_y = h \cdot T_{mean, \, surface} \sim 1.2 \cdot 10^7 \; [W/m^2]$$

Since $\Delta x/\Delta y = 3$, the heat flux out of the plate is ~ 30 times larger than the heat flux into the same area from the uncooled part of the plate. The heat transfer can therefore be regarded as one-dimensional. The equation has been reduced to:

$$C_p \cdot \rho \cdot \frac{\partial T}{\partial t} = k \frac{\partial^2 T}{\partial y^2} + \frac{\partial k}{\partial T} \left(\frac{\partial T}{\partial y}\right)^2 \tag{7.8}$$

When quenching steels, the term $k\, \partial^2 T/\partial y^2$ is two orders of magnitude larger than $\partial \lambda/\partial T\ (\partial T/\partial y)^2$ and the equation is reduced to:

$$\frac{\partial T}{\partial t} = \frac{k}{C_p \cdot \rho} \frac{\partial^2 T}{\partial y^2} \tag{7.9}$$

To determine the heat flow and the temperature at the surface, the temperature only has to be measured at one point, which should be close to the surface. In other methods, several elements at different depths are used and a temperature gradient in the body is created by fitting a curve through these points. This gradient as a function of time can then be used to calculate temperature and heat flow (see 7.2.1).

When using only one measurement point, it is not possible to calculate the heat flow and surface temperature directly. The reason is that a small disturbance in the heat flow at the surface is dampened as the distance from the surface increases. On the other hand, small errors in the measurements give rise to large disturbances in the calculated heat flow at the boundary. This can be overcome by first solving the heat flow equation in the inner part and then extrapolating the solution obtained to the surface. This method converges.

The boundary condition for the equation is then:

$$T\,(t) = T_1 \qquad \text{at} \quad y = d \tag{7.10}$$

$$\frac{\partial T}{\partial y}\,(t) = 0 \qquad \text{at} \quad y = 0 \tag{7.11}$$

The second condition means that the heat flow at the uncooled surface is zero. This is a good approximation since the heat flow due to radiation and convection in air is 100 times lower than the heat flow on the actively cooled side.

The heat conductivity equation can then readily be solved with Finite Difference methods. The solution can then be extrapolated to the surface by using a Taylor extension. It can be shown that the heat-flow out of the surface calculated by:

$$\dot{q} = h \cdot (T_s - T_e) \qquad\qquad (7.1)$$

is relatively insensitive to errors in the positioning of the thermocouple. This is important since many of the interesting events appear early in the cooling process, which implies that the temperature should be measured close to the surface where a small absolute error in the positioning gives a large relative error.

7.3.3 Influence of Scale Formation on the Heat Transfer

During austenitizing, a scale-layer is formed on the surface which has completely different thermal properties than the base material, Table 7.1.

	Low-carbon steel	Iron oxide
Heat conductivity k [W/m°C]	28	2
Specific Heat C_p [J/kg°C]	650	836
Density ρ [kg/m^3]	7800	5700
Thermal Diffusivity a [m^2/s]	$5.5 \cdot 10^{-6}$	$4.2 \cdot 10^{-7}$

TABLE 7.1

Typical thermal data for a low-carbon steel and iron oxides.

The base material will experience a heat transfer coefficient:

$$\frac{1}{h_{steel}} = \frac{1}{h_{scale}} + \frac{t_{oxide}}{k_{oxide}} \qquad (7.12)$$

where t_{oxide} is the thickness of the scale layer and h_{scale} is the heat transfer coefficient applied on the scale. Figure 7.14 shows how the scale formation influences the heat transfer coefficient experienced by the steel plate. Now, one of the purposes of high pressure water quenching is to break up the scale. It can, however, be seen that even thin layers influence the heat transfer considerably. In steels that contain Cr and Ni, very thin (a few µm) scale-layers are quickly created, which then provide protection against further oxidation so that the problem is largely avoided.

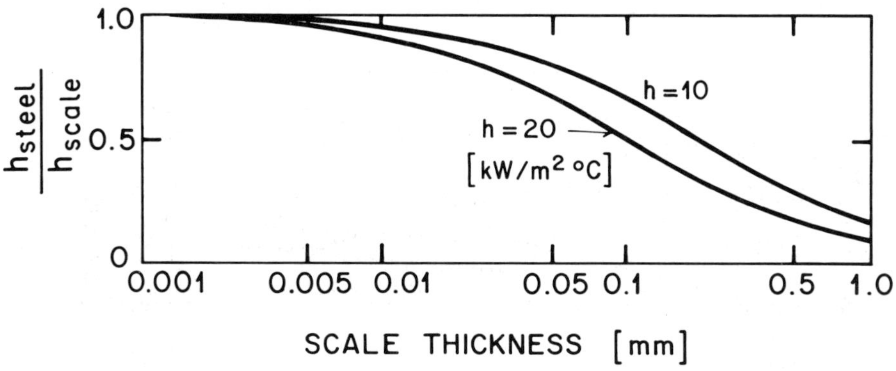

Figure 7.14: Influence of scale thickness on the heat transfer coefficient experienced by the steel.

7.3.4 Experimental Technique

Laboratory equipment was built to study the parameters that influence quenching [7.10]. A 400 x 100 x 12.5 mm^3 plate was equipped with two thermocouples 1 mm from the surface. The width of the plate was chosen so as to avoid influence from cooling of the corners. The plate was heated to 890°C and then extracted from the furnace at a constant speed by a thyristor-regulated

motor and subsequently quenched. The water was sprayed on the
plate through a slit nozzle which created a well-defined water
curtain. The following parameters were varied:

- Angle of attack   15 - 45°
- Water pressure   1.7 - 4.7 atm
- nozzle width   0.4 - 1.0 mm

7.3.5 Material Properties

For the experiments, a carbon steel was used (German material
designation number 1.0566) with 0.15 % C, 0.28 % Si, 1.40 % Mn
and 0.07 % V. The material data are dependent on T and $\partial T/\partial t$
which means that the equilibrium values cannot be used. Instead,
the steel was assumed to be austenitic down to 400°C. Figure 7.15
shows the equilibrium values as well as the values used in the
calculations.

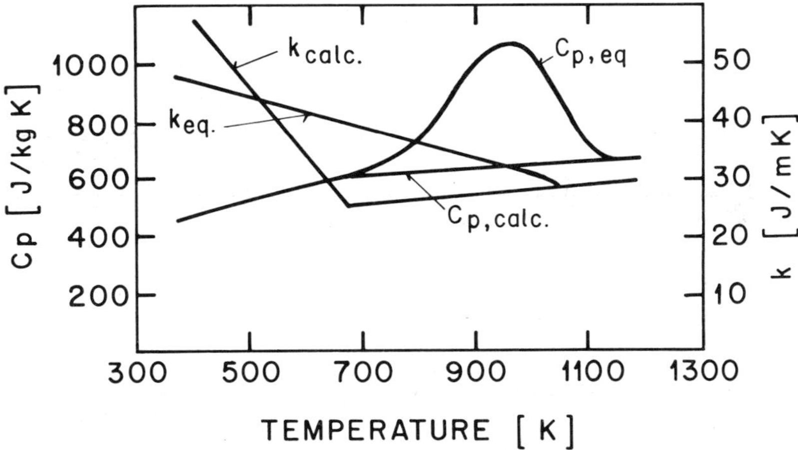

Figure 7.15: Equilibrium values of k and c_p compared to the
values used in the calculations when the quenching
experiments were evaluated.

7.3.6 Experimental Results

A typical calculated temperature distribution in the plate is

shown in Figure 7.16, together with the calculated heat transfer
coefficient.

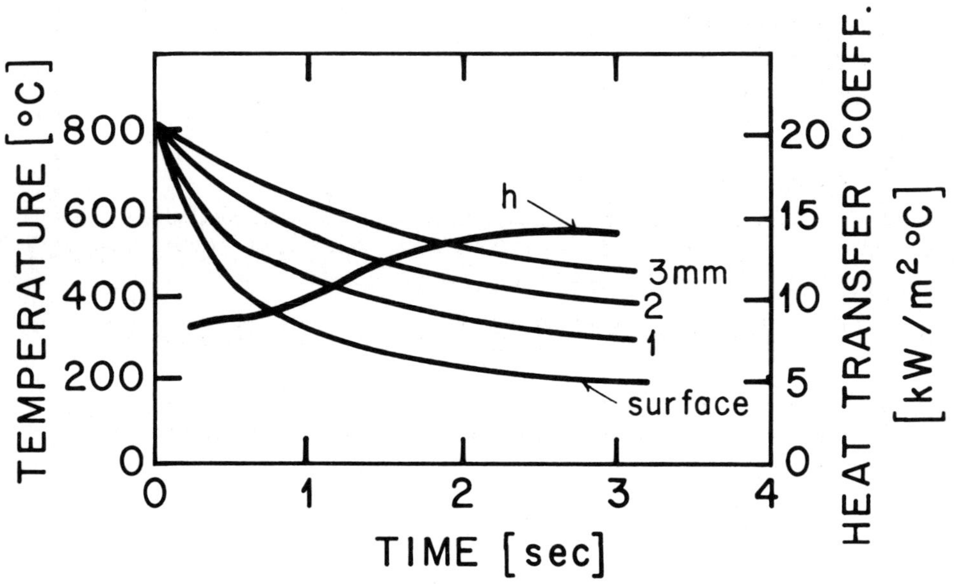

<u>Figure 7.16</u>: Typical calculated temperature distribution in the
plate together with the calculated heat transfer
coefficient.

It was shown that in all the experiments, the water speed perpendicular to the surface was high enough to break up the vapor
film on the surface during the first second of the quenching
process, Figure 7.17. Other work has shown [7.11, 7.12] that
above a certain pressure, there is no effect on heat transfer
coefficient of increasing pressure.

As can be seen in Figure 7.18, the angle of attack influences the
totally transferred energy during the quenching. The time used
in the calculation of the totally transferred energy was taken
as the time from the start of the quenching until the plate had
passed the nozzle and the temperature in the thermocouple started to rise due to conduction in the plate. The reason for the
increase in heat transfer at the lower angles of attack is that

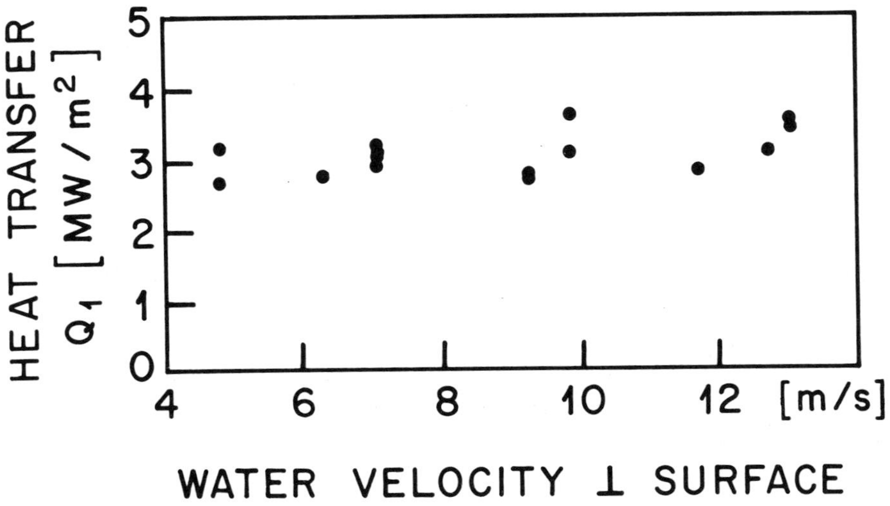

Figure 7.17: Heat transfer during the first second of the quenching, Q_1, vs the water speed perpendicular to the surface.

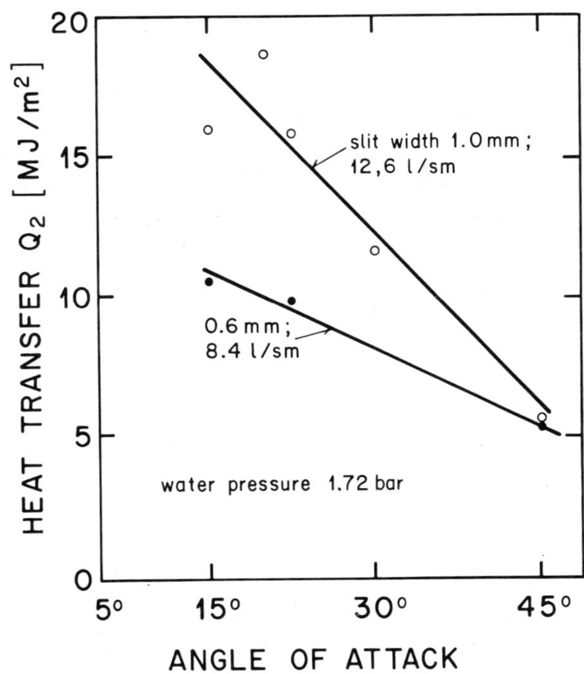

Figure 7.18: The overall heat transfer during the quenching, Q_2, vs the angle of attack.

the water flows along the plate causing a turbulent layer of
water flow along the plate, which enhances the heat transfer. At
the higher angles of attack, the water is deflected from the surface. Similar results are reported by other authors [7.11]. The
influence of water flow on the totally transferred energy is seen
in Figure 7.19. The effect of water flow rate on heat transfer rate has been studied in continuous casting [7.13] and in
spray cooling and in cooling using laminar jets [7.8, 7.9, 7.14].

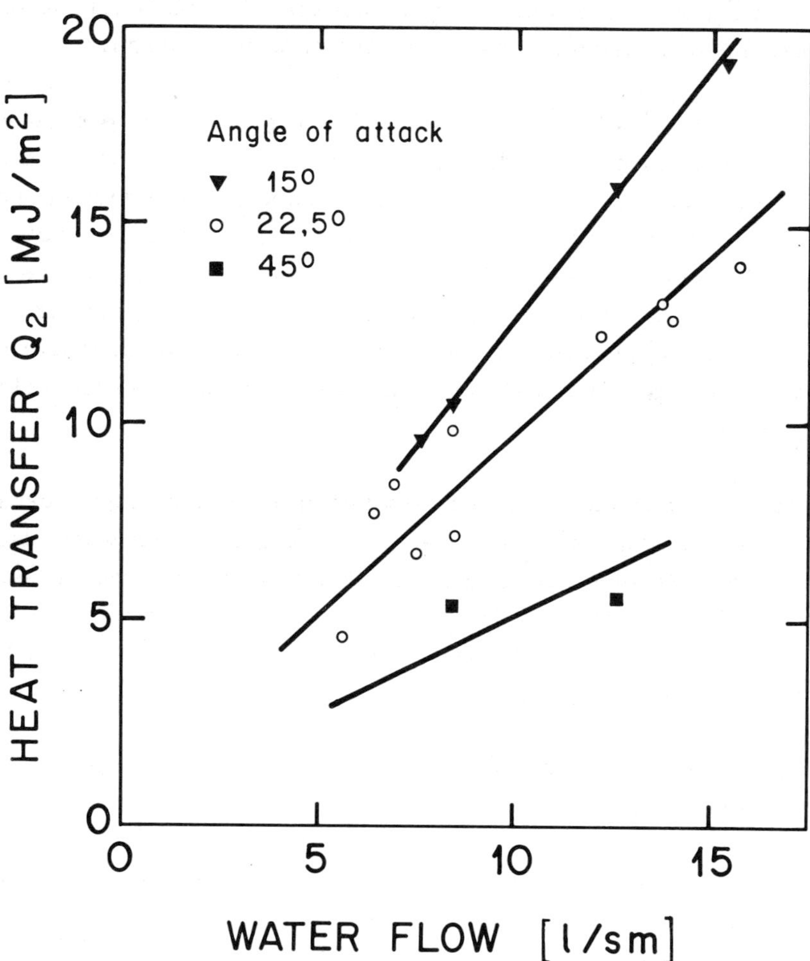

Figure 7.19: The overall heat transfer during the quenching, Q_2,
vs the water flow.

378

7.4 Modelling of Quenching of Complex Parts

7.4.1 <u>Introduction</u>

When a part undergoes a fast quench, the temperature differences
in the part often produce permanent deformations which may
distort the geometry to unacceptable levels. Simultaneously,
residual stresses are produced, which may or may not be desirable. To control and restrict the distortions is a matter of
experience in dealing with quenching methods. Except for very
simple geometries, the only way to assess the residual stresses
so far has been to measure them in destructive testing.

The purpose of this section is to show that it is possible to
numerically simulate the quenching process i.e. to obtain temperature histories at all points of a complex industrial part,
as well as distortions and to obtain residual stress distributions at the end of the quench. The computer programs used are
commercially available packages, so that no development is
necessary. The emphasis is put, rather, on the procedure of
creating the computer model and choosing the appropriate options
in the programs.

It has to be stressed from the beginning that the input data is
not straightforward and therefore will require, in most cases,
some experimental data acquisition. The results are quite sensitive to the heat transfer data. It should also be mentioned that
further changes in the internal stress distribution occur during
later steps in the production process.

The ultimate objective of the development of procedures like the
one described is to study:

- influence of changes of quenching conditions on stress distributions

- influence of changes of geometry on stress distributions

Given a quenching method and a material for which experimental
heat transfer data has been acquired, it should be possible to

predict what happens during the heat treatment. The same data
would be used for calculations in differently shaped parts. On
the other hand, a change of material or a change in the quenching
method requires the acquisition of new data.

7.4.2 Quenching of an Aluminum Impeller

Figure 4.54 shows a picture of an aluminum impeller used in
turbomachinery. After forging, the impellers are heated to 530°C
and quenched in boiling water. Distortions, as in Figure 7.20,
have been observed after quenching in several instances. The
material used for the impellers is Al 2618[*]. Extensive experimental quenches were made with small parts in which five thermocouples were inserted, as shown in Figure 7.21, and the corresponding cooling curves recorded. In Figure 7.22 the cooling
curves typical of quenching in boiling water are shown. These
curves can be interpreted with the help of the pictures of Figure
7.23 a-e. The times at which the pictures were taken are marked
in Figure 7.22.

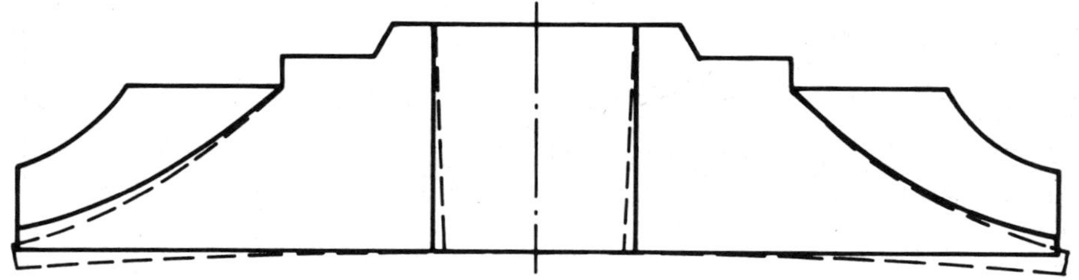

<u>Figure 7.20</u>: Typical distortion pattern in impeller that arise
during quenching.

---

[*] Al 2618: Al, 2-2.6 Cu, 1.2-1.8 Mg, 0.9-1.4 Ni, 0.9-1.4 Fe,
     0.2 Si, 0.1 Ti

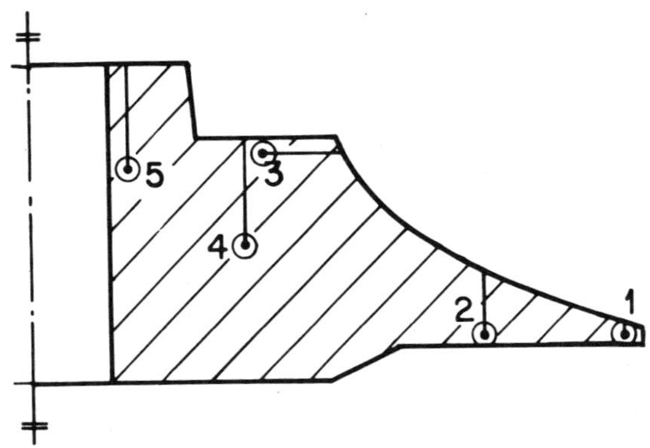

Figure 7.21: Placing of thermocouples in test piece.

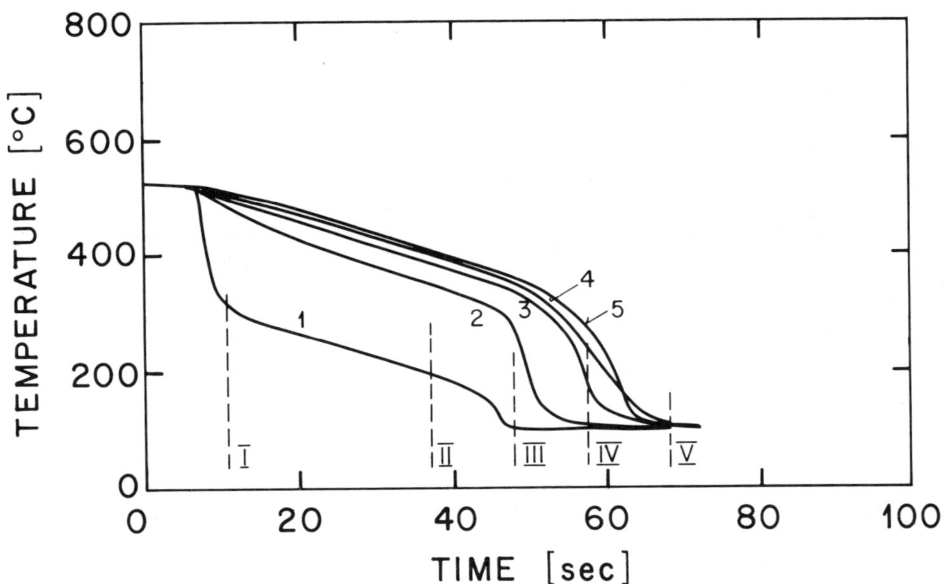

Figure 7.22: Typical cooling curve for the impeller shown in Fig. 7.20 with the thermoelement-placing as in 7.21 when quenched in water at 100°C.

When the impeller is immersed, the heat transfer coefficient is very high because of the high temperature of the contact surface and liquid wetting that surface. This leads to the rapid creation of bubbles that coalesce in a vapor layer. In Figure 7.23a, immediately after immersion, the full impeller is covered by a a vapor film. As soon as the film is formed, the heat transfer rate decreases dramatically, as does the cooling rate at the surface. Because this transitory situation happens very quickly (it could only be detected by thermocouple 1 which has 3 surrounding surfaces dissipating heat), it can be seen that a fast cooling rate is followed by a sharp slowdown. The film is thinner in the outer rim because there is more water surrounding it, less aluminum supplying heat into it, and turbulence around the impeller trying to break it. The situation is steady for a while (Fig. 7.23b). This contributes to the breaking-up of the film, first in that region as seen in Figure 7.23c. When the film breaks up, liquid again comes into contact with a hot surface, vaporizes, and the resulting bubbles float away. This process tends to be rather violent and the heat transfer coefficient is very high. Therefore, around thermocouple 1, the temperature falls abruptly to the water temperature. In the rest of the impeller there is still a film. Towards the end, Figure 7.23d, there is no film boiling at all, but only violent bubbling as the surface cools down, receeding more and more towards the center which is the bulkiest region, i.e. where the residual heat is still stored, Figure 7.23e.

This alloy undergoes no phase changes during quenching. Thus, all internal stresses are caused by thermal stresses. The temperature gradients produce strains and these stresses. At the beginning, while the temperature difference between the interior and the surface increases, the surface tends to shrink faster. The surface is therefore under tension and the interior under compression. Towards the end of the quench, when the temperature gap is decreasing, it is the center that tends to shrink faster. If, at any point, the flow stress of the material is reached, plastic deformation occurs and at the end, residual stresses remain. The flow stress of the material varies strongly with temperature, Figure 7.24, which makes it more difficult to perceive at which points there is plastic deformation. In principle, a cooling

Figure 7.23 a: Picture taken at the time marked with I in 7.22.

Figure 7.23 b: Picture taken at the time marked with II in 7.22.

Figure 7.23 c: Picture taken at the time marked with III in 7.22.

Figure 7.23 d: Picture taken at the time marked with IV in 7.22.

<u>Figure 7.23 e</u>: Picture taken at the time marked with V in 7.22.

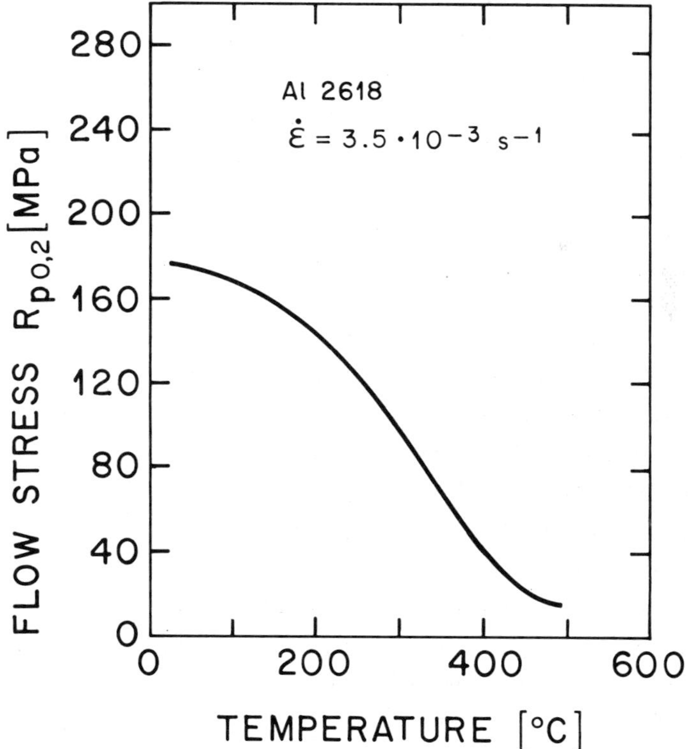

<u>Figure 7.24</u>: Flow stress of Al 2618 vs temperature during quenching.

process that creates greater temperature gradients at the beginning, and/or for more extended periods of time, is one that will produce higher residual stresses.

The geometry of the part, the boundary conditions, and the material properties altogether defy any simplified analysis based on common sense and basic principles. A calculation procedure that accurately picks up spatial and temporal variations is required in order to be able to make a reliable prediction.

7.4.3 <u>Simulation</u>

It was decided to try the Finite Element method using the programs ADINA and ADINAT, both commercially available. The problem was separated into two parts: first the temperature distribution was calculated during the quench using ADINAT, and secondly based on the temperature history, the residual stresses were calculated using ADINA. The problem may be considered uncoupled since the amount of plastic deformation is so small that no internal heat generation has to be taken into account.

As Figure 7.20 clearly shows, the problem is not fully axisymmetric because of the fins. However, the relatively large number of fins and their regular distribution produce distortions that are virtually axisymmetric. On the other hand, the critical areas during later operation are found around the central bore and are therefore far away from their influence. These considerations, together with the large increase in cost that a full three-dimensional analysis would produce, led to the simplification of the model to an axisymmetric two-dimensional analysis. In order to achieve that, it was assumed that the main effect of a fin is to subtract heat away from the area where it is connected to the impeller. Section 7.4.3.1 describes, in part, how an approximate equivalent heat flow was directly applied to the base of the fins. This results in the temperatures and stresses in such areas being averages in the tangential direction. The high heat conductivity of aluminum is an advantage here, since it rapidly homogenizes the temperatures.

A simulation was made of the quenching of an impeller with 300 mm diameter, for which cooling curves under several conditions were measured.

7.4.3.1 Temperature Calculations

With ADINAT, a transient heat transfer analysis was performed. The formulation it uses is identical to the one presented in section 3.4 [7.15]. Figure 7.25 shows the mesh used, consisting of eight-node isoparametric axisymmetric elements. All nodes were given an initial temperature of 530°C. The most sensitive parameter in these calculations is the heat transfer coefficient at the surface, as defined in section 7.2. The experimentally acquired and compiled values were introduced as input data. The specific heat and heat conductivity values according to Table 7.2 were put in.

TABLE 7.2

Materials Properties as a Function of Temperature

T	E	v	α	H	k	ρC_p
100.	59000.	0.35	0.0000216	100.	142.	$2.54 \cdot 10^6$
300.	26600.	0.35	0.0000242	100.	142.	$2.54 \cdot 10^6$
400.	8000.	0.35	0.0000252	100.	142.	$2.54 \cdot 10^6$
500.	3400.	0.35	0.0000257	100.	142.	$2.54 \cdot 10^6$
530.	2800.	0.35	0.0000259	100.	142.	$2.54 \cdot 10^6$
532.	2800.	0.35	0.0000259	100.	142.	$2.54 \cdot 10^6$

T - temperature (°C)

E - modulus of elasticity (N/mm^2)

v - Poisson's ratio

α - thermal expansion coefficient $(°C^{-1})$

H - work hardening rate (N/mm^2)

k - heat conductivity (W/m°C)

ρC_p - specific heat times density $(J/m^3°C)$

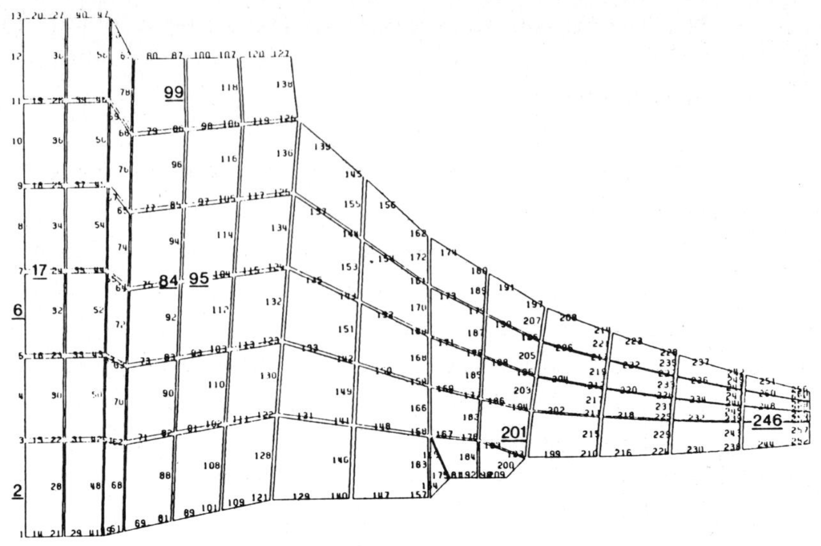

Figure 7.25: FEM-mesh used in the transient heat transfer analysis.

The problem of simulating the fins was approached in the following manner. Firstly, it is easily seen that the heat a fin has stored is negligible compared with the amount of heat it conveys between body and water:

$$Q_f = \rho C_p \cdot V \cdot \Delta T \tag{7.13}$$

where:

ρC_p = specific heat per unit volume ($2.54 \cdot 10^6$ J/m^3 °C)
V = volume (m^3)
ΔT = temperature difference between initial and final temperature (°C)
Q_f = stored heat in a fin

A typical fin 40 mm long, 10 mm high and 5 mm thick stores 2200 J. The heat flux it transmits changes with time, but for an average surface temperature of 350°C, for example, it can be taken as:

388

$$\dot{Q} = h (T_s - T_e) \cdot A \qquad\qquad (7.14)$$

$$A = \text{surface area}$$

for h = 2500 (W/m^2°C) (average), $\dot{Q}$ = 500 W with typical cooling times of 100 s.

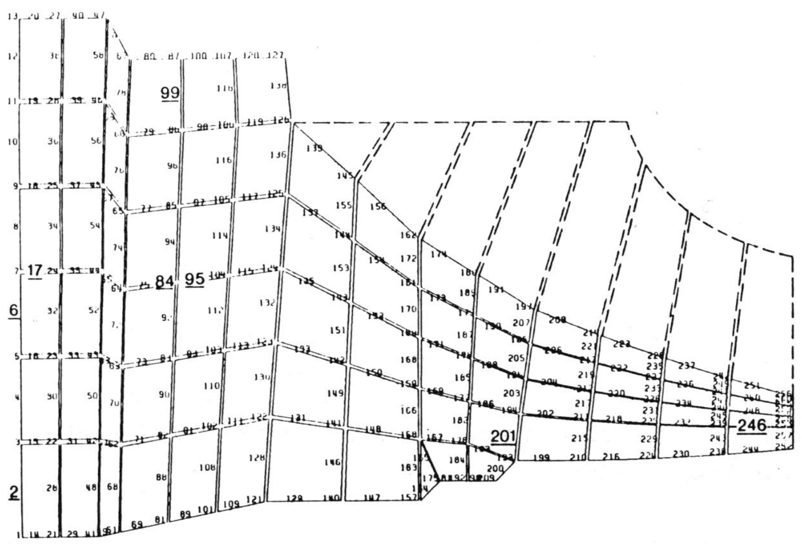

Figure 7.26: Division of the fins for simplified simulation.

From this, it is acceptable to consider at all times a fin in a steady-state situation for which there are theoretical solutions. Each fin is divided, as in Figure 7.26, according to the element division and it is assumed that there is no heat conduction between the strips. As in [7.16], the heat flux that is conducted in steady-state through a fin root, is given by:

$$\dot{Q} = k\,A\,m\,(T_h - T_e)\,\tanh\,mL \qquad\qquad (7.15)$$

$$
\begin{aligned}
T_h &\quad \text{temperature of the root}\\
k &\quad \text{heat conductivity}
\end{aligned}
$$

$$m = \sqrt{\frac{hP}{kA}}$$

L   length of fin

A   cross-sectional area

P   perimeter of area

h   heat transfer coefficient

This allows the calculation of an average equivalent heat transfer coefficient in the surfaces where the fins are placed to:

$$h_{eq} = k\ m\ \tanh mL \qquad\qquad (7.16)$$

For each strip, m was calculated with the average h = 2500 and from that h_{eq}. These values were entered into ADINAT.

With regard to the time integration scheme, an implicit scheme was chosen, using the β method as in equation (2.172), with a value of $\beta = 0.75$. In this way, the integration is unconditionally stable, and reasonably large time increments can be used without loss of accuracy. Small time increments were taken at first, as large temperature gradients are formed initially, and then increased gradually as the temperature variations decrease. In this way, temperature distributions were calculated at all time increments, and stored.

The heat transfer coefficients for water temperatures near and at boiling conditions (see 7.1 - 7.3), and for surfaces at different inclinations were used. Boiling water was simulated, starting by using the heat transfer coefficient in each impeller surface corresponding to its inclination. The cooling curves produced did not even approximately match the experimental ones. The reasons for such behavior can be attributed to the following:

a) Heat transfer coefficients are determined for relatively large surfaces, while in the impeller many of the surfaces have sizes of the order of magnitude of the bubbles formed during quenching.

b) In the impeller, there is interaction between the bubbles produced by one surface and those produced by another surface. For example, bubbles formed in the fins are carried

away by other bubbles formed in the upper surface of the impeller instead of being carried away by fresh water.

Both of these factors certainly produce a change in the bubble
convection regime around the impeller with respect to the regime
of a flat surface obtained when quenching a plate. It should be
mentioned here that in quenching regimes in which bubbling is not
so important (oil quench, polymer quench, or water quench below
60°C), these effects are probably minimized.

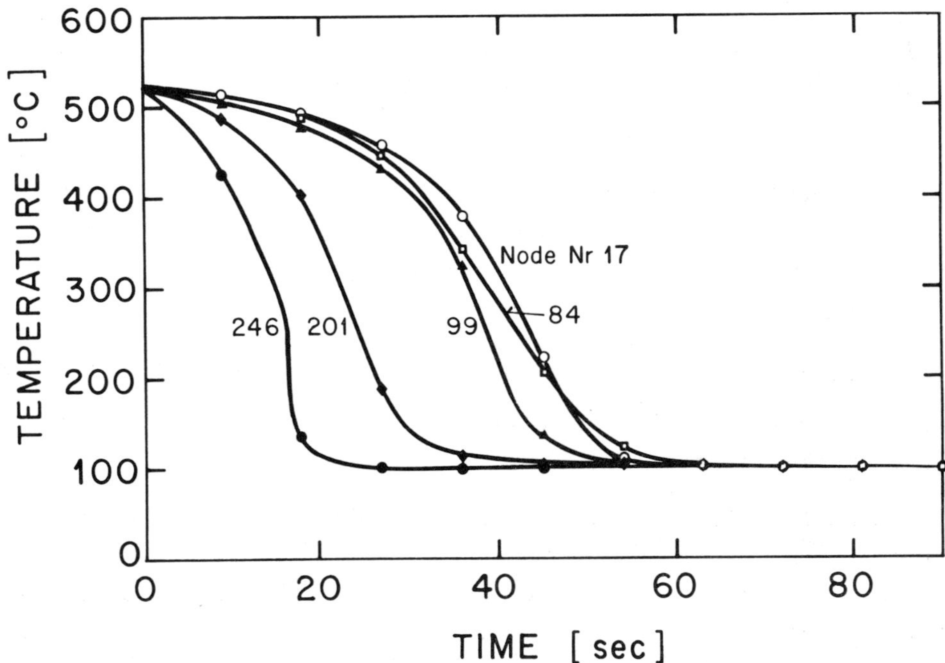

Figure 7.27:   Calculated cooling curves of selected nodes. The
heat transfer coefficient used was that of a surface facing downwards (180°) in water at 100°C.

Having understood this, and recalling from the pictures of Figure 7.23b that a blanket of vapor formed around the impeller is
stable for long periods of time, it was decided to use the heat
transfer coefficient measured for a horizontal surface facing
downward for the whole impeller. The cooling curves produced
(Fig. 7.27) appeared reasonable and were quite close to the ex-

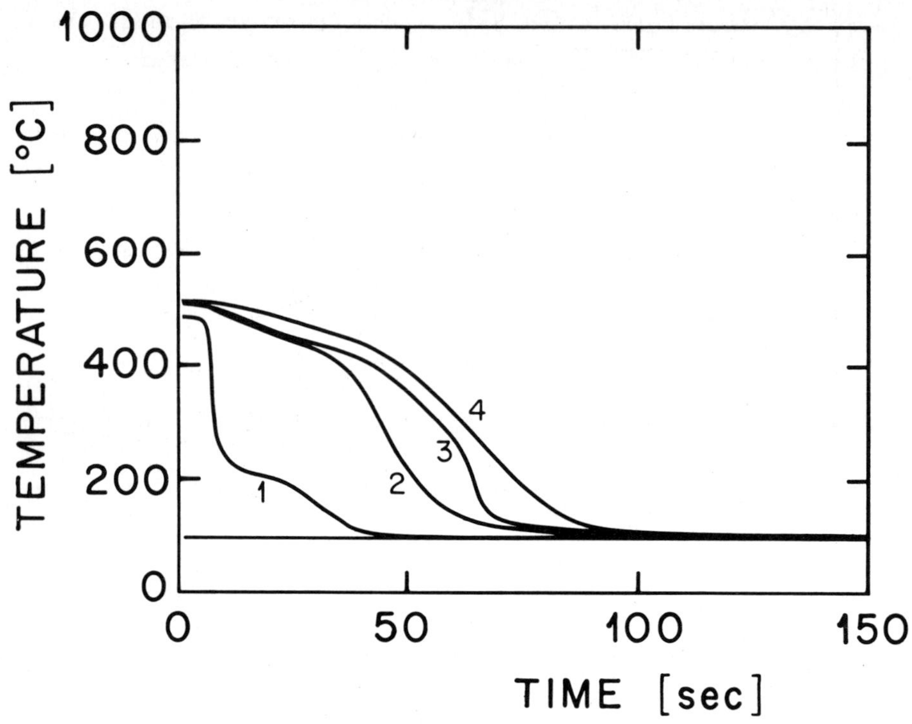

Figure 7.28: Measured cooling curves during quenching in water at 95°C.

perimental cooling curves obtained in quenching in water at 95°C (Fig. 7.28). This shows how difficult it is, in an unstable regime like boiling, to predict the heat transfer coefficient corresponding to a given configuration. However, given the accuracy obtained, including the correct cooling curve shapes, it would not be difficult to iteratively reach correct ones. Figures 7.29 and 7.30 show the temperature distributions after 15 and 30 sec. respectively. It can be seen how the thinner outside cools much faster than the inside. It can also be seen that the inside bore does not have a surface large enough to produce a considerably faster cooling than the central regions.

7.4.3.2 Residual Stress and Distortion Calculations

For residual stress calculations, ADINA was run with the thermo-

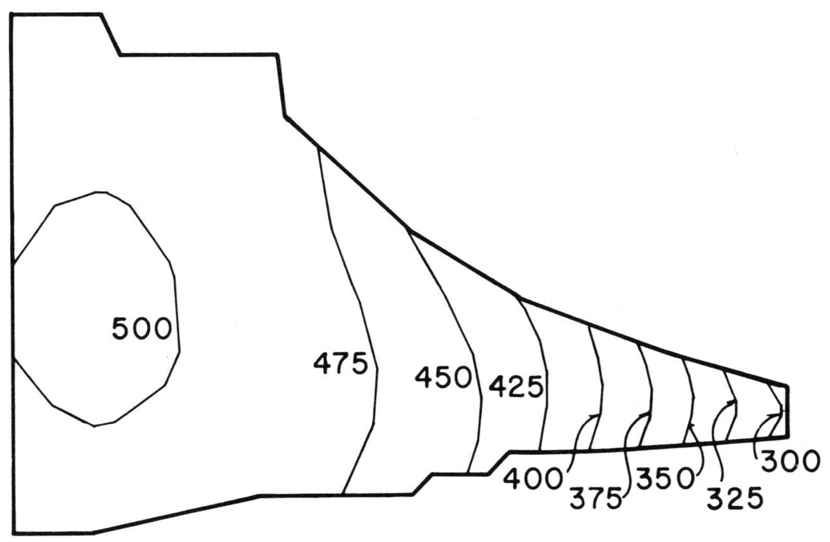

Figure 7.29: Calculated temperature distribution 15 sec. after immersion.

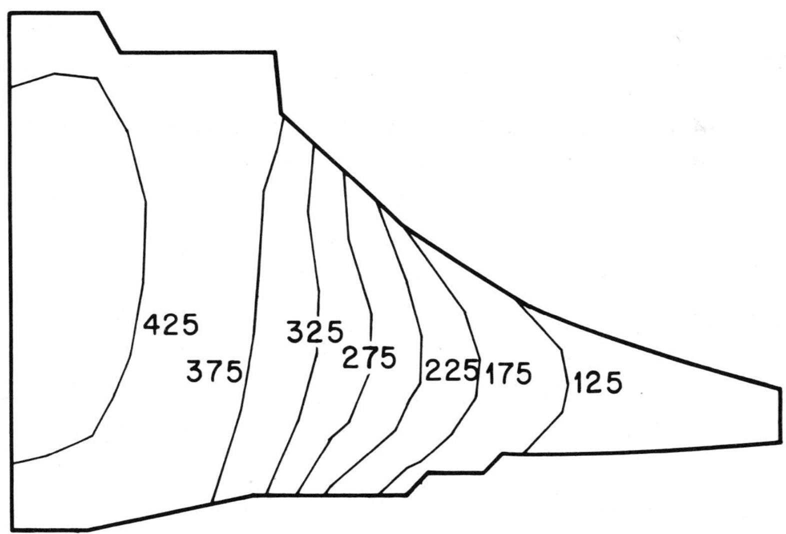

Figure 7.30: Calculated temperature distribution 30 sec. after immersion.

elasto-plastic and creep model [7.17]. The same mesh was used. This model permits the use of the temperature file previously calculated as input. To prevent rigid body motions, the node on the bottom left was fixed in the vertical direction. No other loads other than the step by step temperature distributions were applied. All creep constants were equated to zero (no creep effects), and a yield stress function of temperature was used, as in Figure 7.24. The flow stress of the material during quenching cannot be measured, so the values are for equilibrium conditions. The other material properties are given in Table 7.2. The zero stress strain configuration is the initial one, with a uniform temperature distribution of 530°C. For postprocessing purposes, the stresses calculated were extrapolated to node locations.

With the 8 node isoparametric type used, it was necessary to subdivide the time increments into 0.1 sec to obtain convergence at the start of plastic deformation. Afterwards, no more numerical difficulties were encountered. The calculation procedure with this material model is such that increments in strain are calculated at each time increment [7.18]. The total increment in strain is divided in additive parts:

$$d\underset{\sim}{\varepsilon} = d\underset{\sim}{\varepsilon}^{e} + d\underset{\sim}{\varepsilon}^{p} + d\underset{\sim}{\varepsilon}^{c} + d\underset{\sim}{\varepsilon}^{th} \tag{7.17}$$

$d\underset{\sim}{\varepsilon}$ - total increment in strain tensor
$d\underset{\sim}{\varepsilon}^{e}$ - elastic   "   "   "   "
$d\underset{\sim}{\varepsilon}^{p}$ - plastic   "   "   "   "
$d\underset{\sim}{\varepsilon}^{c}$ - creep   "   "   "   " (= 0 in our calculations)
$d\underset{\sim}{\varepsilon}^{th}$ - thermal   "   "   "   "

and each one is related to stress. Infinitesimal plasticity theory is used in these relations. "Total" strains are updated from increment to increment. The version of ADINA with which the calculations were done, produced stresses and total strains at every time increment. As a result, displacements and stresses are the correct ones. Because plastic strains are not additive, however, total strains should be seen as a measure of displacement gradients, unrelated to the stresses that are calculated.

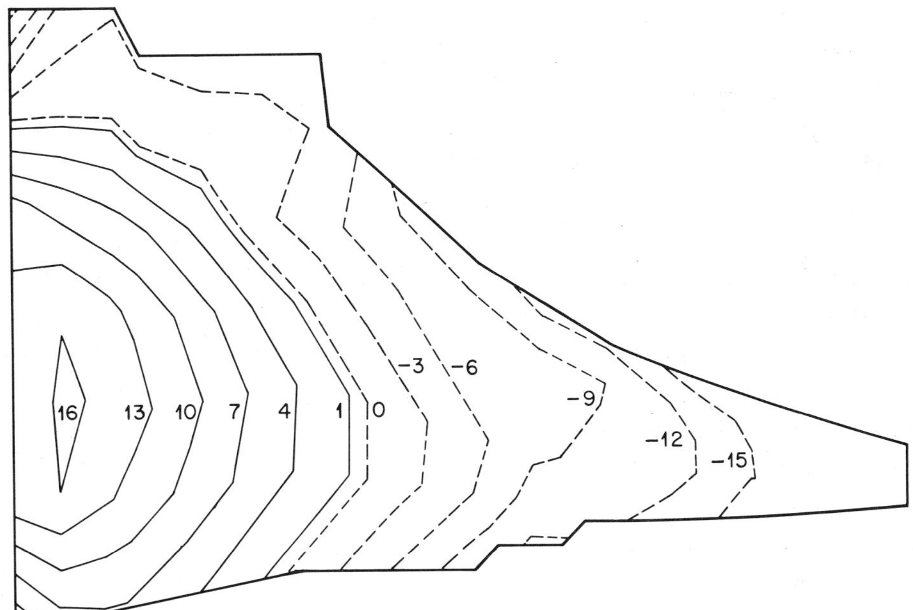

<u>Figure 7.31</u>:   Calculated tangential residual stress distribution.

Figure 7.31 shows the final tangential residual stress distribution. The large, compressive, tangential (and von Mises) stresses at the outer rim show that at the end of cooling the material yields there. They correspond to the large central bulk trying to shrink and the cold outer rim preventing it. Therefore, stresses are tensile in the center, through the mid regions of the axial bore. Both axial and radial stresses are less important. The stress concentration effects of concave corners and the "stress free" effects of convex corners can be observed. The mesh is not detailed enough for the values to be absolutely correct at these corners.

Figure 7.32 shows the deformed profile of the impeller. In order to see the distortions, a uniform contraction has to be subtracted. This contraction is equal to:

$$\frac{\Delta L}{L} \; = \; \int_{530}^{100} \alpha \, (T) \; dT \; = \; 0.01043 \tag{7.18}$$

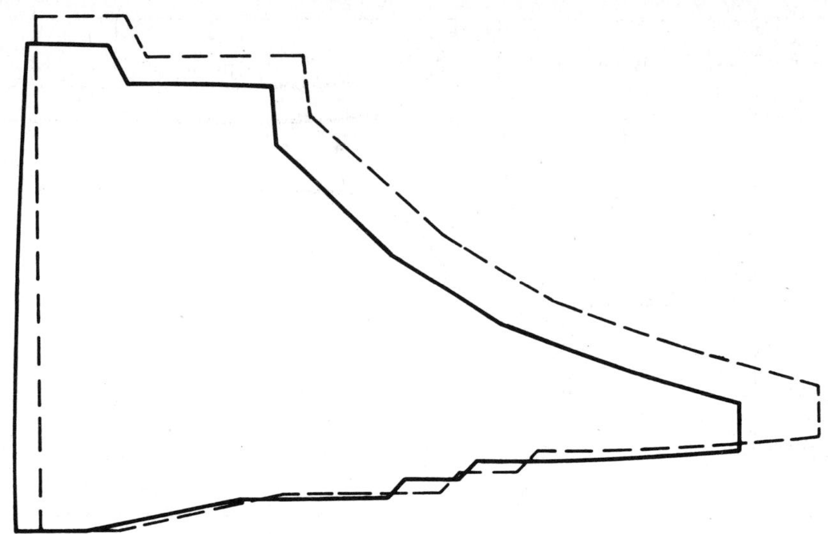

<u>Figure 7.32</u>:   Deformed profile after quenching. Calculated shape.

This results in a distortion like the one in Figure 7.21.

In Figure 7.33 stress histories are plotted for four specific nodal points: number 2, in the bore, lower side, corresponding to
the most highly stressed region in operation; number 6, in the
middle of the bore, where high stresses appeared; 95, in the center of the piece; and 254 in the outer rim. Shear stresses are
virtually negligible. Comparison between von Mises stresses and
tangential stresses show that the latter dominate the stress
tensor.

Through the tangential stress picture, it can be seen how, around
18 seconds after immersion, maximum tensile stresses are produced
on the outside, and compressive stresses on the inside. After
that, the signs reverse and become compressive at the surface and
tensile in the interior. The curves are jerky at some points,
probably due to numerical problems of sudden onset of plastic
deformation.

7.4.4 <u>Comparison with Experiments</u>

Actual residual stress measurements were made on another 300 mm

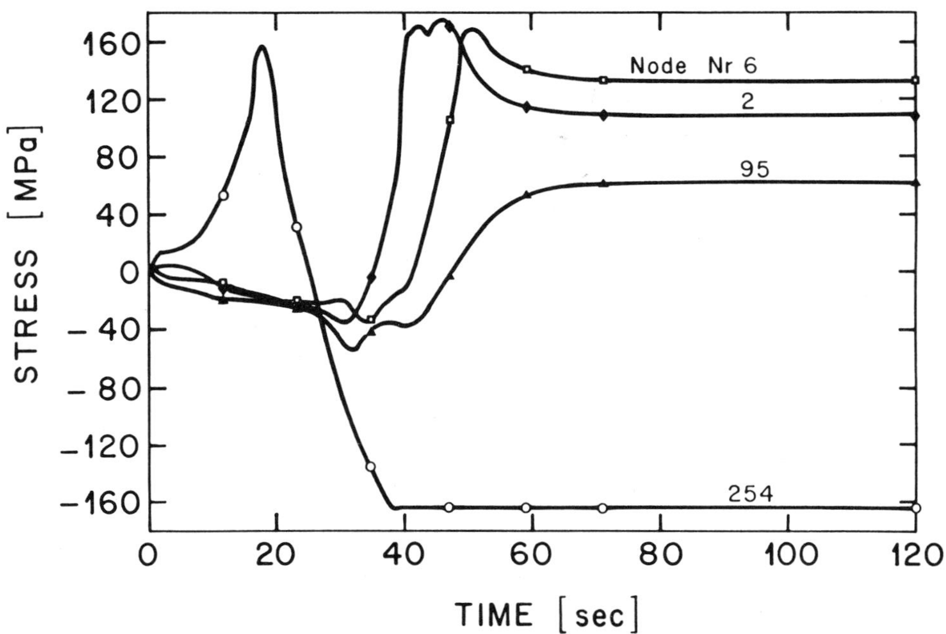

Figure 7.33: Stress histories at selected nodes; tangential.

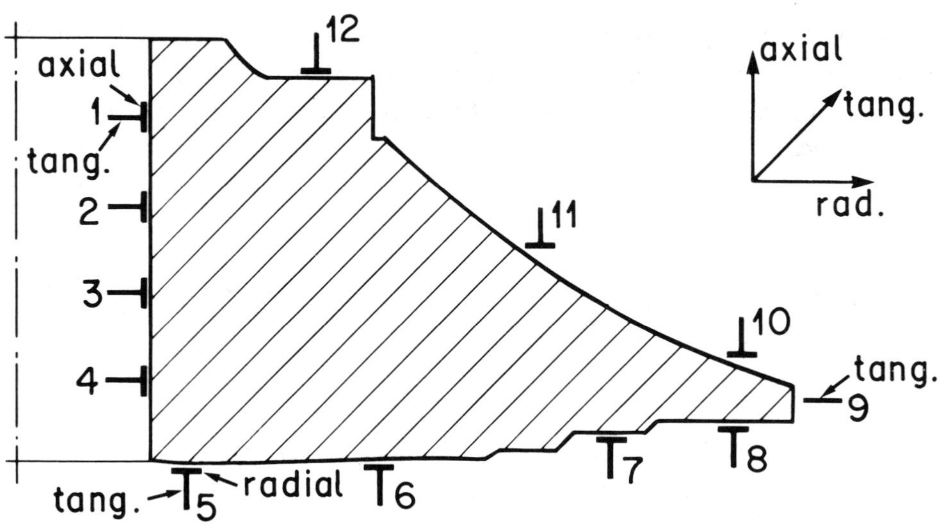

Figure 7.34: Location of strain gauges in the experimental measurement of the residual stresses.

diameter impeller. Strain gauges were glued as indicated in
Figure 7.34. Material was then cut away around each strain gauge,
relieving the strains. Measured values are shown in Figures 7.35
and 7.36. The calculated values are also shown.

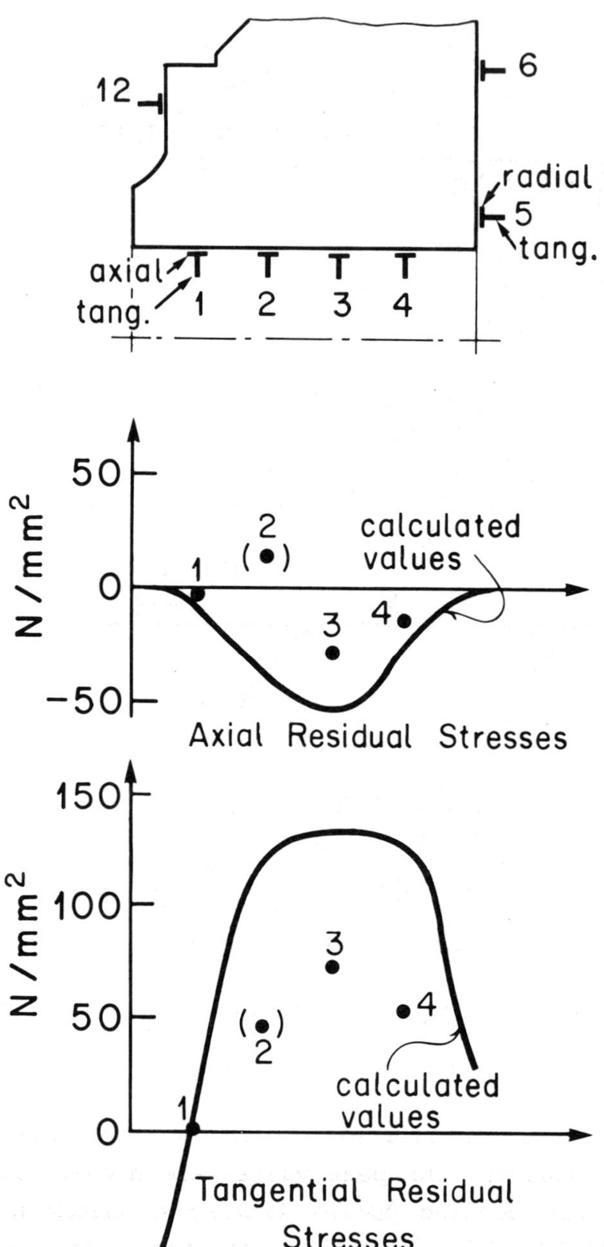

Figure 7.35: Residual stresses in the bore.

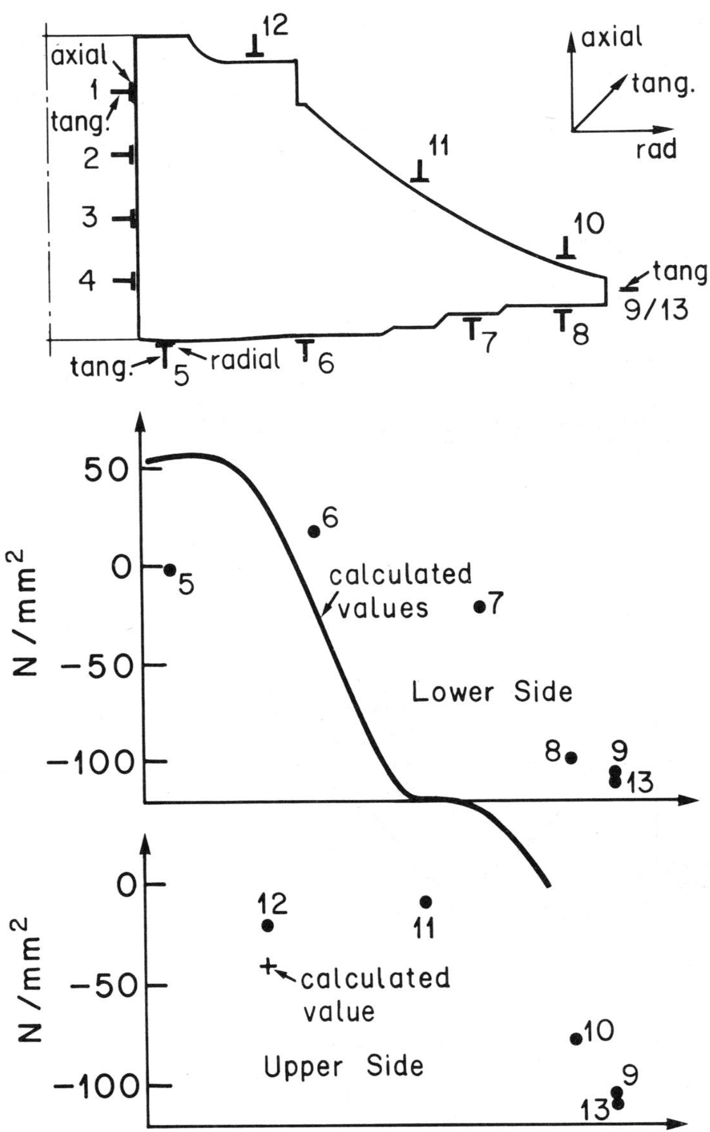

Figure 7.36: Tangential stresses in the upper and lower surfaces.

The stress distributions calculated are in good agreement with the ones measured. However, the peak values are higher. This is in agreement with the cooling curves fitting a quench at 95°C instead of one at 100°C, which would lead to slower cooling.

7.4.5 <u>Conclusions</u>

The success (or non-success) of an analysis, as described here,
depends essentially on how accurate the input data is. The results presented here, obtained with "reasonable" data, show that
the computational means available can produce results that are
in good agreement with reality. Therefore, the feasibility is
demonstrated and attention should be devoted to acquiring the
appropriate input data.

As far as temperature calculations are concerned, to obtain heat
conductivity and specific heat data usually poses no problem. It
is the heat transfer coefficient for the quenching media used
that causes trouble. This coefficient is a function of many parameters, as discussed in section 7.1. For instance, in the case
of water quenching it is not enough to consider the medium but
also its temperature, container restraints, and speed of immersion. Given the quenching conditions, it is still at least a
function of material temperature and surface inclination and
surface condition. It is conceivable that it might be necessary
to consider how many surfaces losing heat are in the neighbor-
hood, and so on.

As far as residual stresses are concerned, thermal expansion is
normally obtainable as well as its elastic constants as a function of temperature. The flow stress and work hardening rate as
a function of temperature need to be determined. It may prove
to be quite difficult to do so because, for instance, in Al 2618,
the flow stress of the material in a solid solution state is
needed. At temperatures above, for example, 400°C and below
150°C, the material can be regarded as being in the stable condition but in between, the material is in a transient state,
therefore the curves have to be approximated.

Finally it should be emphasized that all calculations have been
done in the "worst possible conditions", as boiling water is
the most unstable quenching media that can be thought of. Therefore, when treating other situations the expected errors are
smaller.

7.5 Modelling of Heat Treatment of Large Forgings

This particular application refers to the quenching of massive parts, namely turbine rotor forgings. The size limit up to which a certain steel can be used without serious reduction of critical properties, such as yield strength, creep strength or toughness, must be known in order to decide up to what sizes monoblock rotors can be employed and where other techniques, e.g. welding, provide an alternative. Emphasis of the investigation was on fracture toughness.

A monoblock rotor is machined from one single forging; this requires forgings of huge dimensions - for this application a forging weight of more than 500 tons. The strength requirement - a yield strength of over 700 MPa is typically needed in LP-rotors - calls for a quenched and tempered steel, which has to have a very good hardenability to obtain even properties (strength) throughout the forging. A second important requirement, besides strength, is fracture toughness, especially in the center of the forging, where the service stresses are highest. Good hardenability is provided by a quenched and tempered steel with 3.5 % nickel, which is consequently used for monoblock LP-rotors.

The mechanical properties are governed by the materials microstructure - in the case of quenched and tempered steels mainly by the carbides. The important parameters are carbide size and shape, the carbide spacing, the volume fraction of carbides and the homogeneity of their distribution. With a given carbon concentration, and thereby fixed volume fraction of carbides, these parameters are governed by the thermal history of the material - the decomposition reactions of the high-temperature austenitic phase during quenching and the annealing temperature. Ideally, the microstructure of quenched and tempered steels consists of tempered martensite: finely dispersed carbides in a ferritic matrix with the typical morphology of martensite. In the austenitic high-temperature phase, the carbon is in solution; upon quenching, martensite starts to form spontaneously below the M_s-temperature. The martensitic transformation is diffusionless in nature; it is a shear-transformation of the face-centered cubic austenitic lattice.

As a consequence, martensite can be formed only above a critical
cooling rate, which is specific for a given material and influenced mainly by the chemical composition of the material. With
decreasing cooling rate, the amount of diffusion controlled
transformation increases, and consequently other decomposition
products are formed at the expense of martensite. The first product to appear (with decreasing cooling rate) is bainite, a
microstructure showing characteristics of both diffusion and
shear.

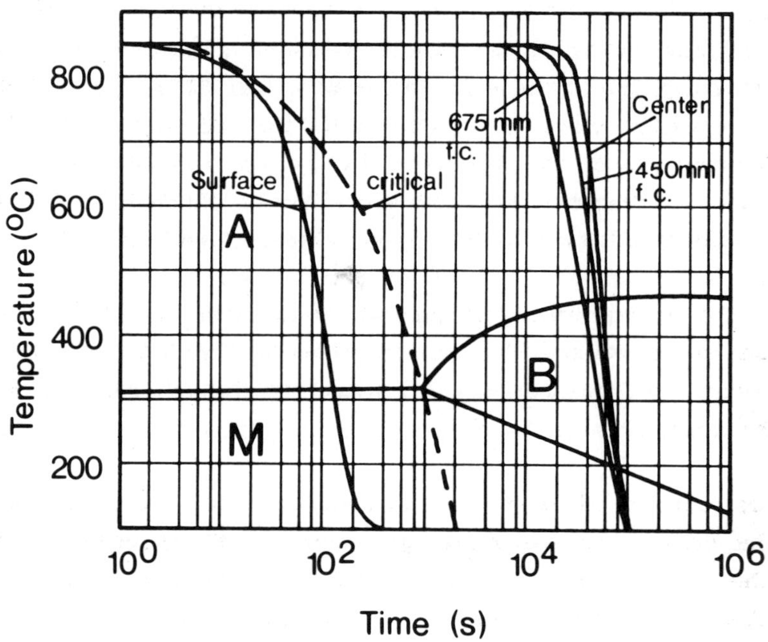

<u>Figure 7.37</u>:   T-T-diagram for 3.5 % ni-steel with calculated
                    cooling curves for a diameter of 2700 mm at va-
                    rious distances from the center (full lines) along
                    with the "critical cooling curve" (dashed)
                    A = Austenite, B = Bainite, M = Martensite.

The transformation reactions of a specific material are summarized in time-temperature-transformation (TTT)-diagrams, which are
shown in Figure 7.37 for the 3.5 % nickel-steel. The martensitic
transformation starts at temperatures below 320°C. It is limited
by a critical cooling rate, shown as a dashed line. Every cooling
rate lower than this critical one will inevitably lead to an

increasing amount of bainite in the microstructure. Some cooling curves are superimposed upon the TTT-diagram of Figure 7.37 as solid lines and will be discussed later.

The cooling rate is a result of many parameters: the nature of the coolant (air, oil, water), the heat flow from the surface of the material into the coolant, which is expressed by the heat-transfer coefficient, the thermal conductivity and the specific heat of the material, and finally the geometry, i.e. the size of the body to be cooled. This leads back to the heat treatment of large forgings. The cooling rate in the center of large forgings is very insensitive to the quenching conditions - it is mainly governed by the rate of the heat flux from the center to the surface, which in turn is given in the physical properties thermal conductivity and specific heat. So, in summary, the cooling rate at the center of large forgings is primarily controlled by the geometry of the forging i.e. diameter and length.

This leads to the conclusion that the microstructure after quenching is controlled by the size of the forging; consequently the forging size has an influence on the mechanical properties of the material.

It was mentioned above that the fracture toughness, especially in the center of the forging, is of great interest with respect to the safety of the component. The fracture toughness of a quenched and tempered steel is controlled mainly by the carbide distribution. It correlates indirectly - and inversely - with the strength of the material. The inverse relationship between strength and toughness is well known and widely observed for most engineering materials. Untempered martensite is very strong, but also very brittle. The tempering treatment precipitates the carbon dissolved in the martensite lattice in the form of carbides. Primarily, the purpose of the tempering treatment, following the quenching, is to adjust the materials strength. Higher tempering temperatures decrease the strength of the material via the carbides. In summary, the mechanical properties, strength and toughness, result from two major parameters: the microstructural phases as a result of the transformation during quenching, and the tempering treatment after quenching.

For reasons of economy it is very attractive to study the correlation between geometry, cooling rate and microstructre using
simulation techniques. In order to analyze the influence of
several parameters in a problem like this rapidly, a completely
interactive program was created. It simulates the cooling of a
cylinder, subjected to convection boundary conditions. Convection
is determined by a temperature-dependent heat transfer coefficient, which can be different at each surface of the cylinder.
These heat transfer coefficients are supplied by the user as data
files, and the material's conductivity and specific heat are
supplied as a user subroutine.

The formulation used is again the one described in section
3.4.4., and the resulting Finite Element equations are as in section 3.4.5, with all the coupling terms with deformation re-
trieved.

In order to automate data input, a standard mesh is mapped into
the dimensions given by the user, and cooling curves at points
again selected by the user are given in a format ready for plotting. In this particular application, time increments are con-
stantly adapted in order to produce a maximum temperature variation between 3 and 6 %, provided numerical accuracy is ensured.

An example of the result of a calculation of the cooling rate is
shown in Figure 7.37. The calculation simulates the cooling of
a cylinder of a 3.5. % nickel-steel with a diameter of 2700 mm
and a length of 6000 mm, which is quenched by water-spraying from
a temperature of 850°C. The calculated cooling curves for various
locations (center, 405 mm and 675 mm from center, surface) are
superimposed as solid lines onto the TTT-diagram of the 3.5 %
Ni-steel. The fact that the cooling curve for the surface lies
to the left of the critical cooling curve (dashed line) indicates that the surface region is transformed completely into martensite. This finding is verified by the metallurgical investigation using transmission electron microscopy.

When the cooling rate falls below the critical rate, i.e. with
increasing distance from the surface, an increasing volume
fraction of bainite is formed. A microscopic investigation re-

404

vealed a microstructure consisting of upper and lower bainite.

The upper bainite is characterized by an inhomogeneous distribution of coarse, often platelike carbides, which are located pre-
ferentially at grain boundaries, whereas the carbides in lower
bainite are finer and homogeneously distributed. The carbide distribution in lower bainite is very similar to that of tempered
martensite. This leads to the discussion of the influence of the
tempering treatment on the carbide distribution and the resulting
mechanical properties.

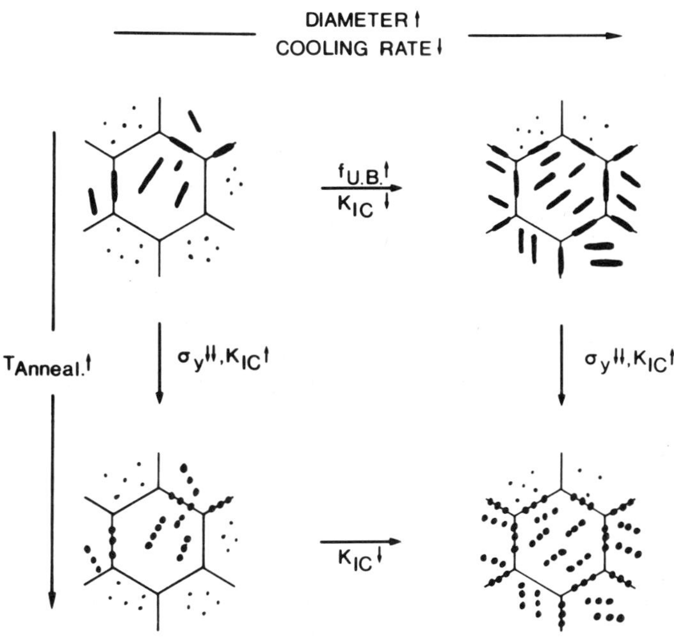

Figure 7.38: Dependence of yield strength and toughness on the
diameter and tempering temperature based on the
carbide distribution (schematic).

The tempering treatment leads to a precipitation of carbides in
the martensite and to a coarsening of carbides in the bainite.
This decreases the yield strength and improves the fracture

toughness. The mechanical properties respond very sensitively
to changes in the tempering temperature; e.g. a difference of
only 40°C (630°C and 590°C, resp.) decreases the yield strength
by almost 200 MPa. It was found that the lower temperature does
not affect the shape of the carbides in the upper bainite; they
remain platelike and consequently the fracture toughness is low.
Higher temperatures lead to a spheroidization of the carbide
plates, and concomittantly the fracture toughness increases. In
summary, the fracture toughness at the center of large forgings
is a result of two physical parameters: the cooling rate, determined mainly by the dimensions of the forging and the heat flux
in the material, and the tempering temperature after quenching.

The cooling rate controls the volume fraction of the phases: increasing forging diameter, i.e. decreasing cooling rate, will
lead to an increasing volume fraction of upper bainite with
coarse, platelike carbides and low toughness.

Increasing tempering temperature coarsens and spheroidizes the
carbides, thereby decreasing the yield strength and improving
the toughness.

These findings are summarized schematically in Figure 7.38 and
described in detail in [7.19].

References Chapter 7

[7.1]   Hsu Y.Y., Graham R.W., Transport Processes in Boiling and Two-Phase Systems including Near-Critical Fluids, Hemisphere Publishing Corporation/McGraw-Hill Book Company, 1976.

[7.2]   Bergles A.E., Colier J.G., Delhaye J.M., Hewitt G.F. and Mayinger F., Two-Phase Flow and Heat Transfer in the Power and Process Industries, Hemisphere Publishing Corporation/McGraw-Hill Book Company, 1981.

[7.3]   Rohsenow W.M., Hartnett J.P. (eds.), Handbook of Heat Transfer, McGraw-Hill Book Company, 1973.

[7.4]   Mitsutsuka M., Fukuda K., Boiling Phenomena and Effects of Water Temperature on Heat Transfer in the Process of Immersion Cooling of a Heated Steel Plate, Transactions ISIJ, 19 (1979) 162-169.

[7.5]   Price R.F., Fletcher A.J., Determination of Surface Heat-Transfer Coefficients During Quenching of Steel Plates, Metals Technology (May 1980) 203-211.

[7.6]   Jakob M., Heat Transfer I, 6th ed., John Wiley & Sons, 1958.

[7.7]   Paschkis V., Stolz G., Quenching as a Heat Transfer Problem, Journal of Metals (Aug. 1956) 1074.

[7.8]   Yanagi K., Makihara K., Eguchi S., Hirai S., Hashimoto R., Nakamura Y. and Sakamoto J., Study on Cooling of Hot Steel Strips with Two-dimensional Laminar Water Jets, Mitsubishi Heavy Industries Technical Review, (October 1983) 273-279.

[7.9]   Sasaki K., Sugitani Y., Kawasaki M., Heat Transfer in Spray Cooling on Hot Surface, Tetsu-To-Hagane (Jan. 1979) 90 (in Japanese).

[7.10]   Alfredsson H., Rydstad H., Investigation of Heat Transfer by Quenching Heavy Steel Plate through Water Flushing, Technical Reports from Royal Institute of Technology, Stockholm, Sweden, TRITA-MEK-76-06 (Sept. 1976) (in Swedish).

[7.11]   Kadinova A.S., Kheifets G.N., Factors Affecting Heat Exchange in Spray Cooling with Water, Met. Sci. Heat Treat., 16/1-2 (1974) 12-15.

[7.12]   Jensfelt P.N., Påskyndad svalning av varmvalsade produkter, Jernkontorets Annaler, 149 (1965) 505-603.

[7.13]   Müller H., Jeschar R., Untersuchung des Wärmeübergangs an einer simulierten Sekundär-Kühlzone beim Stranggiessen, Arch. Eisenhüttenwesens, 44/8 (1973).

[7.14]   Anman P.M., Griffiths D.K. and Hill D.R., Hot Strip Mill Runout Table Temperature Control, Iron and Steel Engineer (Sept. 1967).

[7.15]   Bathe K.-J., A Finite Element Program for Automatic Dynamic Incremental Non-linear Analysis of Temperatures, Report 82448-5, M.I.T. (1977).

[7.16]   Kreith F., Principles of Heat Transfer, 2nd edtn. Int. Textbook Company, 1967.

[7.17]   Bathe K.-J., A Finite Element Program for Automatic Dynamic Incremental Non-linear Analysis, Report 82448-1, M.I.T. (1978).

[7.18]   Snyder M.D., Bathe K.-J., A Solution Procedure for Thermo-Elastic-Plastic and Creep Problems, Nuclear Engineering and Design, 64 (1981) 49.

[7.19]   Albrecht J., Bertilsson J.E. and Scarlin R.B., The Fracture Toughness of Actual and Simulated Large Rotor Forgings made from 3.5 % Nickel Steel, ASTM-STP "Steel Forgings", to be published (1985).

8 Outlook

The examples in this book show that process models, in conjunction with the present generation of computers, can already supply solutions to many problems in manufacturing technology. In some cases, this can be achieved more economically than with tests in the laboratory or in production facilities. With process models, localized predictions can be made of temperatures, stresses and deformations in the workpiece or tool, which cannot be achieved with the most modern measuring techniques.

However, barriers to the use of current computer process models are continually met. One aspect is the economic barrier for the modelling of problems which require many hours of CPU time on large computers. The widespread introduction of process models is being pursued in conjunction with the introduction of CAE/CAD in industry, whereby both are dependent on the further development of more powerful, cheaper computers.

Further impulses can be expected from the development of expert systems in the field of information science. Such expert systems offer the potential of accumulating the experience of experts, regulations, data banks and the results of computer simulations. These can be made available with short access times/computing times.

With increasing precision of the process models, there is a need for far more accurate and more extensive physical materials properties, such as heat transfer coefficients within wide ranges of temperature and pressure as well as thermal conductivity when phase transformations occur. The constitutive laws for the description of plastic behavior are often too inaccurate, particularly for anisotropic materials or in situations where reversed deformations occur. In thermomechanical treatments such as recrystallization, the correlation is practically unknown between local parameters (calculated using process models, e.g. reference values for stress, deformation and deformation rate) and their influence on dynamic recrystallization. A great deal of basic and detailed interdisciplinary research work must also be performed in this area in the next few years.

"

No new basic mathematical-numerical method is currently anticipated for the modelling of elastic-plastic materials behavior. The elementary plasticity theory and the extremal principles will be used for optimization and for on-line process control. Sophisticated Finite Element Methods will be applied on three-dimensional non steady-state problems.

A further area for process modelling appears in connection with the further development of control technology for process control. Simplified process models can be applied for process control in situations where the measurement of key parameters is technically difficult (e.g. cooling of castings within the mold). As a beginning, this development has already been made for continuous casting, rolling and welding. Also in process control, process models and expert systems can open completely new possibilities in the future.

Finally, process models will become more important in new, rapidly-developing technologies, such as laser machining. The increasing importance of manufacturing systems, particularly flexible manufacturing systems (FMS), should be mentioned. Of primary importance here is the simulation of the system and not the process. Process models for technologies will be the building blocks of a system which will bring us closer to computer-integrated manufacturing (CIM).